P9-EEP-982

SOLID-STATE MOTOR CONTROLS

SOLID-STATE MOTOR CONTROLS

BY JOHN A. KUECKEN

FIRST EDITION

FIRST PRINTING—JUNE 1978
SECOND PRINTING—MAY 1980

Printed in the United States
of America

Library of Congress Cataloging in Publication Data

Kuecken, John.
Solid-state motor controls.

Includes index.
1. Electric motors—Automatic control. 2. Electronic control.
I. Title.
TK2851.K83 621.46'2 78-5065
ISBN 0-8306-8929-X
ISBN 0-8306-7929-4 pbk.

Cover illustration by Robert O. Barg.

Foreword

Beginning with the invention of the transistor in 1948, solid-state devices began to displace electronic functions long held by vacuum tubes, metallic rectifiers, electromechanical relays, photosensitive chemicals, and other similar devices. The new solid-state packages not only displaced these old standbys but also went far beyond their capabilities—in certain applications. In receiving functions, transistors soon became the favorite active element of design engineers. The rigorous demands of transmitting tubes, however, precluded the use of transistors in all but "flea powered" units for many years. While today the output stages of many commercial high-powered transmitters use transistors, tubes still hold the market for the really high power stations.

The metallic rectifiers and photochemical devices of the past have largely given way to solid-state products. With rectifiers it is relatively easy to build junctions able to carry high currents and to withstand large reverse voltages. The photoproperties of semiconductor junctions proved to be much more useful than the meager photodevice field of pretransistor electronics.

But when solid-state devices were first investigated as control replacements for industrial motors, generators, relays, and other heavy duty machines, engineers soon learned that the harsh environment of industrial equipment required far greater ranges of parameters than the new technology could provide. Thus, solid-state control of these products had to wait for the technology to advance, to provide devices capable of handling large currents and voltages, extremes of temperature and vibration, and multiple control functions and versatility.

Today, absolute maximum ratings for power transistors and thyristors have reached the point where these devices are able to displace the electromechanical controls on many of the large-horsepower motors in common use. With the cost of these control devices being competitive with older control methods, and with the proven high reliability and performance of solid-state controls, the designer of any new control system would do well to investigate the latest offerings of the manufacturers before deciding whether to go mechanical or solid state. Specifications continue to improve, and costs continue to fall for solid-state control devices.

The author provides a complete course on motor controls—from the principles of control to the use of microprocessors for obtaining maximum efficiency in complex installations.

The three basic types of motors—ac, dc, and universal (ac/dc)—are treated in detail so that the designer has a complete understanding of all parameters before deciding on the best approach to a control system. Supporting these discussions of motors are chapters reviewing principles of dc and ac circuits. Inductor, transformer, and capacitor circuits illustrate many of the principles used in motors.

With an understanding of magnetic flux density, magnetic potential gradient, poles, and other characteristics, a control designer gains an understanding of the inner workings of motors, enabling him to understand the reasoning behind control techniques given later in the book. Ultimately the hope is that the reader will learn to appreciate the superiority of solid-state motor controls in many applications and will design and use them in his own applications.

John A. Kuecken

Contents

1
Principles of Control

The progress of civilization is measured, at least in part, by the means man has devised to control and harness its forces. Abraham rode a saddle ass (Genesis 22:3) and the Pharaoh of Egypt could summon up 600 chariots to chase the Children of Israel (Exodus 14:6—8). By 1500 B.C. the Egyptians were sailing 70-foot papyrus ships with a single sail and 30 oars for propulsion and two for steering. The significance of the 15 pairs of propulsive oars is that they were not only for propulsion but also for control. The wind would move the ship very well but not always in the direction the captain wanted her to go; therefore, the oars.

CONTROL CATEGORIES

Our world is populated with a tremendous variety of controls, devices intended to harness and direct the use of energy. Of these the simplest, fastest, and most flexible are usually electrical. This text intends to explore some of the major categories of these controls and to discuss the means by which solid state devices can be applied.

In the discussions to follow, controls will be divided into two broad categories:

1. Open-loop controls
2. Servo (closed-loop) controls

Some very fundamental differences exist in these categories which tend to influence the design of the components and the principles of operation. Indeed, each category is populated with a number of subcategories, which will also require definition. For the sake of simplicity, we shall try to employ some rather commonplace examples in the definitions.

OPEN LOOP CONTROLS

Open loop controls are defined here as those controls in which there is no direct sensing of the accomplishment of the end goal of the system by the system itself.

For example, the vacuum cleaner is turned on and off at the discretion of the person using it. There is nothing built into the mechanism to tell whether the rug is actually clean. This is a simple on/off control and will be referred to as a *manual control*. Other manual controls can be a good deal more sophisticated and still remain in this category. For example, a blender, a cake mixer, or a fan may be equipped with a variable speed control (the speed control itself may contain a closed-loop servo). However, in each case, the unit is turned on and off and the speed is regulated only at the discretion of the operator and not by having the device sense that the job is accomplished.

A second type of open-loop control is the *automatic open loop control*. This type generally takes the form of a *sequencer*. A good example of this is the automatic dishwasher and the automatic clothes washer. These machines generally have been built to operate in a time sequence. Fill, wash for 10 minutes, pump out, rinse for 5 minutes, pump out, and so forth. Here again the device does not sense the cleanliness of the clothes or dishes but rather assumes that if all of the steps in the sequence have been adequately performed, the clothes and dishes will be clean. In the past these sequencer controls have consisted of some rather elaborate clockwork schemes to turn on pumps, valves, and motors. If the device is relatively simple and has only one set sequence, this sort of thing works out relatively well. However, more-modern washers tend to be supplied with a choice of cycles for various wash and wear fabrics, delicate lingerie, sensitive cotton cloths, and other washable items. Provisions for a flexible cycle selection tend to multiply the complexity of the sequencer. It is particularly in these areas that the use of an electronic sequencer is attractive. A very modest microcomputer or microprocessor chip combined with the simplest of read only memory (ROM) and random access or read/write memory (RAM) permits an unprecedented amount of flexibility in the cycle control and in a very simple package. Ultimately, the cost of the solid state or electronic sequencer should be smaller than that of the simplest mechanical unit.

The use of the open-loop system is by no means confined to the relatively simple applications described. One of the most important and sophisticated members of this family is the *direct numerical control*. This is the system commonly found in numerically controlled (NC) machine tools. In these machines, *stepper motors*, which produce a precise angular shaft rotation for each input stimulus, are used to position the table, carriage, or tool. For example, stepper motors are available that provide a shaft rotation of 7.2° for each input pulse. It requires 50 input pulses to make the shaft rotate through a full revolution. If the machine is provided with a 20-thread-per-inch lead screw, the tool or carriage propelled by the lead screw will advance 0.001 inch with each input pulse.

The convenience of this arrangement in digital programming of the machine tool is relatively obvious. The sequencer need only provide a train of pulses numerically equal to the desired travel in thousandths of an inch. Most machines of this type read sequencing instructions from a punched paper or Mylar tape. The tape sequentially addresses the various table motion or tool motion drives and the machine produces the piece with little or no operator intervention. The original tape can be punched out by a programmer from the drawings.

One of the reasons for the popularity of this system for machine control over the closed-loop system is the fact that the system can be easily made to operate with essentially zero overshoot. This is particularly important in machine tools and plotters. If the drill or mill were to go a little too far and then move back, the damage would be done because the taking-off tools work better than the putting-back-on tools. On the plotter, the ink would be on the paper beyond the point where the line should have ended.

SERVO CONTROLS

The term *servo* stems from the Latin servus—meaning servant. The dictionary definition indicates that the term is applied to any device permitting control of large forces with negligibly small ones. In technical usage, though, this definition is too broad since it could be applied to nearly any valve, switch, or amplifier. The common technical usage of the term *servo system* is usually reserved for devices in which the control input is construed by the system as an error and the machine acts to remove the error.

Perhaps the simplest example available is the refrigerator. The operator sets in a desired temperature in the control dial and the thermostat senses the actual temperature within the box. If this temperature is more than a few degrees warmer than the set level, the motor starts and operates until the temperature is slightly less than the set level. The machine then shuts down and waits until the temperature again rises to the upper limit whereupon the cycle repeats. The system is called a closed loop because it actually senses the accomplishment of the task at hand.

The refrigerator cooling system and thermostat form a unidirectional closed loop servo. The system is unidirectional in the fact that the compressor system is able only to cool the box and not to warm it. The warming action, of course, comes from leakage through the walls and the requirement to freeze or cool the contents. In other closed-loop systems, this unidirectional property would not be acceptable.

An important property can be defined in terms of the refrigerator example. The system possesses an important property termed *hysteresis*. This property is carefully engineered into the product, forms an important part of the operating principle, and is very necessary to ensure the life of the machine.

Most refrigerators are equipped with single-phase induction motors, which require the use of a starting winding. During the starting operation the torque of the motor is very limited, and the motor cannot start the compressor against the back pressure of the refrigerant. You may have had the experience of having the refrigerator stopped by a brief power interruption during a storm. When the power came back the motor, perhaps, attempted to resume the refrigeration cycle, but couldn't start against the charged compressor cycle, so the unit groaned until the thermal overload switch kicked out due to the excessive current being drawn. If the thermal overload had not interrupted the current, the motor would have burned out. Sometime later the thermal overload re-closed and the motor tried again. If the refrigerant had an opportunity to pass from the cooler into the evaporator during the idle period, the back pressure was reduced and the unit started and ran successfully.

From the example, we can see that it is neither possible nor desirable to have the refrigerator attempt to hold the temperature absolutely constant. The practical mode of operation is to have the unit run until the box is cooled to some low temperature, for example 34°F, then shut off. The thermostat then waits until the box warms to some higher temperature, for example 36°F, before attempting to restart the compressor. This gives some time for the refrigerant to pass from the cooler through the evaporator and relieve the back pressure. The difference between the 34° shutoff point and the 36° turn-on point is termed hysteresis. An important aspect of hysteresis is that it is directional; the thermostat switches off at 34° with a falling temperature and switches on at 36° with a rising temperature.

There are two other terms that we may define here. When the motor shuts off at the 34° level, there is still a certain amount of refrigerant left in the cooler. As this passes into the evaporator, the cooling process continues so that the box temperature will actually fall below the turnoff point to perhaps 33°. This is termed *overshoot*. A similar case of overshoot takes place on the warming cycle. As the box warms beyond 36°, the compressor starts; however, it takes a little while before the cooling process begins. The box temperature will actually overshoot and get to be warmer than 36°, perhaps to 37°, before the cooling process takes over. In this example the system would have a hysteresis of 2°F, with a 1°F overshoot at each limit. A good quality refrigerator would have much tighter limits, but the principle is the same, and some amount of overshoot will be noted, and a measurable hysteresis will be built in.

It is a general rule that switching mechanisms should cleanly toggle on and off and not be allowed to halt at a halfway point.

THE BIDIRECTIONAL CLOSED LOOP SERVO

Many servo systems have the requirement that they operate in both directions. For our example of this type system we shall examine an oversimplified power steering system shown in Fig. 1-1. As the discussion progresses, we shall see some of the properties of the system and learn why the system is oversimplified.

In our illustration we see that the system consists of a steering wheel which is attached to a driven gear with a springy insulated joint. The gear carries two batteries, A and B, with opposite poles grounded and a pair of contact pins attached to the floating, or "hot," battery terminals. A permanent magnet electric motor is connected with a slip contact to the switch rod. Now the property of permanent magnet dc motors is that they reverse their direction of rotation when the polarity is reversed, and we can see that this is accomplished with the directional switch.

We arrange the motor such that when the steering wheel is twisted clockwise (cw), the switch rod swings over and touches the B contact, which causes the motor to drive the driven gear also in the clockwise direction. This rotates the gear and pulls the B contact

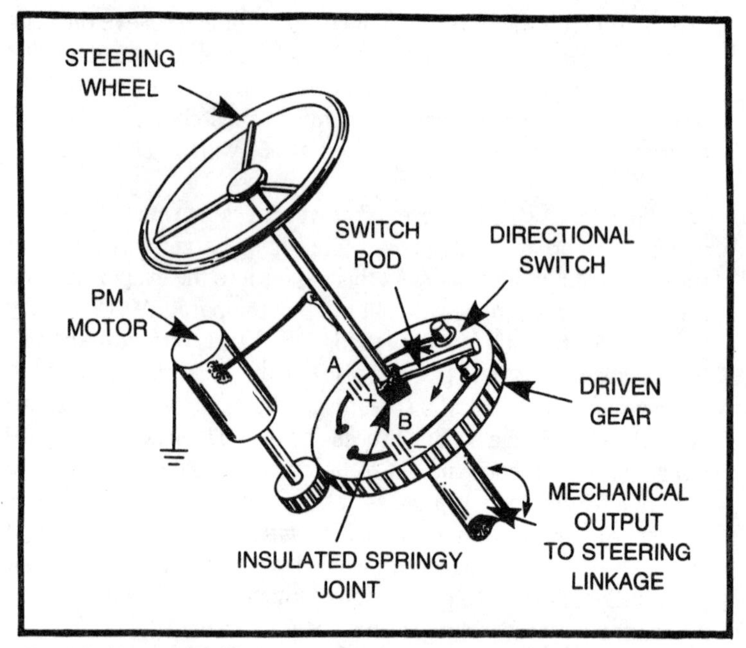

Fig. 1-1. An oversimplified power steering system.

away from the rod. If the motor coasts a little, and the system has the right amount of friction, the driven gear rotates through the same angle that the steering wheel was turned and coasts to a stop with the switch rod not touching either the A or B contacts. If the wheel is turned counterclockwise (ccw), the switch rod touches the A contact and the motor drives the gear in the ccw direction.

This is obviously a full closed-loop servo system. The operator need only supply enough torque to the steering wheel to keep the switch rod in contact with the pin and the output torque is supplied almost entirely by the motor. The output shaft can be made to rotate through the same angle as the steering wheel with essentially no effort on the part of the operator. The system is said to have *negative feedback* in that the action of the system tends to cancel the effect of the input (by rotating the contacts on the driven gear away from the switch rod).

Now note the fact that it is always necessary to have some small angle where the steering wheel may be moved without touching either contact. This *deadband* is necessary so that the rod cannot contact A and B simultaneously since this would short circuit the two batteries. In addition, there are other factors which necessitate some amount of deadband.

In the initial discussion we observed that if the motor coasts a little and if the system has the right amount of friction, the system would coast to a halt with the switch rod half way between A and B at the new angle. However, this implies a rather sluggish performance on the part of the system. Suppose that after turning through some clockwise angle the driven gear pulls contact B away from the switch rod but the motor coasts past the midpoint and, in fact, coasts to the point where contact A touches the switch rod. This would give the motor an impulse to drive in the counterclockwise direction. Now, depending upon the response rate of the motor, the driven gear will do one of several things. If the amount of "zip" in the motor system is relatively small compared to the friction and inertia, the system will back up just enough to center the switch rod, and the system will come to rest. However, if the motor system is really "zippy" the system will overshoot until B is contacted again and the system sent back toward A.

The curves of Fig. 1-2 illustrate the four principal classes of performance that may be obtained from the system, as established by different degrees of damping. The critically damped system with a single overshoot cycle is the preferred one for most situations since it reduces the error to the minimum level in the shortest time and achieves a small error much faster than the overdamped response. It should be noted, however, that in our power steering example the presence of any significant amount of overshoot would be unacceptable at high speeds, and underdamping would be disastrous.

It should also be noted that the degree of damping in a practical system is somewhat of a function of the size of the input. For a small steering correction the motor might just barely get moving and thus have little tendency to overshoot. The same system with a large angle to turn through could overshoot badly once the angle has been passed beyond which the motor could accelerate to full speed. In a practical power steering system, some damping mechanism adjustment with motor speed probably would be required.

The damping is related to the width of the deadband of the system. If the space between contacts A and B is increased the *input error* rises; however, the damping requirement is relaxed since the driven gear can coast farther before contacting the opposite contact. This is usually undesirable because it makes the system "loose," or "sloppy." The steering wheel must be rotated through some larger angle before anything happens. A preferable method is to make voltages A and B adjustable so the system velocity can be controlled.

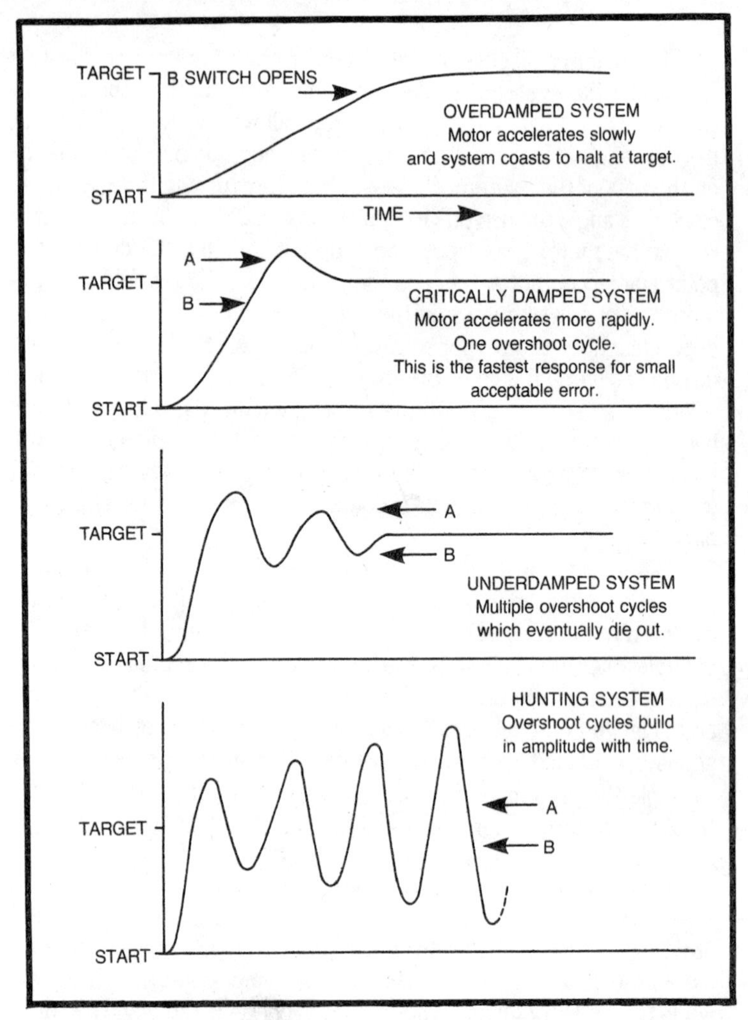

Fig. 1-2. The effects of system damping. The A and B notation refers to the switches shown in Fig. 1-1.

In a system with a significant amount of friction loading, some level of drive can be found where the existing damping will suffice to provide the critically damped performance.

To define the terms deadband error and input error we will refer to Figs. 1-3 and 1-4. Figure 1-3 illustrates that as the switch rod (Fig. 1-1) is rotated there is a deadband area where nothing happens. Beyond switch points A and B, however, the motor receives some fixed input voltage, which causes it to rotate.

For purposes of illustration, Fig. 1-4 supposes that the operator takes the steering wheel and starts it rotating at a constant rate at time T_0. Until the switch rod contacts A or B, at T_1, nothing happens. However, after contact is made the motor receives full voltage and begins to accelerate up to speed, but this takes some time (T_2). As a result the driven gear can lag still farther behind the switch rod, and the input error is larger than the deadband error. In Fig. 1-4 the assumption is made that the switch rod is rotated at a rate equal to the maximum driven gear rate. The steering wheel obviously can be turned either faster or slower.

Figure 1-5 illustrates the behavior of a critically damped system in which the switch rod moves at a rate lower than the maximum driven gear rate. In this case the driven gear can catch up to and coast past switch point A or B, and its path will oscillate about the switch point.

In Fig. 1-6 the switch rod simply runs away from the driven gear, and the input error increases with time. This would be a particularly bad performance in a power steering system since the system would continue to run after the input had ceased in order to

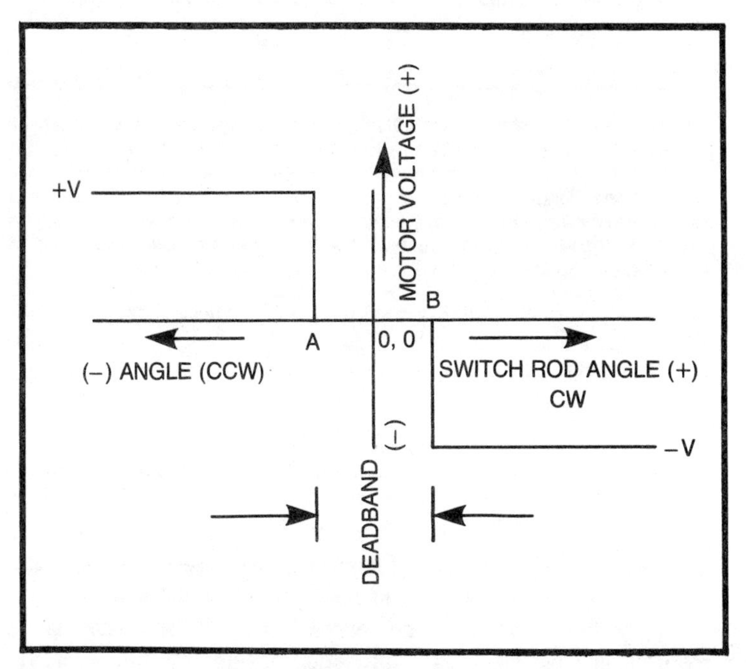

Fig. 1-3. The switch rod (Fig. 1-1) moves through a slight deadband area where no power is applied to the motor. Once the deadband area is passed, however, full voltage drives the motor.

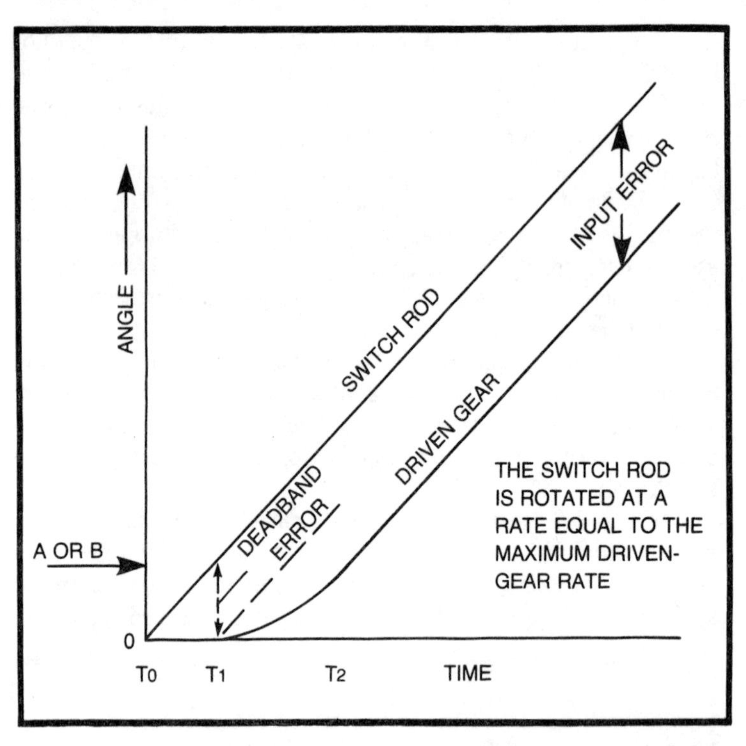

Fig. 1-4. Here the switch rod is rotated at a rate equal to the maximum driven-gear rate. At time t_0, the rod begins to move at a constant rate. At t_1, the rod makes contact to energize the motor; the angle associated with the time from t_0 to t_1 is the deadband error. From t_1 to t_2 the motor builds up speed, adding error to the deadband error. The sum of the two is the input error. Beyond t_2 the driven gear tracks the switch rod (because we decided to have the switch rod move at the maximum driven-gear rate).

crank out the accumulated input error, which would produce dangerous oversteering.

From this we can state that the system will track with minimal input error up to the maximum driven rate.

Our scheme of adjusting the motor drive voltage in order to match the existing frictional damping has a few rather significant faults. First and foremost is the fact that it accomplishes freedom from overshoot oscillation or hunting by slowing the maximum driven rate. As we have seen in the power steering example, this rate must be higher than the greatest input rate if reasonable input errors are to be obtained. A second reason is that the frictional loading inherent in most systems is a function of temperature, wear, the condition of lubrication, and other variables. Since these may change, for reasons beyond the control of the designer, they tend to

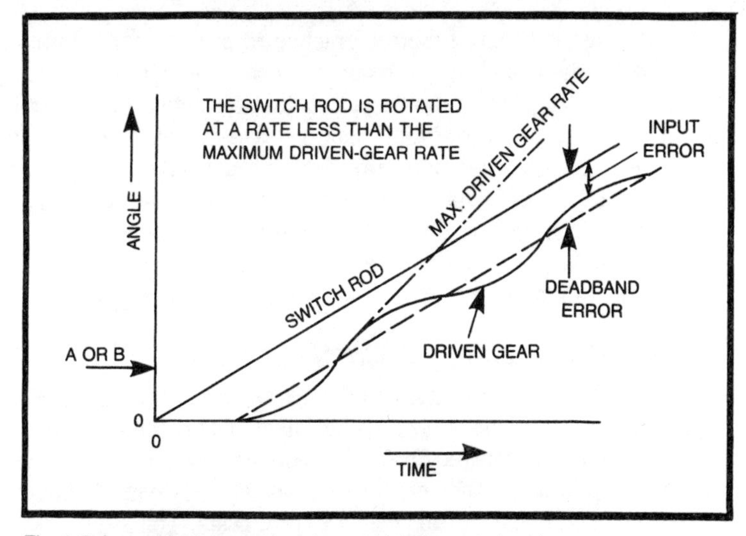

Fig. 1-5. In a critical damped system, which the switch rod moves at a rate less than the maximum driven-gear rate, the gear can catch up to and coast past the switch rod movement, thus establishing an oscillation about the switch point.

present a hazard to system stability. For instance, a system that was critically damped or overdamped and sluggish at cold temperatures could become underdamped to the point of hunting at higher temperatures or after lubrication.

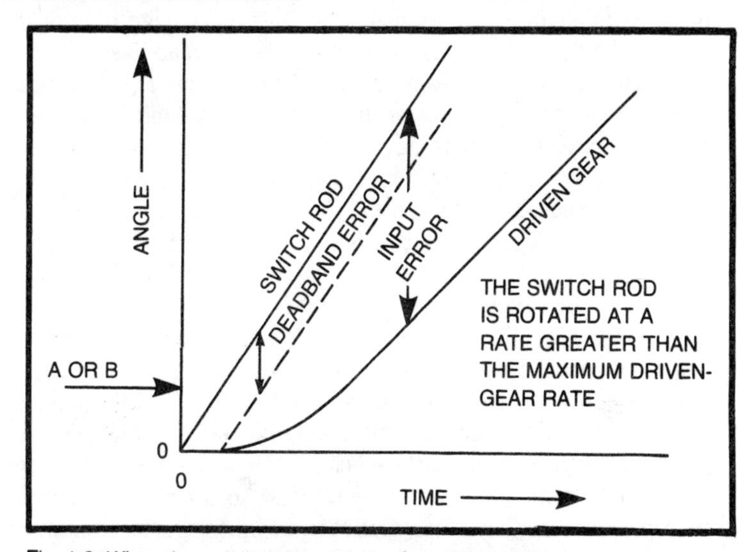

Fig. 1-6. When the switch rod runs away from the driven gear, the system will continue to run after the rod stops so that the accumulated input error can be cranked out.

It is usually a much better practice to provide some form of braking for the damping. For example, a solenoid operated friction brake can be applied to the output shaft. Connecting this brake in series with the motor to automatically release when the motor drew current and to set when the motor did not draw current permits far faster and "tighter" system operation with very good damping. A second alternative is to short circuit the motor, making use of its dynamic braking action whenever the error is within the deadband. The ramifications of dynamic braking will be discussed later.

LINEAR VERSUS NONLINEAR SERVOS

All of the systems discussed thus far fall in the category of nonlinear servos. These used to be called *bang-bang* systems because they were operated on an all or nothing basis. The motor was turned on or it was turned off—with nothing in between. Whenever the system is running it runs full bore, or it rests. This seems like a rather violent way to accomplish a given task so we might take the time here to examine the properties of a *linear servo*.

For our example, let us make an alteration in our oversimplified power steering system so that it qualifies as a linear servo. If we refer to Fig. 1-7 we see that this is accomplished by replacing the bidirectional switch with a potentiometer. This potentiometer will be presumed to give us a motor driving voltage that is linearly proportional to the input error. (Actually this isn't really quite true as we shall see later, but it will suffice for now.) We shall also assume that the motor *acceleration* is proportional to the drive voltage. This boils down to the fact that the bigger the input error, the more the motor will accelerate. In order to see what happens with this system we will have to get a little more mathematical.

Mathematically, linear servo systems are very elegantly solvable by means of differential equations. For the benefit of those more mathematically inclined readers, I would recommend section 4, page 55 of Trimmer's book, listed in the bibliography.

Conversely, the solution of bang-bang servo problems by means of pencil, paper and slide rule tended to be a horrendous fit-and-file type of exercise since no single simple set of equations could be written for the system. The response of the system depended upon previous history—such as, "How fast was the system going when it hit the switch point?" and so forth. This reason, perhaps more than any other, was responsible for the fact that most of the original precision electronic servos were *semilinear*. (The "semi" will be explained shortly.) The SCR-584 radar used to direct

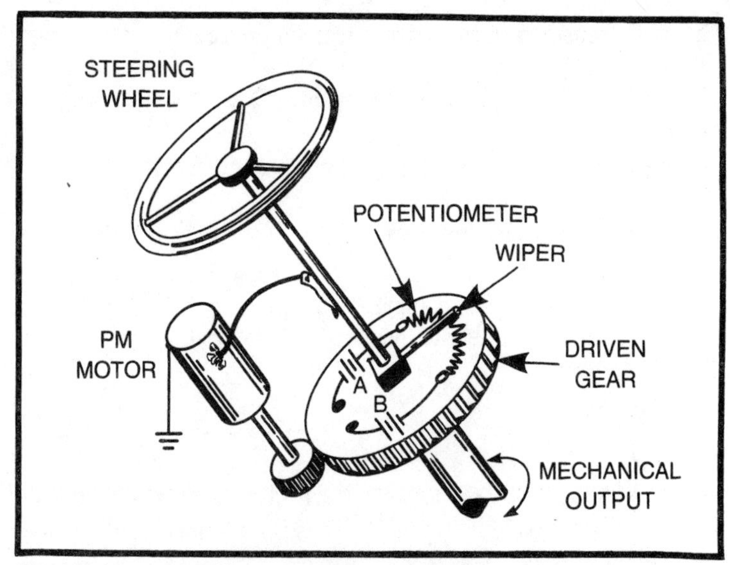

Fig. 1-7. An oversimplified power steering system.

antiaircraft fire during World War II made use of semi-linear servos for antenna positioning. A pair of 6L6 beam tetrodes supplied the field excitation for an amplidyne (in the General Electric version) or a motor/generator set (in the Westinghouse version). The latter machine weighed perhaps 200 pounds and was required to amplify the signal power up to the level required to run the antenna positioning motors. Today, the entire control could be a solid state package weighing but a few pounds.

If our ability to control these drives has advanced in the past thirty-odd years, however, so has our computational power. The once mind-boggling differential equations and the tedious step-by-step solutions for bang-bang systems have bowed to the power of the programmable computer with its ability to remember past data and make conditional jumps based upon data. Accordingly, we shall base our study on solutions that can be performed with a relatively modest desk-top computer. Most of the studies in this and later chapters have been run on a Monroe 326 Scientist; however, most are within reach of even a pocket programmable if the programmer can write "tight code."

For the linear system we will set up the problem as follows:

Let:

Δt = time increment (in seconds)

Sw = the position of the wiper (in degrees)

ΣS = the position of the driven gear (in degrees)
Then:

$$\text{Input Error} = S_W - \Sigma S \qquad (1.1)$$

And

$$\text{Acceleration} = \frac{S_W - \Sigma S}{km} \Delta t \qquad (1.2)$$

Then current velocity becomes:

$$V = V_{n-1} + \text{Acceleration} - K_2 V_{n-1} \text{ (deg./sec.)} \qquad (1.3)$$

where we will designate damping with K_2 and acceleration with A. The term $\Delta t/km$ represents the inertia of the system in response to the motor torque.

And finally:

$$\Sigma S = \Sigma S_{n-1} + V\Delta t \qquad (1.4)$$

The computer will of course solve the problem in a repetitive set of steps, and the designation (n-1) refers to the solution for the particular value on the previous step.

Before going to the results, let us consider the physical meaning of a few of the terms. The use of the damping as a multiplier times current velocity implies *viscous damping*, that is, the kind of damping one would get if the system were stirring heavy-oil or corn syrup. When the output is moving fast the damping load is high, and when the output is moving slowly there is little damping. In our example we shall assume a value of $K_2 = 0.5$ which means that if the drive power were suddenly interrupted the system would burn off speed so that it would be going only half as fast one second later and one quarter as fast two seconds later, and so forth.

If we take our interval for Δt as one second we can drop it from the calculation since it would simply be multiplying the argument by one on each step.

The term km relates to the response of the motor to input errors. In this example we shall vary the term (km) in order to determine the conditions for stability for the system with a fixed level of damping. For the first of the plotted curves, where km = 1, this implies that for an input error of 1 degree, the supply voltage is high enough to accelerate the load at the rate of 1 degree per second. If we were to halve the supply voltage or double the load inertia (m) the acceleration would also be cut in half.

Now it is noteworthy that each of these terms is directly measurable in a physical system, so the calculation is set up in a manner that permits application to real systems with easily measurable properties.

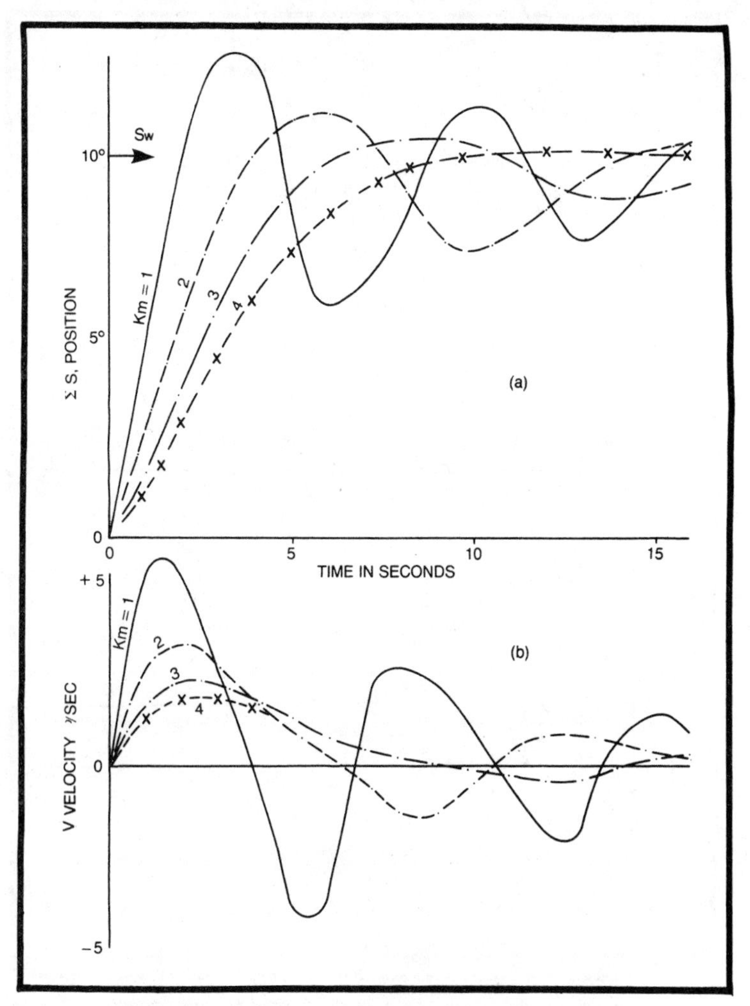

Fig. 1-8. The curves at (a) show the value in a linear system of S with time for various values of km. The value of damping (k_2) = 0.5. The curves at (b) show the value of V with time for the conditions at (a).

If we refer now to the curves of Figs. 1-8 and 1-9 we will see the results of the calculations for a step function input of 10°. This is as if we had held the system power off and turned a 10° error into the wiper. The results are really not a whole lot different from the results shown in Fig. 1-2; however, we observe that the curves are very much distorted from a decaying sine wave. This is due to the presence of the damping. When the system has overshot, both the damping and the motor are trying to stop the load, but when the

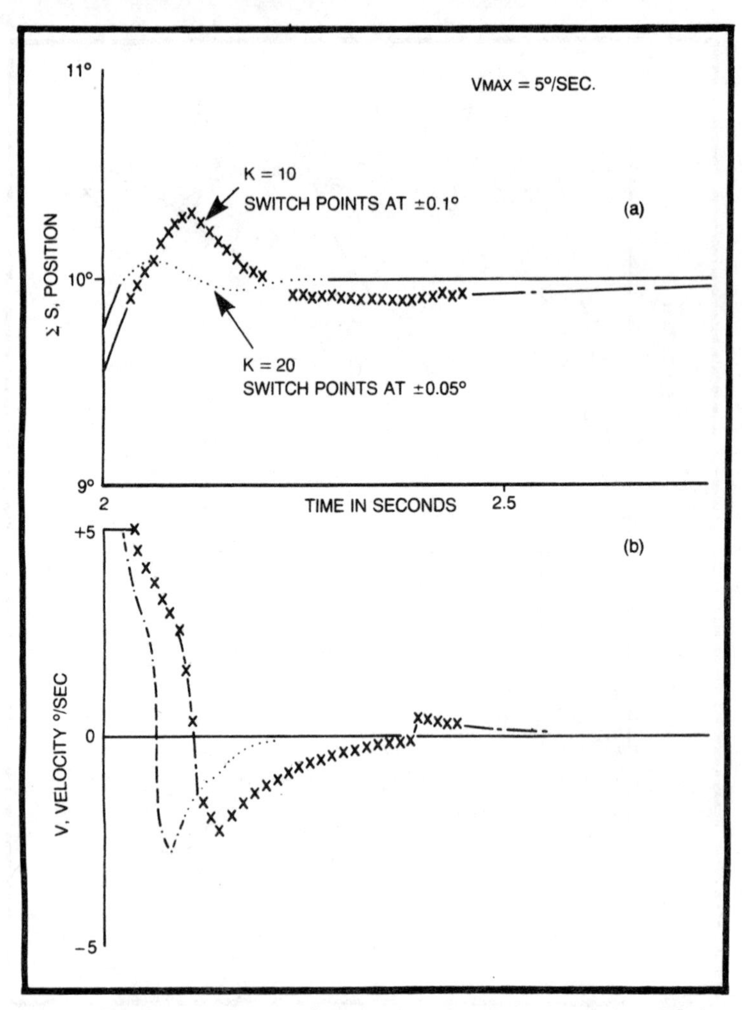

Fig. 1-9. These curves for a nonlinear system show the response to a 10° step input error.

motor is trying to get the load back to the target position it must overcome both the inertia and the damping.

From the curves we see that the system is critically damped somewhere between km = 3 and km = 4. With lower values of km the system is badly underdamped and actually spends a great deal of time trying to run in the wrong direction.

Now there are several points which should be noted. The peak system velocity for the truly linear system varies linearly with the size of the step function input. If the initial step input had been set at

20° both the motor speed and the motor acceleration would have been twice as great. In any practical system there is a very real limit on how much voltage (and current) a motor will handle without burning out and how much torque the motor will handle without twisting the shaft off. Any real system must, then, be limited in torque and acceleration. In practical systems, the linear performance is restricted to very small input errors. Beyond this point the system limits and becomes nonlinear or saturated. A second limitation stems from the fact that a real motor will not develop enough torque to overcome its own internal friction at very low voltages. For this reason the linear servo is not truly linear in the small-error region. This accounts for the note earlier about "semilinear" systems. Because of the small-error limitation a real semilinear system is usually designed to produce very large error-voltage for small errors and consequently the system saturates at any significant departure from the small-error condition.

Another logical question concerns the effect of changing the damping. Suppose we wanted to increase the value of km to 2; what level of damping would be required to restore stability? We find that it would be necessary to increase the value of K_2 to $\sqrt{0.5}$, or 0.707, in order to restore the critically damped condition. In this direction the system would have so little *coast* and would take so long to get up to speed that it would require 18 seconds rather than the 9 required on the example of Fig. 1-8 to reduce the input error to 0.1°. In the other direction, if we were to increase the value of km to 16 (thereby making the system only one-fourth as fast on acceleration) we would still have to hold K_2 at a level near 0.33 for critical damping. This would be a relatively free coasting system, but because of the sluggishness of the acceleration it would require 22 seconds to reduce the input error for a 10° step to less than 0.1°.

Generally speaking, if we halve the acceleration on a stable linear system we must nearly reduce the damping to the square root. Conversely if we double the acceleration we must nearly square the damping. The stable system will be just about the fastest when the deceleration due to damping is close to the acceleration from the drive.

NONLINEAR SYSTEMS

Next let us perform a competitive numerical examination of a roughly comparable *nonlinear system*. We shall again have the system jump from an initial input error of 10°. In this case we shall assume that the unit will accelerate along the curve given by:

$$V = V_{max}(1-e^{-kt}) \quad (1.5)$$

This continues until it hits the switch point. This is the same curve followed by the current in an inductance-resistance circuit or the voltage across a capacitor in resistance-capacitance series circuit. This is actually a pretty good approximation of a typical motor/load speed curve. It implies that the speed will reach 63% of full speed when $kt = 1$, 87% of full speed when $kt = 2$, and 95% of full speed when $kt = 3$.

In between switch points A and B we shall assume that we have removed the power and have short circuited the motor so that it acts as a dynamic brake (more about dynamic braking later). In this case we will presume that the motor follows a discharge curve such that:

$$V = V_{n-1}\,(e^{-kt}) \quad (1.6)$$

On this curve V has fallen to 37%, 14%, and 5% of the value with which it crossed the switch point at $kt = 1$, 2, and 3.

When the driven unit hits the overshoot switch point we shall assume that:

$$V = V_{n-1} - V_{max}\,(1-e^{-kt}) \text{ For } V_{n-1} > 0 \quad (1.7)$$

It may be seen that the use of the motor of both driving and braking provides an effective doubling up on the capabilities. When V falls to zero in the overshoot, the calculation reverts to equation (1.5).

For the example we have shown in Fig. 1-9 the case for $k = 10$ and $k = 20$. Note that the time scale has been magnified compared to the linear example. Only portions of the $k = 10$ curve are shown to avoid confusion. Compared to the linear system, it is apparent that the system is much faster, even though the motor did not get up to as high a speed as it did for the same jump in the linear system. This is because the lack of continuous damping allows the motor to be running at full speed about a half second after starting, and it continues to run full tilt until the system is within 0.1° of the goal. This is not to say that the operating procedure is easy on the machinery. The high speed reversal at 2.3 seconds places very large stresses upon the motor brushes and shaft if the system is slightly underdamped, not to mention the large torque loads placed upon the shafts and gears. It might be necessary to limit the motor current for a variety of reasons to prevent premature mechanical failures. However, the point remains that a given motor can reduce the input error much faster in a bang-bang or nonlinear (*saturating*) system.

It is important to note that this system must be capable of very rapid switching. The calculation made no allowance for switching time and the motor was considered to be either driving or braking. For the parameters used it was necessary to consider time increments of 0.01 seconds, and a switching time as large as 0.001 second would change the results considerably. If this system were built of practical components, it would be necessary to use solid state switching components to obtain the calculated performance since ordinary relays are rather difficult to drive open or closed in less than about 0.011 second without obtaining excessive bounce. Small vacuum relays can be obtained with bounceless switching times on the order of a millisecond; however these are very expensive components.

In general, the nonlinear system will be critically damped if the system velocity is reduced to about half of the maximum speed by the damping during the coast across the deadband.

This, of course, leaves us with the question of determining the damping rate in a practical system. While this is not too difficult to do in a well equipped laboratory it does require some fairly fancy equipment. However, a relatively simple method is available. If the system is allowed to run at full speed and then tripped into the damped condition and the time and distance to coast to a halt are measured, the damping constants can be determined. For free coasting systems:

$$\frac{\text{Time to halt (in seconds)}}{4.6} \cong \text{kt}$$

For systems with higher friction that jerk to a halt:

$$\frac{\text{Time to halt (in seconds)}}{3.5} \cong \text{kt}$$

The velocity of the system will be halved in about half of the total coasting distance when kt = 0.7. While the determination of kt is not terribly accurate from the above, the distance rule generally works out to be accurate within about 5%—except in very sticky systems.

Table 1-1 can be used to find other damping values. The figures in the table are printed in triads showing:

- Normalized velocity, V/V_{max}
- Damping time, kt
- Cumulative position, $(k \Sigma S)/V_{max}$

Table 1-1. Decay Curve For $V = V_{max} (e^{-kt})$.

0.9139 0.1000 0.0956	0.3715 1.0000 0.6352	0.1510 1.9000 0.8546	0.0614 2.8000 0.9438	0.0249 3.7000 0.9801
0.8269 0.2000 0.1821	0.3362 1.1000 0.6704	0.1366 2.0000 0.8689	0.0555 2.9000 0.9497	0.0225 3.8000 0.9825
0.7482 0.3000 0.2604	0.3042 1.2000 0.7023	0.1236 2.1000 0.8819	0.0502 3.0000 0.9549	0.0204 3.9000 0.9846
0.6770 0.4000 0.3313	0.2752 1.3000 0.7311	0.1119 2.2000 0.8936	0.0455 3.1000 0.9597	0.0184 4.0000 0.9866
0.6126 0.5000 0.3954	0.2490 1.4000 0.7571	0.1012 2.3000 0.9042	0.0411 3.2000 0.9640	0.0167 4.1000 0.9883
0.5543 0.6000 0.4534	0.2253 1.5000 0.7807	0.0916 2.4000 0.9138	0.0372 3.3000 0.9679	0.0151 4.2000 0.9899
0.5015 0.7000 0.5059	0.2039 1.6000 0.8021	0.0829 2.5000 0.9225	0.0337 3.4000 0.9714	0.0137 4.3000 0.9913
0.4538 0.8000 0.5534	0.1845 1.7000 0.8214	0.0750 2.6000 0.9303	0.0305 3.5000 0.9746	0.0124 4.4000 0.9926
0.4106 0.9000 0.5956	0.1669 1.8000 0.8388	0.0678 2.7000 0.9374	0.0275 3.6000 0.9775	0.0112 4.5000 0.9938
				0.0101 4.6000 0.9949

Key	Example
V/V_{max} Velocity	0.9139
time (kt)	0.1000
Distance $\frac{K\Sigma S}{V_{max}}$	0.0956

In terms of our previous example, where $V_{max} = 5°/sec$, the last triad entry in our table gives us a $V/V_{max} = 0.0101$. Therefore:

$$V = 0.0101 \times 5°/sec = .051°/sec$$

Actually, only a very free rolling system would continue to coast to this slow a velocity. A system with more friction would have stopped at something more like 5% of V_{max}.

The above velocity would have obtained for a damping-time product $kt = 4.6$. For the example with $k = 10$ this would be at a time:

$$t = 4.6/10 = 0.46 \text{ seconds}$$

From the time of the braking application the driven gear would have coasted through an angle:

$$\frac{k \Sigma S}{V_{max}} = 0.9949$$

Therefore:

$$\Sigma S = (0.9949 \times 5°)/10 = 0.4975°$$

Now the half velocity point is to be found at $kt = 0.7$ with $V = 0.5015 V_{max}$. At this point we see that:

$$\Sigma S = (0.5059 \times 5°)/10 = 0.2530°$$

This is equal to the whole deadband, so switch points A and B would be ±0.1265° for a stable, critically damped system. This is just a bit broader than our example with the switch points set at ±0.1° which may be seen from the curve of Fig. 1-9(a) to have been just a bit under damped. As noted earlier, the damping performance could be improved by making the value of V_{max} about 20% lower, or by slightly widening the deadband.

In subsequent chapters we shall see some of the factors of physical motors and equipment that will affect our choices in the trade offs involved in a usable and stable system.

2

Solid-State Switches

The most fundamental part of nearly all electrical controls is the switch or switches used to regulate the flow of current in the circuit. These switches may be manually operated mechanical devices or electrically operated mechanical devices, as they are in the oldest controls. In a somewhat more modern control, they may be thermionic triode or multigrid tubes, magnetrons, ignitrons, thyratrons, or saturable reactors. In the most modern type of controls, most of these older control systems have given way to solid-state switches and amplifiers. The use of solid-state switching in applications requiring substantial power handling capabilities is relatively recent and is perhaps worth a little explanation.

DESIRABLE SWITCH PROPERTIES

To begin with, let us consider the question of the desirable properties of a switch; what do we have to look for? Some of the simplest properties can be summarized as follows:

- **Voltage rating.** How much voltage will the switch safely handle? In the open condition there is some maximum voltage that the switch will "stand off" without arcing or otherwise failing.
- **Current handling capability.** The amount of current the switch will carry without overheating.
- **Current interrupt capability.** The amount of current that the switch can safely interrupt without being de-

stroyed. This will later be shown to be a function of the type of load and current.

- **Surge current rating.** The amount of current that can be handled on a brief surge. The surge may occur when starting a motor or perhaps lighting an incandescent lamp load.

For switches that are operated by means other than simple manual switching, a few other parameters must be known:

- **Switching time.** How long after the open/close signal does the switch respond?
- **Rise/fall time.** What is the transition period required for the current to rise/fall.

While these are by no means the only items worth considering, they shall suffice for the time being. We are concerned here with solid-state switching devices; however, these devices frequently are driven by or eventually drive electromechanical switches. The properties of these solid-state switches must be considered.

SEMICONDUCTOR BACKGROUND

The use of semiconductor materials in junctions for rectification dates back at least to the experiments of Hertz and Boze in the 1860s. The Branley coherer and cat-whisker-galena detector were all in use long before the turn of the century in radio communications. These and the electrolytic and the copper oxide rectifiers all represent crude "solid state" devices which antedated the Fleming valve (thermionic diode) detector by many years. However, the principles of semiconductor rectification were but poorly understood and a usable semiconductor amplifying device awaited the invention of the transistor in 1948 by John Bardeen, Walter H. Brattain and William Shockley. It was nearly a decade later before transistors capable of switching a few amperes and holding off a few tens of volts began to appear on the market, and at that, the devices were too expensive for most applications. For this reason, commercial and comsumer use of solid-state controls did not begin to appear before the early 1960s, and these were mainly silicon controlled rectifier devices (SCRs). By comparison, a current General Electric catalog carries a page on inverter SCRs in the 700 to 1000 ampere range with voltage ratings of 500 to 1300 volts.

The D56W transistor is rated at 1400 volts on the collector and at 5 amperes of collector current (absolute maximum ratings) along with a dissipation rating of 50 watts at a case temperature of 70°C. It is fairly obvious that the art has advanced to the point where very

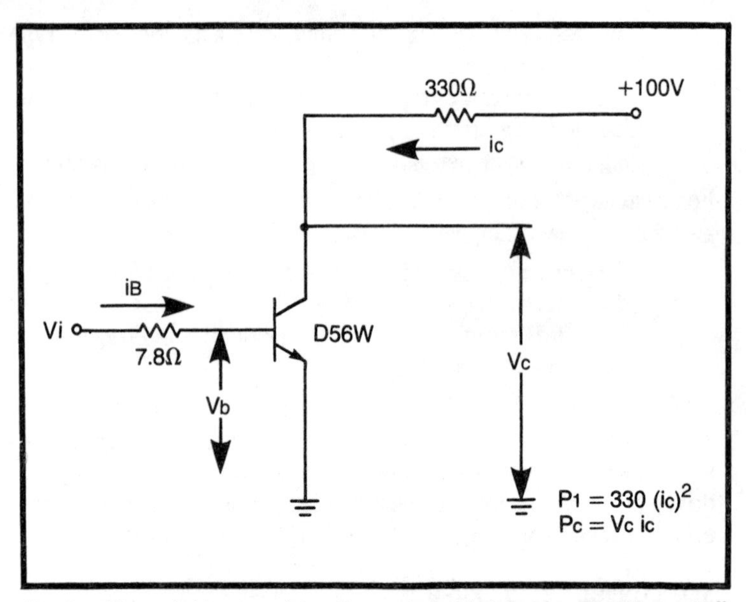

Fig. 2-1. The switching transistor. Parameters tabulated for saturation, cutoff, and linear biasing show that maximum dissipation occurs with the linear biasing. Values for the parameters were taken from catalog data sheets for the General Electric D56W transistor.

substantial currents and voltages can be handled by solid-state devices.

The emphasis to this point has been on the subject of switching and switching characteristics rather than on amplification. Also, the rating on that D56W transistor will seem a bit odd to someone used to building linear hi fi amplifiers. After all, 1400 volts times 5 amperes is 7,000 watts, not the 50 that the transistor is rated for. Let's take a look at the large-signal characteristics of this device to see how we might be able to use it sensibly.

The D56W is an NPN high-voltage mesa power transistor designed to operate in monochrome and color TV horizontal deflection circuits directly from the rectified 117 V ac line. In this case let us consider what we can do with the unit to control the current through a 330 ohm resistor. The circuit is shown in Fig. 2-1, with the voltage and current labels defined. You will note that we have backed down from the absolute maximum ratings of the device. At an initial temperature of 25° C for the case we obtain the data shown in Fig. 2-1 from the unit data sheets. It would seem from the data that the device should well be able to handle either the on, or saturated condition, or the off, or cutoff condition. The linear condition, however, would destroy the unit in a matter of seconds. As a matter of fact, with a collector voltage of 100 volts continuously, the unit is rated as safely handling only 0.18 amperes, or 18 watts. One of the catalog curves indicates that the safe operating region for any current should not be extended above about 600 volts in a dc condition. If the supply voltage were reduced to 600 volts and the load resistance reduced to 197 ohms, the unit would be operating entirely within the safe operating region when either on or off and would be switching 1770 watts, provided, of course, that the switching transient power did not overrate the unit and that the collector temperature could be held below 70° C by the heat sinking.

Now obviously, in order to pass from $V_i = 0$ to $V_i = 5$ volts, we must pass through the linear portion of the curve. The time taken for the voltage to fall from the open circuit level of 600 volts to the on level of 10 volts is given as 1 microsecond, maximum, for the transistor. The maximum instantaneous dissipation in the transistor will take place when the value of V_c is approximately half of the supply voltage and the current is half of the saturated current. This works out to be 300 V × 1.5 A = 450 watts. Now during the time that the voltage is falling, the current is rising, and the average power in the transistor collector works out to be about 68 percent of the peak power, when averaged over the switching time; therefore, the total transient pulse in the transistor is:

$$450 \text{ watts} \times (0.68 \times 10^{-6} \text{ sec.}) = 3.06 \times 10^{-4} \text{ watt seconds, or joules,} \qquad (2.1)$$

per switch transition. This is in addition to the 30 watts lost during the on condition due to the forward drop. If we assume that the device is operating in a square wave condition the 30 watts will exist for only half of the time, giving 15 watts average on-loss. If we further allow an additional 5 watts for switching transient power, which occurs twice per cycle, we find that we can switch from on to off and back

$$\frac{5}{2 \times 3.06 \times 10^{-4}} = 8170 \text{ cycles/sec.} \qquad (2.2)$$

From the above, we can see several points which bear upon the nature of solid state motor controls using transistors:

- A transistor can switch power loads which are in order of magnitude larger than can be handled in linear amplification.
- The forward-drop losses in a transistor are not negligible. Heat sinking provision is generally required.

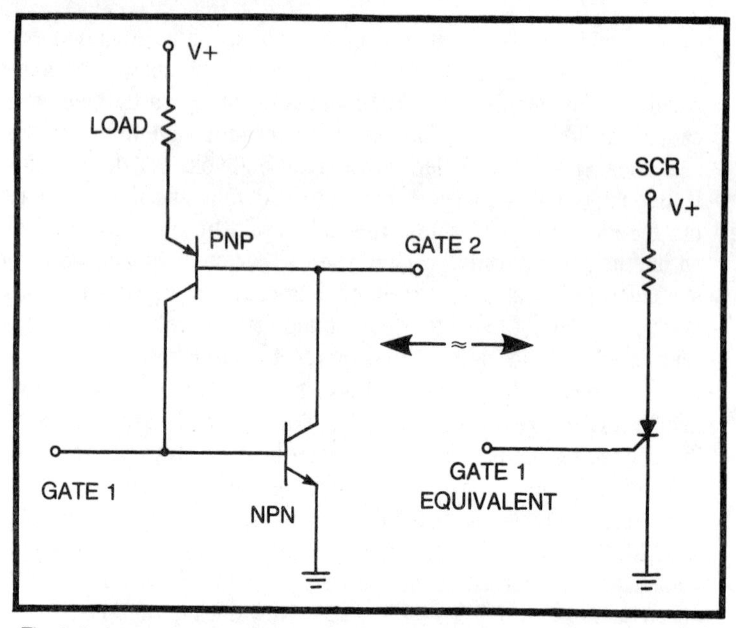

Fig. 2-2. A two-transistor analog for the silicon controlled rectifier. When V+ is initially applied, both transistors are cut off, and no current flows. Biasing gate 1 toward V+ or gate 2 toward ground will send both transistors into a latched, saturated state, and nearly all of V+ will appear across the load.

- Full advantage of the rapid on/off switching should be taken to minimize device dissipation and heating.

Historically, the silicon controlled rectifier was developed for high power, high voltage applications considerably before the high voltage transistor and was introduced into the commercial control market in 1957 by General Electric. Figure 2-2 shows the PNPN two-transistor analog of the SCR. A discussion of the detailed operation of the SCR is beyond the scope of this text, but a comprehensive description of the function and limitations of the SCR and allied devices is found in General Electric's SCR manual.

In later sections we shall deal with SCR and Triac type controllers; however, here we are interested principally in the comparison with the transistor switch. Figure 2-3 illustrates a direct comparison of the SCR with the transistor switch shown in Fig. 2-1. SCRs are available in much higher ratings than bipolar transistors. For our purposes, however, we shall consider the GE type C122 which has a rating roughly comparable to the D56W transistor.

Fundamentally, the SCR behaves like a diode rectifier that can be triggered into conduction by drawing current from the gate. In the two-transistor analogy of Fig. 2-2 this would be accomplished by applying a positive voltage to the base of the NPN transistor or a negative voltage to the base of the PNP transistor (in an ordinary SCR only the NPN base is brought out of the case). Once one of the transistors is pulled into conduction, notice that it pulls the second unit into conduction, and that each tends to keep the other latched on in a saturated state. In an ordinary SCR this ends the control obtainable from the gate, and the device stays turned on until the current is interrupted. Because of the diode nature of the SCR the turnoff in ac supply applications can be accomplished by the normal current reversal. In dc applications, however, a separate commutating mechanism must be supplied. The dc circuit of Fig. 2-4 illustrates the use of shunt and series switching to accomplish this end.

For the circuit of Fig. 2-3, we obtain from the catalog data sheets the following:

i_G	V_G	P_A	V_T	i_L	P_L
0.04A dc	2V dc	2W	+1.5V −400V	1.59A avg	500W

The action is not really the same as with the transistor since the device is halfwave rectifying the current through the load; however,

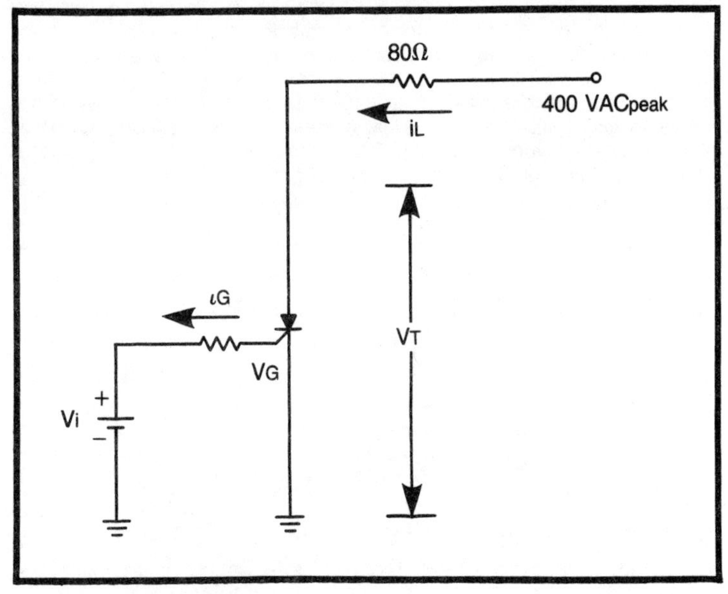

Fig. 2-3. The SCR switch.

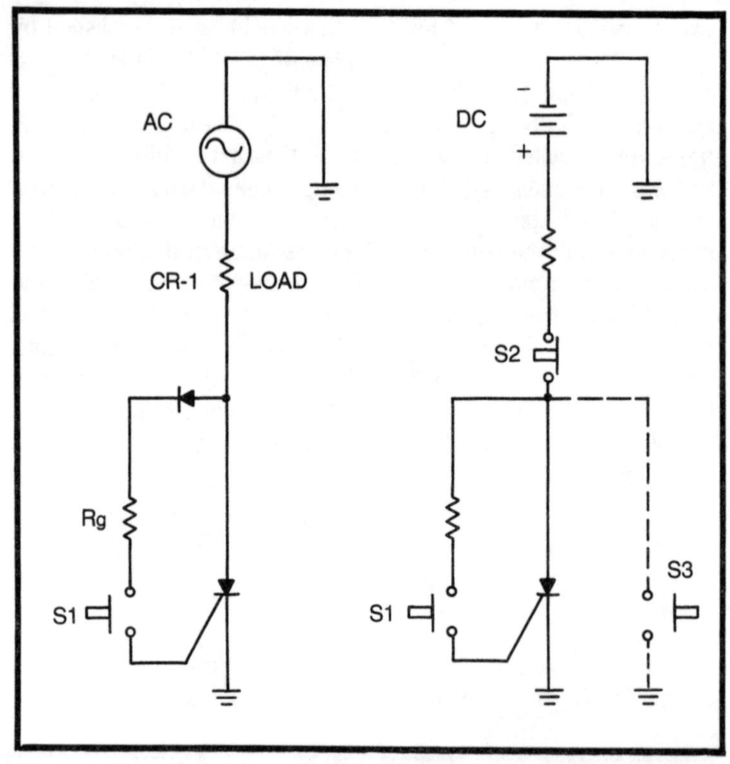

Fig. 2-4. The basic SCR switching circuits for commutation, or turn-off, mechanisms. In the ac case the current direction reversal every half cycle serves to shut off the switch current. Diode CR-1 protects the SCR gate from high negative potentials. Resistor R_g limits the gate current to safe values. In both cases, pushing switch S1 turns the load current on. In the dc case the current will continue to flow until interrupted by S2 or shunted out of the SCR by S3. Note that S2 (or S3) must bear the brunt of interrupting the load current.

certain comparisons with the transistor are possible. We may see that the SCR requires considerably less drive current than the transistor, and that in the forward direction the voltage drop at the anode is considerably lower. This leads to a lower device dissipation. For this reason, it is possible to package the C122 in a JEDEC TO-220AB power tab case whereas the D54W is packaged in the larger and more expensive TO-3.

It is obviously somewhat easier to use the SCR in ac applications because of the self-commutating action, but SCRs have also been used in dc applications—despite the commutating difficulties—because of their large voltage and current capabilities. Fairly recently, though, the development of high voltage transistors combined with the reduction in prices has reduced this advantage considerably in dc applications. Also, the rather extensive measures required to commutate an SCR have begun to throw the net commercial advantage back toward the transistor for dc applications below a few hundred watts.

SURGES AND INTERRUPTS

We had noted earlier that the rating of a switch was characterized by the size of the *surge* that it could handle and also by its *interrupt capability*. Let's look at these properties and see how they arise.

In my laboratory there is a drill press with a 1/3 horsepower induction motor. It also has a lamp to illuminate the work. When the drill press is working very hard the motor can draw up to about 6 amperes for what would seem to be 720 watts which is actually 0.97 horsepower (1 horsepower = 550 ft. lb/sec = 745.7 watts). Most of the time, however, it loafs along at about 5 amperes which seems to work out to 0.8 horsepower. That doesn't seem to track the "1/3 horsepower" on the label too well. The label on the lamp says 100 watts and it seems to live up to its rating a little closer than the motor since the lamp draws 0.82 amperes which works out to be 98.6 watts. However, if you take a transformer, a voltmeter, and an ammeter you will find that at 3.58 volts, the motor draws 0.57 ampere, and the lamp will draw 0.32 ampere. The lamp is dark and cool and the motor is stationary. This works out to an impedance of 6.28 ohms for the motor and 11.1 ohms for the lamp. When a dc ohmmeter is used on the items the lamp checks out at 11.0 ohms and the motor checks out at 1.3 ohms.

Without getting into the vector algebra (which will be briefly reviewed in Chapter 4), we see that the cold lamp is essentially a pure resistance. When hung across the 120 volt line it will initially draw a current of 120V/11.1 ohms = 10.8 amperes. This rapidly falls to the 0.82 amperes that the lamp draws in the hot condition. The *surge* will last for only a fraction of a second.

The motor is a horse of another color. Since the dc resistance and the impedance disagree rather sizeably we conclude that the motor must be largely reactive when it is stopped and in fact looks like a 1.3 ohm resistor in series with a 0.0163 henry inductor. In complex notation this is 1.3 + j6.14 ohms. In polar notation, this would be 6.28/78°. When line voltage is first applied the motor will draw a current of 120 V/6.28 ohms = 19.1 amperes at a phase angle of 78°. The decrease in the motor *starting surge* will be a good deal slower than the lamp surge. On the drill press, the motor will accelerate to 1400 rpm (where the starter drops out) in about 3 seconds. Just before this point the current will be down to about 12 amperes and the starter dropout will account for another 3 amperes. The motor will approach the rated 1750 rpm no load condition in another 3 seconds, or so, and the current will be down to the 5 ampere "cruising" level. If the same motor had been trying to accelerate a tub full of wet wash in the spin cycle of a washer, the whole process could have stretched out to 15 seconds or more.

From the catalog data on the C122 SCR we see that a pair of these units (to get both halves of the sine wave) would tolerate 50 amperes peak or 35 amperes rms for 60 cycles (or 1 second) after having run at full load. This would suffice to *start* 3 or 4 of the lamps even though the SCRs would be capable of handling 8 lamps in the hot condition.

Extrapolating the curve in the catalog gives 30 amperes for 5 seconds and 28 amperes for 15 seconds, so the pair of C122s should suffice to handle the motor even on a slow start. The units would in fact handle both the motor and a single lamp; however, the switch would by no means be over designed and should be carefully checked. Good engineering practice calls for a somewhat more conservative design.

Returning to the original discussion of the ratings, the current and voltage monitoring setup in Fig. 2-5 resolves some of the reasons for the discrepancies between the rated power and the 5 ampere current. An oscillogram reveals that the waveforms for both the lamp current and the motor current are distorted from the sine wave condition. The motor current is lagging the voltage by about 84.5° and the distorted lamp current leads the voltage by about 17°. In the case of the lamp, the waveform distortion is due to the change in filament temperature through the cycle. This, in turn, changes the lamp resistance and brings about the observed phase shift.

In the case of the motor, however, the very large phase shift is brought about by the inductive reactance of the motor. The majority of the current is simply going into charging the magnetic field. The motor could be characterized as a 250 ohm resistor in parallel with a 0.064 henry inductor. The presence of this large inductance has certain consequences on the interrupt rating of the switch.

A magnetized inductor has a certain amount of energy stored in it. This is kinetic energy and is analogous to the kinetic energy stored in a moving mass such as an automobile, a hammer head, or a bullet. If you attempt to stop the moving object rapidly, large forces are developed. For the inductor, the voltage developed by a change in current is given by:

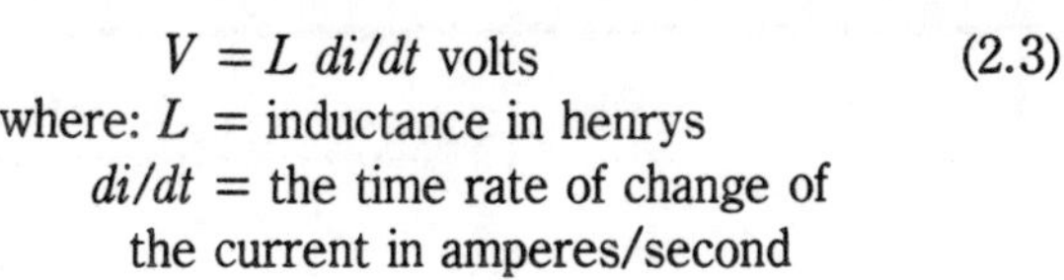

$$V = L\ di/dt \text{ volts} \tag{2.3}$$

where: L = inductance in henrys

di/dt = the time rate of change of the current in amperes/second

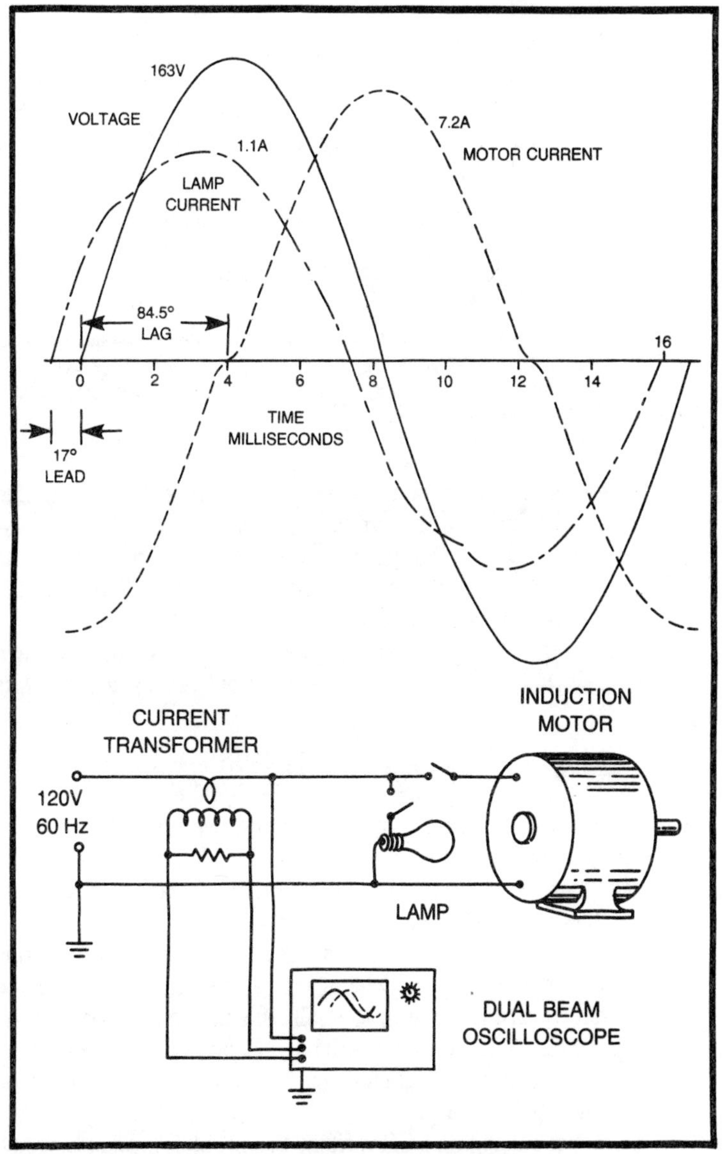

Fig. 2-5. Current and voltage monitoring of an induction motor.

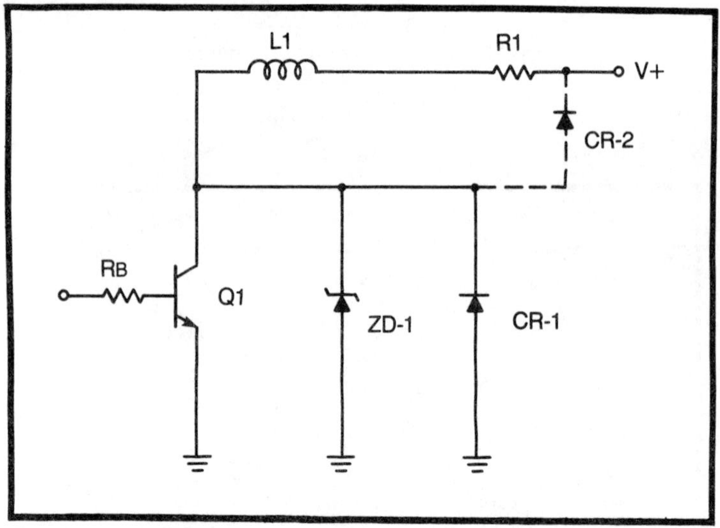

Fig. 2-6. Diode CR1 or CR2 provides freewheeling protection to Q1 by limiting the turnoff rate of the current. Voltage regulating diode ZD-1 provides forward voltage transient limitation.

Now let us consider the possibility of switching off the current in the motor at the peak of the current cycle using the D54W transistor. As we noted on the oscillogram, the peak current in the motor—with the motor at speed and lightly loaded—is 7.2 amperes. We had also noted earlier that the D54W is capable of switching a current from the saturated level to essentially zero in not more than 1 μsec. The time rate of change of the current is therefore not *less* than:

$$di/dt \geqslant -7.2 \text{ amperes}/10^{-6} \text{ sec}$$
$$\geqslant -7.2 \times 10^{6} \text{ amperes/sec}$$

Now for a motor inductance of 0.064 henrys we find that the *interrupt voltage surge* would be not less than –461 kV!!! Obviously this is far higher than the transistor can withstand, especially in view of the fact that the surge is reversed in polarity.

In the mechanical switch actually used with the drill press, this energy is dissipated in an arc which occurs when the unit is stopped well up on the current curve. Eventually this arcing will wear out the switch contacts, and the switch will have to be replaced. However, in a solid state switch, a single internal breakdown of this magnitude will destroy a transistor.

If we were using an SCR switch, the problem probably would not arise since an SCR stays on until the current falls to zero for the ac load excitation. After the gate excitation is removed the SCR will simply stay saturated until the next zero crossing of the current, and a very "quiet" interruption of the current occurs. Actually, in an SCR a small overshoot actually will take place while the carriers are swept up.

If a transistor switch is used, some means must be provided to absorb or reduce this inductive kick, or "hammer blow," effect. Such mechanisms are shown in Figs. 2-6 and 2-7. In the first a *free-wheeling diode* has been added so that the kinetic energy stored in the inductor can be slowly dissipated in $R1$. This action is analogous to taking a car out of gear and waiting for it to coast to a halt. The expression for the rate of decay of the current in the inductor is:

$$i = i_0 \left(e^{-Rt/L}\right) \qquad (2.4)$$

where:

i = current at some instant after opening of the switch
i_0 = current at the instant the switch was opened
R = the loss resistance of the inductor in ohms
L = the inductance in henrys
t = time after opening of the switch in seconds

It may be seen that this is the same as equation 1.6 with suitable substitutions, and the solutions of Table 1-1 may be employed.

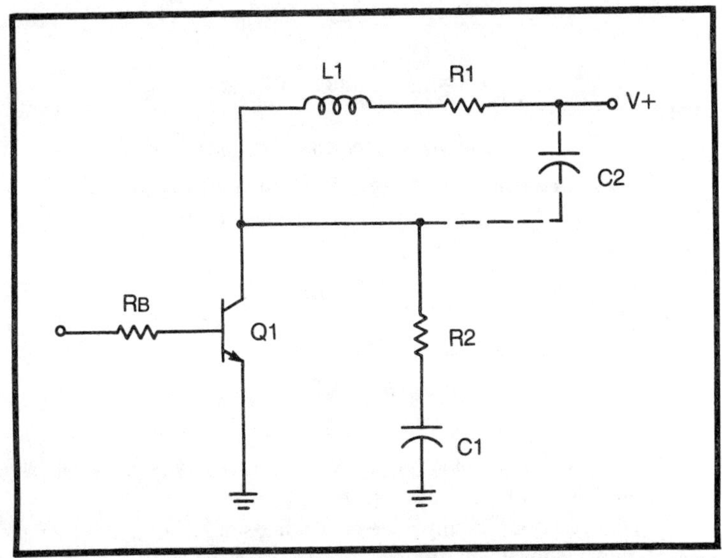

Fig. 2-7. Snubber capacitor C1 or C2 provides resonant voltage protection to Q1. Resistor R2 limits the turn-on surge through Q1.

Depending on the size of R, the decay can be quite fast. The R term is analogous to the braking or friction effect in the automobile. This resistance will always be present in some measure in any real solenoid or motor unless the system is a superconducting type.

The technique shown in Fig. 2-7 is somewhat different in that a *snubber capacitor* has been added to the circuit. This is analogous to having the car coast into a spring. After a little thought, it should be obvious that eventually, with the switch turned off, the collector voltage of Q1 should settle to the level of $V+$. However, we are interested here in the extent to which the capacitor limits the collector voltage to safe values. It turns out that this case is mathematically analogous to the linear system and may be represented with very elegant differential equations or with the summation expressions of equations 1.1 through 1.4. However, there is a very easy way to calculate the not-to-exceed voltage developed across the capacitor. The energy stored in the magnetic field of the inductor is given by:

$$W = \frac{Li^2}{2} \text{ joules, or watt-seconds.} \qquad (2.5)$$

Now the expression for the energy stored in a capacitor is:

$$W = \frac{C\,V^2}{2} \text{ joules} \qquad (2.6)$$

where:

C = capacitance in farads

V = the voltage across the capacitor

If we were to assume that all of the energy in the inductor would eventually wind up in the capacitor (at least temporarily), then we could equate 2.5 and 2.6, and obtain:

$$CV^2 = Li^2 \qquad (2.7)$$

and thus:

$$V = i\sqrt{\frac{L}{C}} \text{ volts} \qquad (2.8)$$

This is the maximum value that V can obtain. With any practical value of R the value of V will always turn out lower.

In our previous example we used values of 7.2 amperes for i and 0.064 henrys for L. If we wanted to determine the value of C required to keep V below 600 volts, we find, from equation 2.7,

$$C = \frac{0.064 \times (7.2)^2}{(600)^2}$$

$$C = 9.21 \times 10^{-6} \text{ farads}$$
$$\text{or } 9\ \mu F$$

The actual voltage across the capacitor as a function of time could follow any of the curves shown in Fig. 1-8, depending upon the relationship of R, L and C. For cases where:

$$\left(\frac{R^2}{2L}\right) \geqslant \frac{1}{LC} \tag{2.9}$$

the response is heavily damped and would follow the curves of $km \geqslant 4$. For the converse case where

$$\left(\frac{R^2}{2L}\right) < \frac{1}{LC} \tag{2.10}$$

the unit would be underdamped and follow curves for $km \leqslant 3$.

Put another way, if the car were coasting freely when it ran into the spring, we would expect it to rebound. In a similar manner this system will rebound and oscillate about the value of $V+$. In the very underdamped case, we would expect the circuit to *ring* or *oscillate* at a frequency of:

$$f = \frac{1}{2\,\pi\sqrt{\frac{L}{C}}} \text{ hertz} \tag{2.11}$$

Now note in particular that the voltage developed in the oscillation may be many times as large as the value of $v+$. This fact is taken advantage of in the Kettering ignition system which still is used in automobiles.

In the circuit of Fig. 2-7 a resistance, R2 has been added to C1 in order to limit the turn-on surge. Without this resistance, the current surge through Q1 would be very large.

In this chapter we have examined some of the principal properties which we must provide in a switch for use with either ac or dc circuits and have compared the advantages of transistors and SCRs in both. It should be noted that the inherent *zero current turnoff* of an SCR in an ac circuit is a great advantage, particularly where inductive loads are involved. In subsequent chapters we shall see that certain motor properties also enter into the switch parameters, and must be accounted for.

3

DC Motor Properties

Historically, the invention of the electric motor is generally credited to Michael Faraday. In 1821 Faraday invented the first crude electric motor. A stiff wire, pivoted at the top, hung down so that the bottom end was just immersed in a dish of mercury which rested upon one pole of a large magnet. When a large current passed through the wire, using the mercury and pivot as movable contacts, the wire whirled in a conical fashion. This does indeed demonstrate the principle of the electric motor, but it tends to get mercury all over the lab.

DEVELOPMENT OF ELECTRIC MACHINES

The development of the electric generator was closely tied to early experiments in electrical illumination. In 1802, Sir Humphrey Davy constructed an incandescent lamp using a thin platinum filament sealed in a glass bottle and heated to incandescence by an electric current. This lamp could be used in the explosive atmosphere of a coal mine (with firedamp present) in safety. Unfortunately the only way to produce electricity in those days was with chemical batteries which were expensive and messy, so the Davy electric safety lamp gave place to the Davy gas safety lamp invented in 1815. The electric illumination experiments continued, however, and by 1809 Davy had operated an immensely powerful arc light operating at 3 kV from a string of 2000 batteries.

The high pressure steam engine entered the picture with the development of a 200 lb/in.^{2} engine by Oliver Evans of Philadelphia in 1815 and by the 1820s most of the major laboratories were equipped with line shafts and slapping belts to power tools and experiments. The times were right for the development of a machine that could convert machine power to electricity. In 1831 Faraday demonstrated a dynamoelectric machine which could either produce electricity from mechanical rotation or absorb electricity and produce mechanical power.

The Faraday dynamo is what would be called today a homopolar machine, and it had to be turned far too fast to produce any significant voltage. Therefore, a machine invented in 1832 by a Frenchman, Hypolite Pixii, was to become the father of our modern motors. The Pixii dynamo was a heteropolar, permanent magnet machine with a wire wound core and an iron armature. This machine could be built to produce usable voltage and current levels at a few hundred rpm which could be derived from a steam engine and supported by the bearings of the day.

This machine, in simplified form, seems like a logical place to begin our examination of the properties of dc motors.

Figure 3-1 shows a schematic representation of the Pixii dynamo. This machine, like all other dynamos, is based upon the discoveries by Ampere and Oersted that a current-carrying wire experiences a physical force when immersed in a magnetic field and conversely that a voltage is induced in a wire which is moving through the magnetic field.

Since this is a book about controlling electric motors rather than about designing them, we can afford to be somewhat qualitative about the arithmetic and go only deep enough to see the properties we are interested in. For those wishing to delve deeper I would suggest:

> Alexander M. Langsdorf
> *Principles of Direct Current Machines*
> McGraw Hill, 1940

For the coil as shown in the inset which is rotating with a peripheral velocity V it may be seen that the velocity of the conductors across the magnetic lines is:

$$V \cos \alpha = \omega r \cos \alpha \tag{3.4}$$

It seems likely that the induced voltage should be directly proportional to the length of the wires (l) and the number of wires (nn) and since both sides of the coil are at work we should come up with a net expression that looks something like:

$$CEMF = V_g = 2\, nr\, \omega\, \beta\, l \cos \alpha \tag{3.5}$$

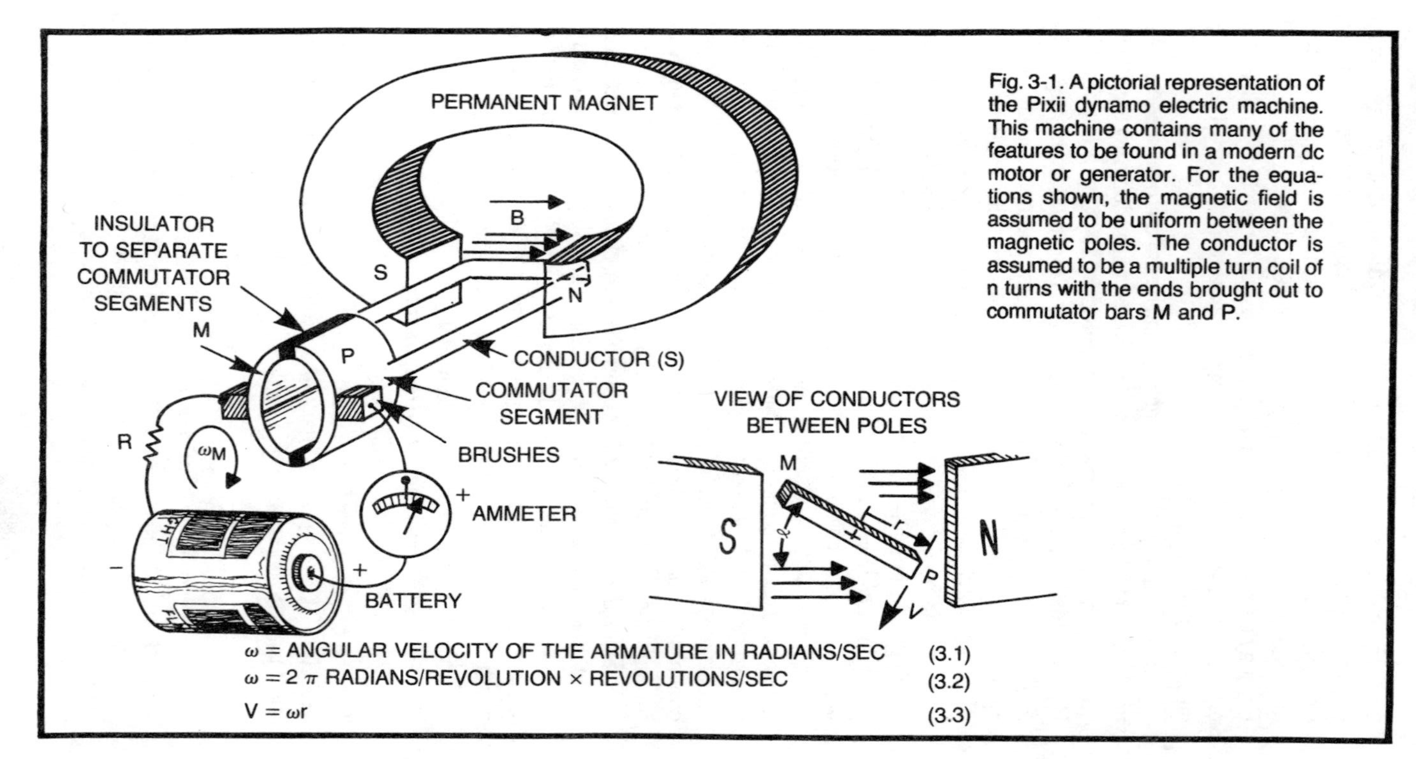

Fig. 3-1. A pictorial representation of the Pixii dynamo electric machine. This machine contains many of the features to be found in a modern dc motor or generator. For the equations shown, the magnetic field is assumed to be uniform between the magnetic poles. The conductor is assumed to be a multiple turn coil of n turns with the ends brought out to commutator bars M and P.

ω = ANGULAR VELOCITY OF THE ARMATURE IN RADIANS/SEC (3.1)

$\omega = 2\pi$ RADIANS/REVOLUTION × REVOLUTIONS/SEC (3.2)

$V = \omega r$ (3.3)

where:

$$\beta = \text{magnetic field strength}$$

$$\text{or } V_g = k\ \omega\beta \cos \alpha \qquad (3.6)$$

where:

$$k = 2\ nr\ l \qquad (3.7)$$

Now perhaps the first thing that we should note about expression 3-6 is that the presence of the cosine term implies that the voltage reverses polarity during the rotation at $\alpha = 90°$ and $270°$. In order to prevent this reversal Pixii invented the commutator $M - P$ so that the brushes could "swap hands" and keep the current flowing in the same direction on the average (although it looks like a full wave rectified sine wave). The truth of the matter is that, in the true sense of the word, the Pixii machine is an alternator and if the commutator were replaced with a set of slip rings the output would be ac. This is actually true of all electromagnetic dynamos with the exception of the homopolar family.

Now by a similar line of reasoning we find that the machine absorbs or delivers a torque given by:

$$\text{Torque} = k\ \beta\ i \cos \alpha \qquad (3.8)$$

where:

K is as defined in 3.8

It seems fairly reasonable that the torque ought to be proportional to current multiplied by the things needed to determine voltage.

Next, let's consider what happens when we connect the machine up to the battery. First of all, the machine is stationary and a current equal to the battery voltage divided by the lumped resistance R would flow. This develops a torque which causes the machine to accelerate, or raise, the value of ω up to some level. Now we see from equation 3.6 that the voltage generated by the machine is proportional to ω and it is in the direction to buck the flow of current from the battery. This in turn will tend to reduce the value of the current and the average torque until an equilibrium point is hit such that:

$$V_{\text{battery}} = ir + V_{\text{generator}} \qquad (3.9)$$

The term $V_{\text{generator}}$ is usually described as the COUNTER ELECTROMOTIVE FORCE OR CEMF. This counter EMF is present in all motors to some degree and is one of the properties which affects the motor performance in a major manner.

Suppose, for example, that the motor were up and running and that we connected another motor to it to make it move even faster than the equilibrium speed. Since the value of the counter EMF is proportional to speed we would find that a speed existed where it just equaled the battery voltage (on the average) and the current in equation 3.9 would have to go to zero. If the speed were increased still further, you can see from equation 3.9 that the direction of the current in the circuit would have to reverse and the machine would be charging the battery. Conversely, if an additional torque load were placed on the machine it would slow down slightly, which would cause the counter EMF to fall, and the machine could draw current from the battery. The existence of the counter EMF is the thing that permits any electric motor to deliver a useful amount of power.

Another factor of interest may be obtained from the equations. If you note equations 3.5 and 3.8 you will observe that increasing the magnetic field strength β simultaneously increases both the counter EMF and the torque, if other things are held constant. However, we see from equation 3.9 that increasing the counter EMF would decrease i. This winds up with a result that tends to come as a surprise to most. The net result of increasing $\vec{\beta}$ is that the motor has to slow down so that it can draw enough current.

Generally speaking, the value of β must be decreased in order to make the motor run faster and increased to make the motor run slower.

Streetcars and electric trains commonly used to employ field weakening as a means of speed control.

Another property of interest in the permanent magnet machine is the fact that a reversal of the current direction in the machine reverses the direction of the motor torque. As a matter of fact, this is the mechanical means by which the motor is able to extract power from the external drive when it is generating. The ability to reverse the direction of rotation of a PM motor by reversing the drive polarity is frequently used in servo mechanisms.

THE MODERN PERMANENT MAGNET MOTOR

The modern permanent magnet motor has benefited from a number of developments but the largest of these has been the development of Alnico *V* and the ceramic magnets. The very high βH product of these materials makes possible the use of short magnetic structures. This point is brought home in a very striking manner if you compare an old fashioned magneto intended for telephone or ignition purposes, with its inverted "U" array of iron magnets, with a modern version of the same machine. In the modern machine the cross section is round and is shrunk to a fraction of the older size. Typically, the field structure may be only half again the diameter of the armature.

Permanent magnet motors and generators are widely used today in fractional horsepower and servo applications where the size, weight, and efficiency advantage over shunt wound dc motors is desirable. The falling cost of the magnets has even brought these motors into use in automobile heater, windshield wiper, and auxiliary control (window lift, seat lift, etc.) applications.

A typical example of a low cost and very modern PM motor is the Siemens 990412052405. This motor is rated at 2.1 watts output at 60 rpm and 6 volts. It is encased in plastic and has a molded-on planetary gear train to reduce the rotor speed of some 14,500 rpm, nominal, to the 60 rpm on the output shaft. Overall, the unit is just slightly larger than a D size flashlight cell.

It is instructive to get a feel for the parameters of a device like this. The motor is a plain bearing or journal-bushing type and in addition has the high reduction ratio planetary gear train. Both of these factors contribute rather significantly to the mechanical losses. These losses can be evaluated in a rather simple set of experiments. The motor was measured with a dc ohmmeter and found to have an armature resistance of 3.7 ohms. It was then operated at 6 volts and the no-load current measured. From this the iR drop can be calculated and the CEMF obtained using equation 3.9. A known torque load was then added and the new current and CEMF calculated. The results were:

for $V = 6$ volts

	i	iR	CEMF	Speed
NO LOAD	0.095A	0.35V	5.65V	60 rpm
5.93 in. oz	0.15A	0.56V	5.45V	57.8 rpm

The speed may be measured or the speed ratio calculated from the CEMF using equation 3.6 and noting that:

$$\frac{\text{CEMF}_o}{\text{CEMF}_1} = \frac{\omega_o}{\omega_1} \tag{3.10}$$

The internal torques may be calculated using the fact that torque is proportional to current from equation 3.8.
Thus:

$$\frac{\text{Torque}_o}{\text{Torque}_o + \text{Torque}_{load}} = \frac{i_o}{i_1} \tag{3.11}$$

From this manipulation we find that the internal torque is approximately 10.24 inch ounces.

It can be seen from the preceding that the internal work done just to keep the bearings moving and churning the grease in the planetary gear train is fairly substantial. In our example we loaded the motor only slightly so that the speed reduction would not be too great. The friction losses are generally dependent upon speed in a more or less linear fashion and should be accounted for if wide speed ranges are considered.

It may be noted that the torque equation is dependent only upon current and not upon speed. The *locked rotor* torque could have been measured at a reduced voltage and a current of 0.095 amperes to determine the torque required to run the motor at 60 rpm with no external load. This technique is very effective with larger motors but is a little difficult to work out with a smaller motor, particularly with the gearhead.

Figure 3-2 illustrates a simple form of *Prony brake* which may be used to provide a measurable dynamic or static torque load. This unit simply operates by pinching and binding the bearing *which must be kept well lubricated*. Also watch out for the heating of the bearing. The bearing is dissipating the entire output of the motor and in larger sizes *it will get hot. The safety stop is an important feature of this unit.* The effort to swing the pendulum is maximum at $\theta = 90°$ and falls thereafter so that if the pendulum swings up near 90° it will latch on and begin rotating at the motor speed.

WARNING

When working with motor controls you are exposed not only to the normal electrical hazards but also to mechanical ones as well. Even a relatively small motor can throw pulleys or slam a Prony brake around fast enough to break or cut off fingers. A necktie, shirt, or shop apron wrapped around a lead screw can decapitate a man. Keep fingers, tools, oil cans, and clothing out of moving machinery.

The cos α term in equations 3.4 through 3.8 is actually replaced in a modern motor with something like:

$$A \cos \alpha + B \cos 3\alpha + C \cos 5\alpha \ldots,$$

However, this is of interest to us only in terms of locked rotor measurements; actually in the so-called locked rotor measurements on small motors, it is better to let the shaft slip around very slowly to account for this torque variation with angle.

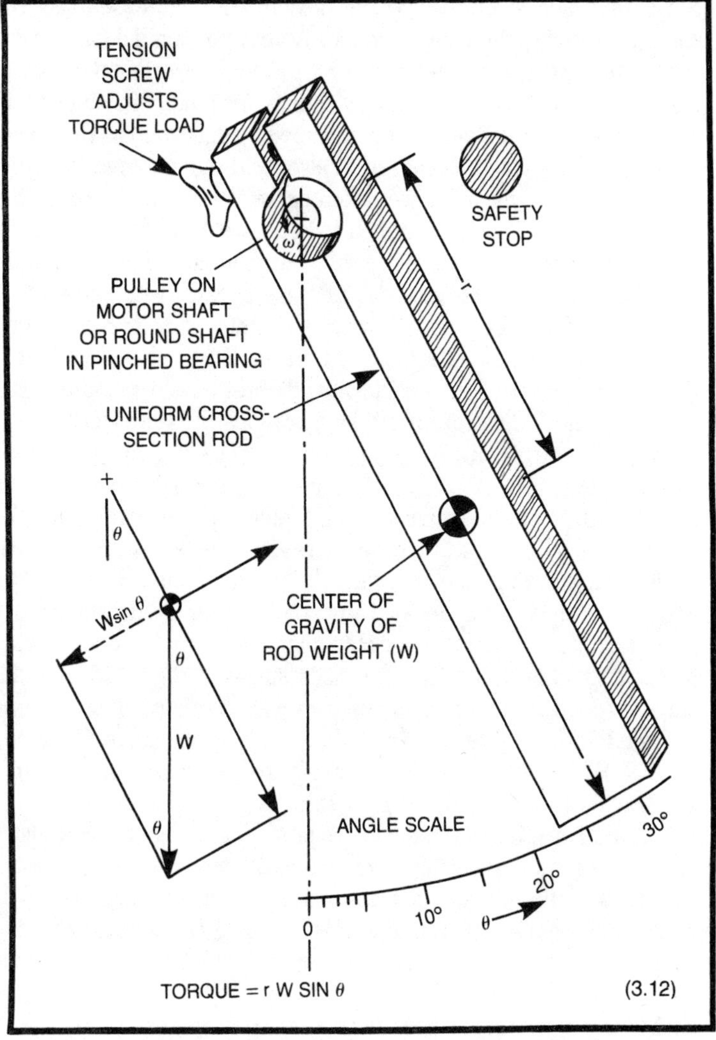

Fig. 3-2. A simple Prony brake apparatus for small motors. The vector diagram shows the resolution of the weight vector into normal and parallel components.

Most of the properties of the PM motor that are of interest to the control designer can be derived while considering the motor as a black box and taking a few simple electromechanical measurements.

THE SHUNT MOTOR

The shunt motor is very similar in properties to the PM motor if operated at a fixed voltage. At any given voltage the field will take on a constant field strength and the device will have essentially the same constant speed characteristic. A very substantial loading will reduce the speed only slightly. This type of motor is found in a great many automotive and marine applications for fans, windshield wipers, and other relatively constant speed applications. However, when the line voltage increases, it also tends to increase magnetic field strength β which, as we noted earlier, tends to decrease the motor speed. In practical shunt motors, the field winding is generally designed to run well up the saturation curve of the iron in the field so that the magnetic field does not increase linearly with line voltage and the motor speed does increase slightly with increasing line voltage; however, the effect is much smaller than with a PM motor.

Several things are significantly different between the shunt motor and the PM motor in servo applications. First and foremost is the fact that reversing the input polarity does not reverse motor direction. A review of equation 3.8 will show that reversing the sign of *both* β and i produces a canceling effect in the equation, and the torque does not reverse. In order to reverse the sense of rotation it is necessary to reverse one or the other, but not both. Shunt motors are very frequently controlled entirely with field current control apparatus. Several popular variable-speed drives in fractional horsepower and integral horsepower sizes supply a regulated constant current to the field, which is supplied with a reversing switch and provided with an intermittent voltage to the armature for speed control. With a constant field, the performance of the shunt motor is indistinguishable from the PM motor.

A particular danger found in this procedure is that an *open* failure in the field supply will bring about one of the most dramatic happenings to be found about an electrical engineering lab. If the field current fails in a shunt wound motor, the field falls toward zero but remains at a finite value because of remnant magnetism in the field iron. From equation 3.5 we observe that this tends to take the CEMF toward zero, and the armature current rises to try to pick up the motor speed. This effect is about optimized in a 5 horsepower motor that tries to attain the nearly infinite speed required for the CEMF to reduce the armature current. The audible effect is very much like a fire siren and is generally terminated either by the line fuse (less dramatic) or by the armature windings being expelled by centrifugal force, whereupon the subsequent arc will take care of the line fuse. The latter condition is the more spectacular.

Good engineering practice calls for the provision for some form of automatic shutdown on the separately excited shunt type motor to prevent overspeeding in the event of loss of field excitation.

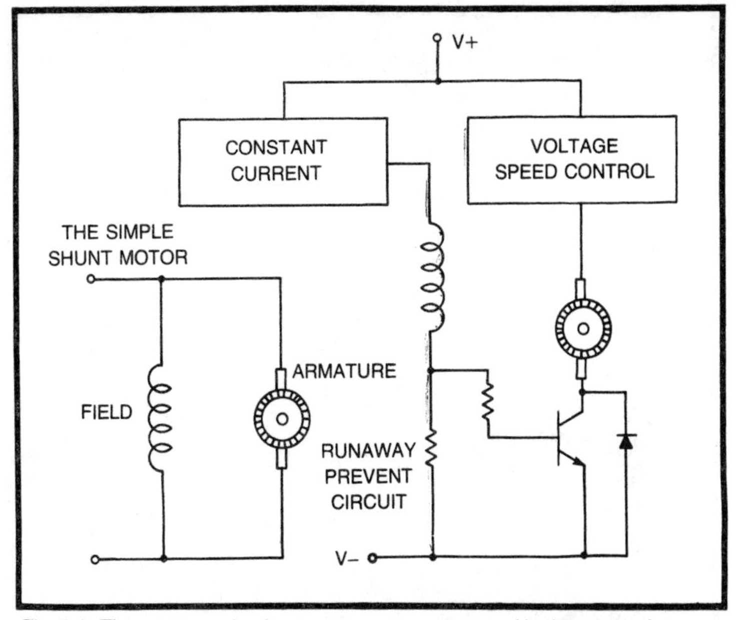

Fig. 3-3. The runaway circuit prevents motor overspeed in the event of an open failure of the field supply. A loss of field current opens the armature circuit because the transistor ceases to conduct. The diode protects the transistor from inductive kick from the motor and from counter EMF.

Figure 3-3 illustrates a shunt wound motor and one form of protection. A transistor is inserted so that loss of field current will open the armature circuit. This may also be handled with an electromechanical relay. In the circuit shown, the presence of the field current saturates the transistor and the motor sees only the forward drop. The diode protects the transistor from the motor inductive kick and from the motor CEMF.

A fuse in the armature circuit might or might not protect the motor. If the failure occurred with the motor at full speed, the inrush of current due to the failure of the field will be less than the normal starting surge and the motor could get to dangerously high speeds before the fuse blew.

THE SERIES MOTOR

The series motor is a horse of quite another color from the PM or the shunt wound motor. Fig. 3-4 shows the schematic connection, but the actual differences between a practical series motor and a shunt motor are substantial and do not show up in the diagram. First of all, since the field winding is in series with the armature, both are

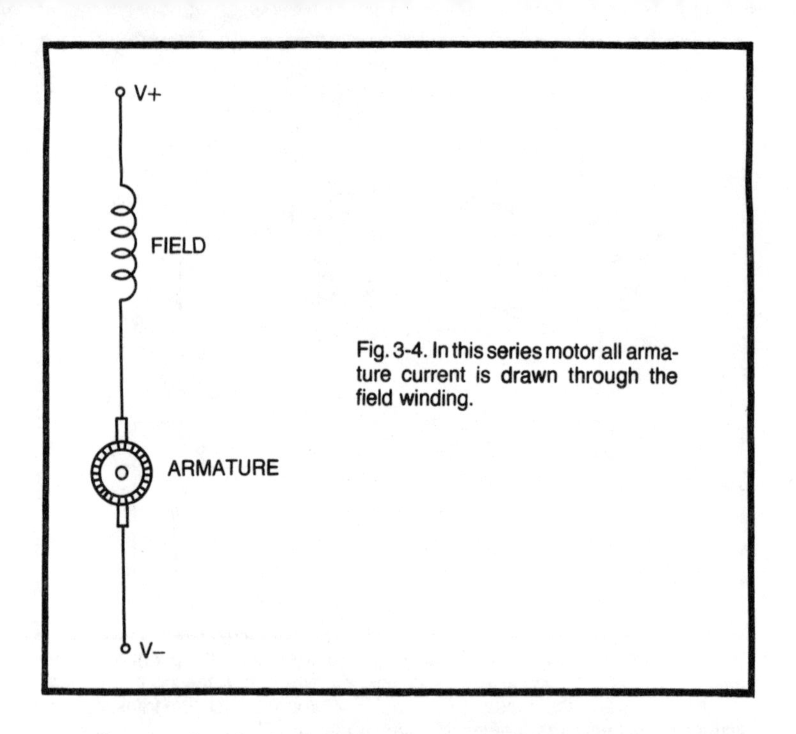

Fig. 3-4. In this series motor all armature current is drawn through the field winding.

usually wound with much lower resistance than is found in a shunt motor. Second, if the field winding is arranged to just barely saturate with the starting inrush (for maximum starting torque), then when the motor gets up to speed, the field is considerably weakened and well below saturation so the speed must rise to very high values. In theory, a frictionless series motor would tend to run at infinite speed since each increase in speed would weaken the field still further through the action of the CEMF in reducing the current.

Langsdorf says,"For this reason, a series motor must always be so installed as to be positively connected to its load, by gearing or direct connection, never by belting, and the minimum load must be great enough to keep the speed within safe limits; such is the case for instance in railway motors, hoists, rolling mills, etc." While today's reader may not be very interested in the design of trolley cars, the advice is as good today as it was in 1940.

Actually, the properties of the series motor are very desirable for traction and hoisting applications. The torque is extremely high at zero speed and falls off rapidly as the speed increases. This gives a trolley car or an elevator a very rapid acceleration away from a stop without moving up to very high speeds too rapidly for control. This was an outstanding characteristic of the old electric automobiles.

When I was a boy, a little old lady in her 70s had an ancient Detroit Electric in our neighborhood and it would show a clean pair of heels to a flathead Ford V8 up to about 35 mph. Furthermore, it did it with only the faint ticking of the brushes and acceleration smooth as silk!

Probably the most common series wound motor is the ubiquitous starting motor developed by Charles Kettering for the 1911 Cadillac. It has the requirement to develop not only enough torque to pull the engine through the compression stroke but also to pick up enough speed to stay with it when the engine begins to fire. Since the starting motor is geared to turn about ten times as fast as the engine crank this means about 5000 rpm for the 500 rpm the engine will turn as it begins to come to life. A typical car starter will have an internal resistance of less than a tenth of an ohm. The internal resistance of the battery and the cable resistance will hold the starting surge down to about 150 amperes. When running no-load with only the brush friction, windage, and the friction of the sleeve bearing on the stub end, the motor will churn up 6 to 10,000 rpm and will draw less than an ampere. It would take a bank of 10 each 2N3772 transistors to handle this job and at that the 1.5 volt forward drop would noticeably affect the starting "zip" of the system.

Nowadays, the series motor is largely confined to starting motor and hoisting applications, except for certain applications which are discussed under the heading of Universal Motors in a following chapter.

THE COMPOUND MOTOR

It seems relatively obvious that a motor could be provided with both series and shunt windings. Depending upon the relative efficiency of each, the motor (or generator) could be made to partake of the properties of either or both. For example, the addition of a series winding will tend to improve the starting torque of a shunt motor. Conversely, the addition of a shunt winding to a series motor will tend to limit the maximum speed the motor will attain under light load conditions. By using a combination of series and parallel windings the motor designer has at his disposal a relatively wide spectrum of characteristics so that a motor may be tailored for any given application.

4
AC Algebra Review

The origin of the ac dynamo actually antedates the dc dynamo. As we saw earlier, Hypolite Pixii found it necessary to invent the commutator in order to turn his machine into a dc machine. As a matter of fact, it was a good many years before alternating currents were useful for much of anything. With alternating current one cannot charge a battery, conduct electrolysis, or perform plating operations—and operation of arc lights and arc welding is more difficult than with dc. Furthermore, the behavior of alternating current circuits was profoundly more difficult to understand for investigators of that time. One of the most puzzling phenomena was the fact that the sum of the currents entering and leaving a junction did not always add up to zero. Also, current would seem to flow where there was no circuit. In order for us to deal effectively with the alternating current devices in this and in subsequent chapters it is necessary for us to have some understanding of these puzzles. For this reason a brief review of ac electricity is included here.

EARLY EXPERIMENTS

Although there are such things as electrostatic motors, practically anything that one thinks of as an electric motor is electromagnetic in its principle of operation. The development of the electromagnet dates back to the discovery in 1820 by the French astromomer Francois Arago that a helical coil of copper wire (which has no magnetic properties of its own) would behave in a fashion indistinguishable from a bar magnet when a current from a battery was flowing through it. This, along with the discovery in the same year by Hans Christian Oersted that an electric current flowing through a wire would deflect a nearby compass needle provided the evidence that electricity and magnetism were somehow linked.

This work attracted the attention of Andre Marie Ampere and in the same year, 1820, he showed that there was a physical attraction, a measurable force, between parallel wires carrying current in the same direction and that there was a repulsive force between the wires if the direction of the current in one wire was opposite to the direction in the other.

In an experiment in 1824 Arago discovered that a whirling copper disc (which at rest had no effect upon a compass needle) would drag the compass needle along in its direction of rotation. Figure 4-1 depicts this experiment. This principle forms the basis for one of the most common type electromagnetic instruments, the drag-cup speedometer. In a good auto model year, some seven million of these instruments will be delivered in new cars. Figure 4-2 depicts this instrument. While it may not be immediately apparent, this instrument is directly related to the induction motor. Faraday correctly deduced that the operation of the whirling disc on the compass needle was caused by the induction of circulating, or eddy, currents in the disc which in turn gave rise to a virtual magnet pole that was responsible for the drag on the needle of the compass.

In 1829 an American, Joseph Henry, conceived the idea of winding insulated wire upon a U shaped iron core and succeeded in producing some extremely powerful magnets. He used these magnets, one of which weighed over 100 pounds, to fashion a rather crude electric motor. Figure 4-3 depicts one of Henry's magnets. He discovered to his surprise that the circuit would close very easily; however, when he pulled the switch open, a brief but extremely intense arc occurred at the switch contacts. He also noted that a

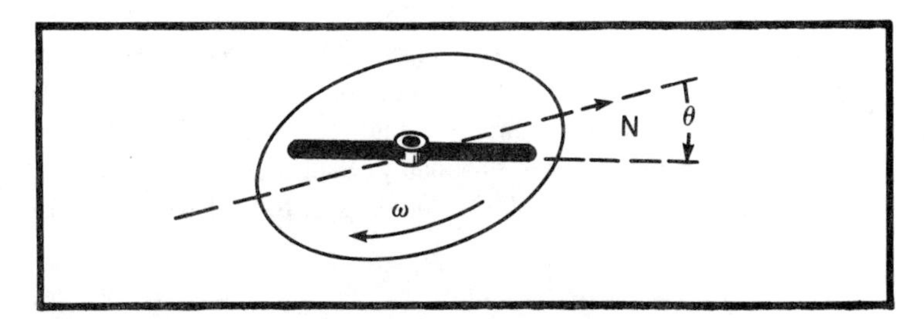

Fig. 4-1. The Arago whirling disc experiment. The nonmagnetic whirling copper disc tends to drag the compass needle in the direction of its rotation.

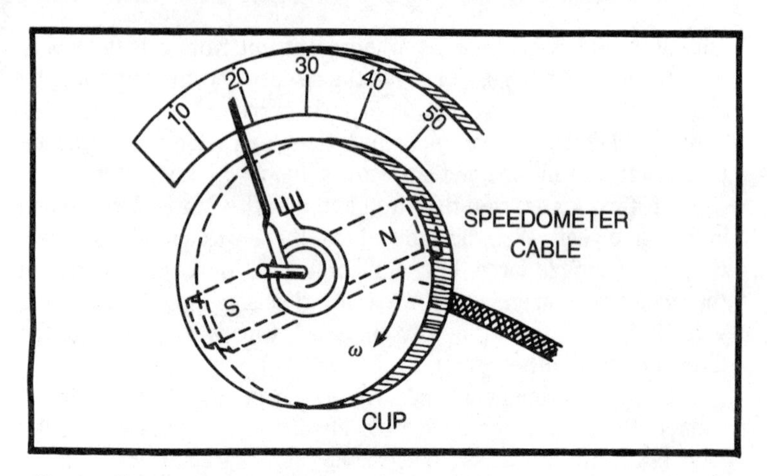

Fig. 4-2. The drag-cup speedometer. A powerful bar magnet is whirled by the cable at a speed related to the wheel speed. The torque induced in the close fitting aluminum cup is opposed by the hairspring. The pointer deflection is proportional to speed.

second winding which was not electrically connected to the circuit would arc as well! Furthermore, the spark was far more intense and energetic than could be obtained by just short circuiting the battery.

In a simultaneous but completely independent effort, Michael Faraday, working in England, found that he could induce a current in a coil of wire by physically moving a magnet through the coil. In another experiment, Faraday found that the galvanometer shown in Fig. 4-3 would kick one direction when the primary circuit was closed and the other direction when the primary was opened. In a steady closed state, the galvanometer showed no current. Credit for the discovery of electromagnetic induction is generally accorded to Faraday who published in 1831. The discovery that the voltage induced in a wire was proportional to the time-rate-of-change of the magnetic field is central to our understanding of the properties of ac motors.

At this juncture we must become a little more mathematical again in order to understand some of the behavior of the devices we are interested in. The treatment is intended to be a brief review, only enough to permit some reasonably quantitive evaluation in later sections. For those interested in delving further, I would recommend:

Stephen S. Attwood
Electric and Magnetic Fields
Dover Publications, Inc.

In parts of the following discussion we shall be dealing with *vector quantities* and they shall be designated with an arrow over the top of the symbol as in the symbol $\vec{\beta}$ for magnetic flux density which was used without explanation in the previous chapter. Now all this means is that a vector has a direction. If someone tells you, "I'm driving away at 35 mph," he has told you nothing about *where* he is going. If on the other hand he says, "I'm driving away, going north, at 35 mph," he has supplied you with a complete vector description of his intended path. You know whence he is departing, you know how fast, and you know the direction he is going. Quantities that have no direction associated with them, such as dollars, watts, population, etc., are known as *scalars*.

When dealing with vectors we often will be using *imaginary numbers*. This is really just a form of mathematical notation and works out rather simply. Referring to Fig. 4-4, suppose that we would like to have a mathematical operation that will rotate a vector 90°. If the vector started out by pointing *east* then after one such operation it would be pointing *north*. However after two such operations it would be pointing *west*. Now 10 miles *west* is the same thing as minus 10 miles *east*; therefore, our operator—which is designated as j in the figure—must be the square root of a minus one. A mathemetician would use the symbol i for this imaginary number,

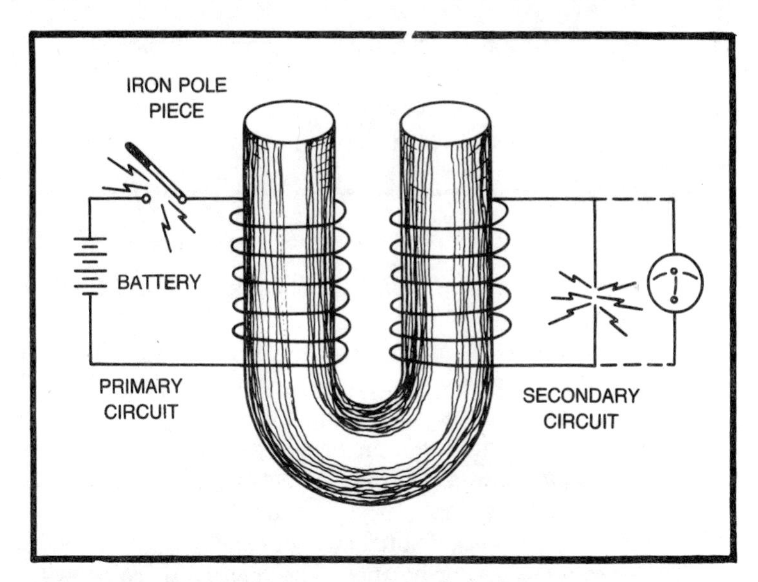

Fig. 4-3. The electromagnet and electromagnetic induction. When the current is established in the primary circuit, energy is stored in the magnetic field. When the current is interrupted, this energy is dissipated in the form of one or more arcs.

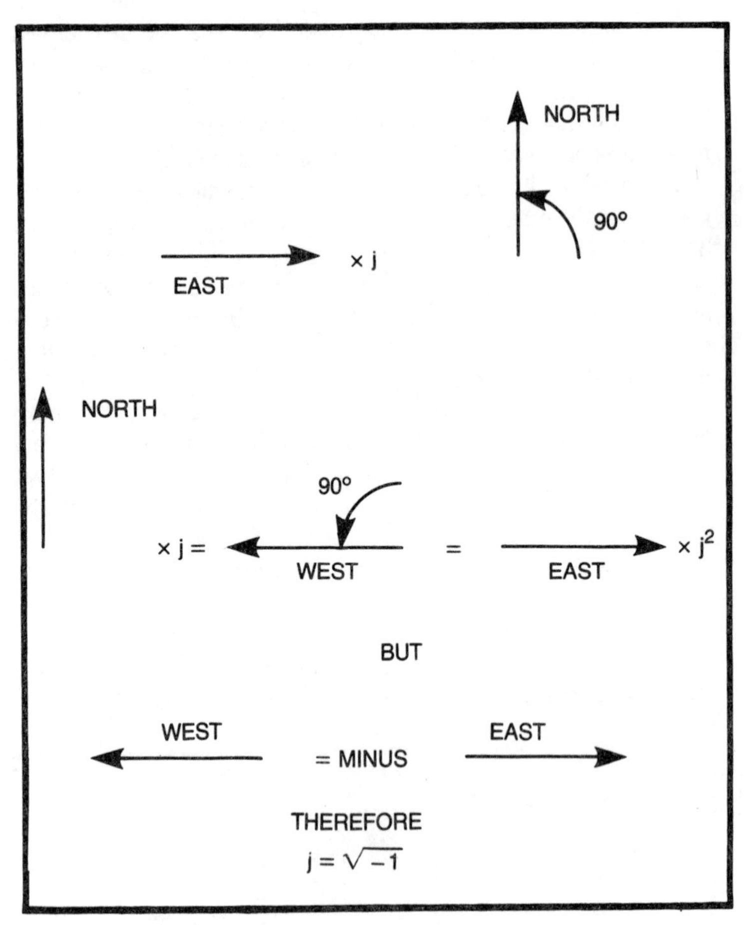

Fig. 4-4. The imaginary operator j.

but electrical engineers are in the habit of using j instead because i is used tor current.

Charles Proteus Steinmetz was in the habit of saying that there is nothing more imaginary about imaginary numbers than there is about the distance between Albany and Schenectady. Referring to Fig. 4-5 we observe that, with the definitions given, we could give the *vector distance* between Albany and Schenectady as $-9 + j12$ miles or as $15/126.86°$ miles. The first of these descriptions is in *Cartesian*, or *rectangular*, co-ordinates, or notation, and the second is in *polar* co-ordinates, or notation. The rectangular co-ordinates are easier to use in addition and subtraction, and the polar co-ordinates are easier to use in multiplication and division.

In Fig. 4-6, the distance 8.6/35.54 easily converts to the distances for X and Y:

$$8.6 \cos \theta = 7 \tag{4.1}$$

$$j8.6 \sin \theta = j5 \tag{4.2}$$

Also related to this figure is another mathematical convenience—Euler's equation:

$$A^{j\theta} = A \cos \theta + jA \sin \theta \tag{4.3}$$

$$A^{-j\theta} = A \cos \theta - jA \sin \theta \tag{4.4}$$

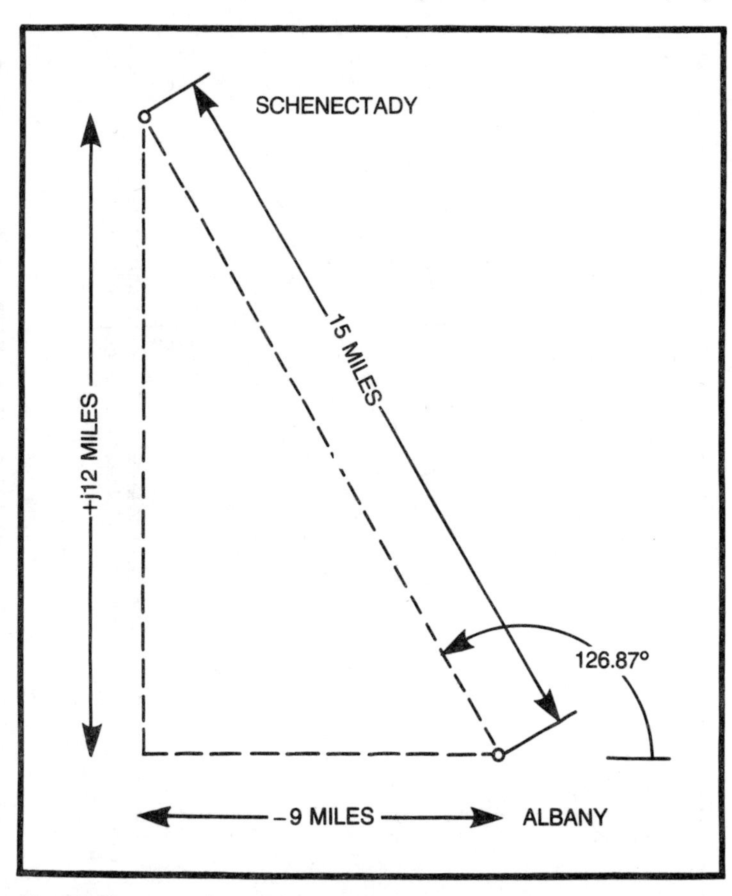

Fig. 4-5. The vector distance from Albany to Schenectady. If we define the plus direction as east and the +j direction as north, then the minus direction is west and the −j direction is south. The path, or vector distance, is −9 + j12 miles, or 15 /126.87°, miles.

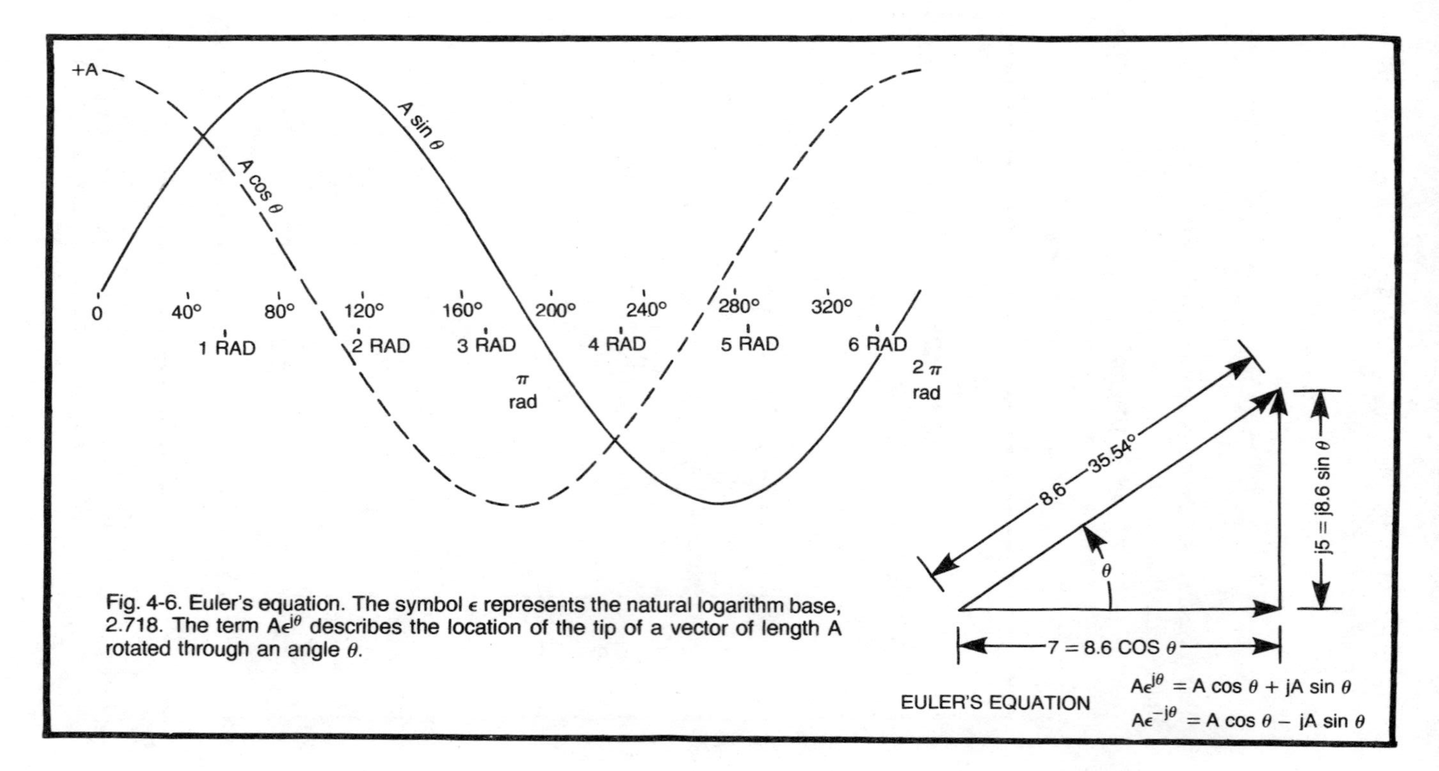

Fig. 4-6. Euler's equation. The symbol ϵ represents the natural logarithm base, 2.718. The term $A\epsilon^{j\theta}$ describes the location of the tip of a vector of length A rotated through an angle θ.

where

$$\epsilon = \text{the natural logarithm base}$$

This is frequently used in electrical engineering to describe the rotating vector. Very often electrical engineering expressions contain the angle θ written in terms of *angular frequency*, ω. Thus the expression $A\epsilon^{j\omega t}$ implies a vector of length A rotating at a rage of ω radians per second. The cosine and sine terms of Euler's equation are simply the projections of the vector on the real and imaginary axes. When the angle is given as the exponent of ϵ this is referred to as *exponential notation*.

Many of the popular electronic "slide rules" are equipped with a rectangular-to-polar and polar-to-rectangular conversion feature which is probably most popular for use in electrical engineering problems. In a reverse Polish Notation machine such as the Hewlett-Packard 21 the operational sequence is:

Write imaginary part
Enter
Write real part
[F]
To polar (radius vector appears in display)
$X \leftrightarrows Y$ (angle appears in display)

The angle will be in degrees or radians depending upon the setting ot the degree-radian switch.

In an algebraic notation machine such as the Texas Instruments SR-50 or the Monroe 326 the sequence is:

Write real part
[F] to Polar
Write Imaginary part
= (the angle appears in the display)
= (the radius vector appears in the display)

Repeated depressions of = produce alternation of the angle and the radius vector. The actual details vary with different calculators.

THE INDUCTOR

Now Faraday's experiments indicated that the voltage induced in an inductor was proportional to the time-rate-of-change of the magnetic flux linked by the inductor.

$$V = L \frac{d\phi}{dt} \text{ volts} \tag{4.5}$$

where

$$\frac{d\phi}{dt} = \text{time rate of change of flux} \tag{4.6}$$

In an inductor excited by a current, ϕ is related to the current i; therefore:

$$V = L \frac{di}{dt} \quad \text{volts} \tag{4.7}$$

where

$$L = \text{inductance in henrys}$$

$$\frac{di}{dt} = \text{time rate of change in amperes per second} \tag{4.8}$$

This will be seen to be identical with equation 2.3. If we now assume that an alternating sinusoidal voltage $V = V_o \sin(\omega t)$ is applied to the inductor we obtain: (4.9)

$$V_o \sin(\omega t) = L \frac{di}{dt} \tag{4.10}$$

then

$$di = \frac{V_o}{L} \sin \omega t \, dt \tag{4.11}$$

Integrating this expression yields:

$$i = \frac{-V_o}{\omega L} \cos(\omega t) + \text{constant} \tag{4.12}$$

For the time being we shall neglect the constant since it pertains to transient conditions and we are interested in the steady state conditions here. Now from the curves at the bottom of Fig. 4-7 we can see that:

$$-V_o \cos = V_o \sin(\omega t - 90°) \tag{4.13}$$

thus

$$i = \frac{V_o}{\omega L} \sin(\omega t - 90°) \tag{4.14}$$

From this we see that the current in the inductor lags the voltage by 90°. Most of the ac problems normally dealt with are steady state problems and we are interested in the values averaged over many cycles. For this reason we may include the sin ωt term implicitly and simply neglect to write it. Also, we noted in Fig. 4-4 that we could effect a 90° counterclockwise rotation by application of the operator j. Making use of both of these conventions we may rewrite equation 4.14 as:

$$i = \frac{V_o}{j\omega L} \text{ amperes} \tag{4.15}$$

The term $j\omega L$ is called the *inductive reactance* of the inductor. It is usually designated as X_L. It is a measure of the current which will be drawn by the device with a given voltage applied. It is a parallel of Ohm's law which states for direct current:

$$i = \frac{V}{R} \tag{4.16}$$

where:

V = the applied voltage in volts
i = the current in amperes
R = the resistance in ohms

or for *ac*

$$i = \frac{V}{Z} = \frac{V}{R + jX} \tag{4.16a}$$

Rather than going through a detailed mathematical proof we will note that the energy involved in the building of the magnetic field is given up on the subsequent quarter cycle; therefore, *no real power is consumed in the imaginary components of an impedance.*

This deserves a little more explanation, and is perhaps most easily visualized in terms of the graphs of Fig. 4-7. We know from Ohms's power law (published by George Simon Ohm at Nurenberg in 1827) that power dissipated in an electrical circuit is given by the product of voltage times current. From equation 4.12 we may obtain the instantaneous value of current and obtain the following:

$$\text{Power in the inductor} = -V_o \frac{\sin \omega t}{\omega L} (V_o \cos \omega t) \tag{4.17}$$

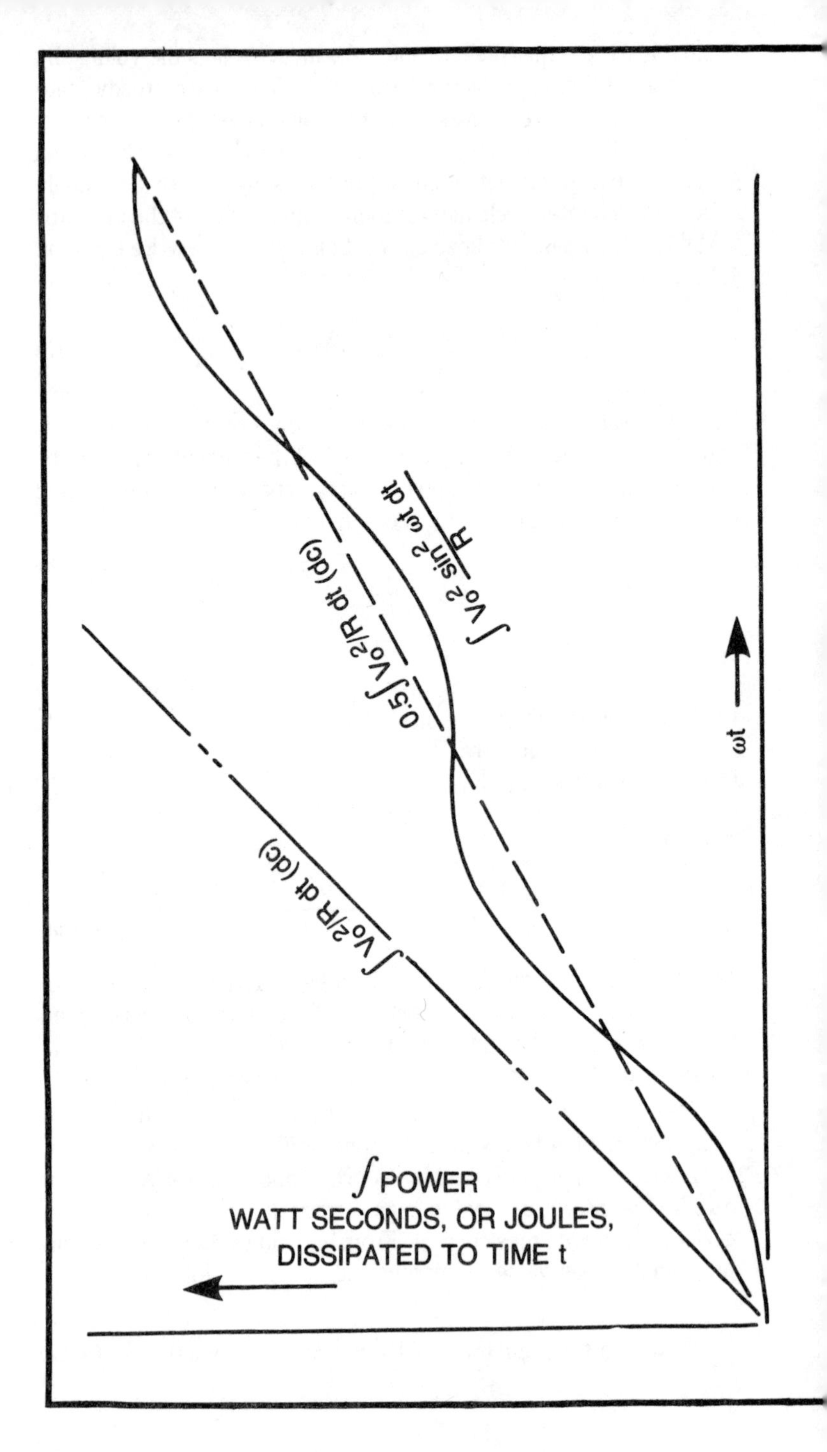

∫ POWER
WATT SECONDS, OR JOULES,
DISSIPATED TO TIME t
ωt
∫ Vo²/R dt (dc)
0.5 ∫ Vo²/R dt (dc)
∫ Vo² sin² ωt dt / R

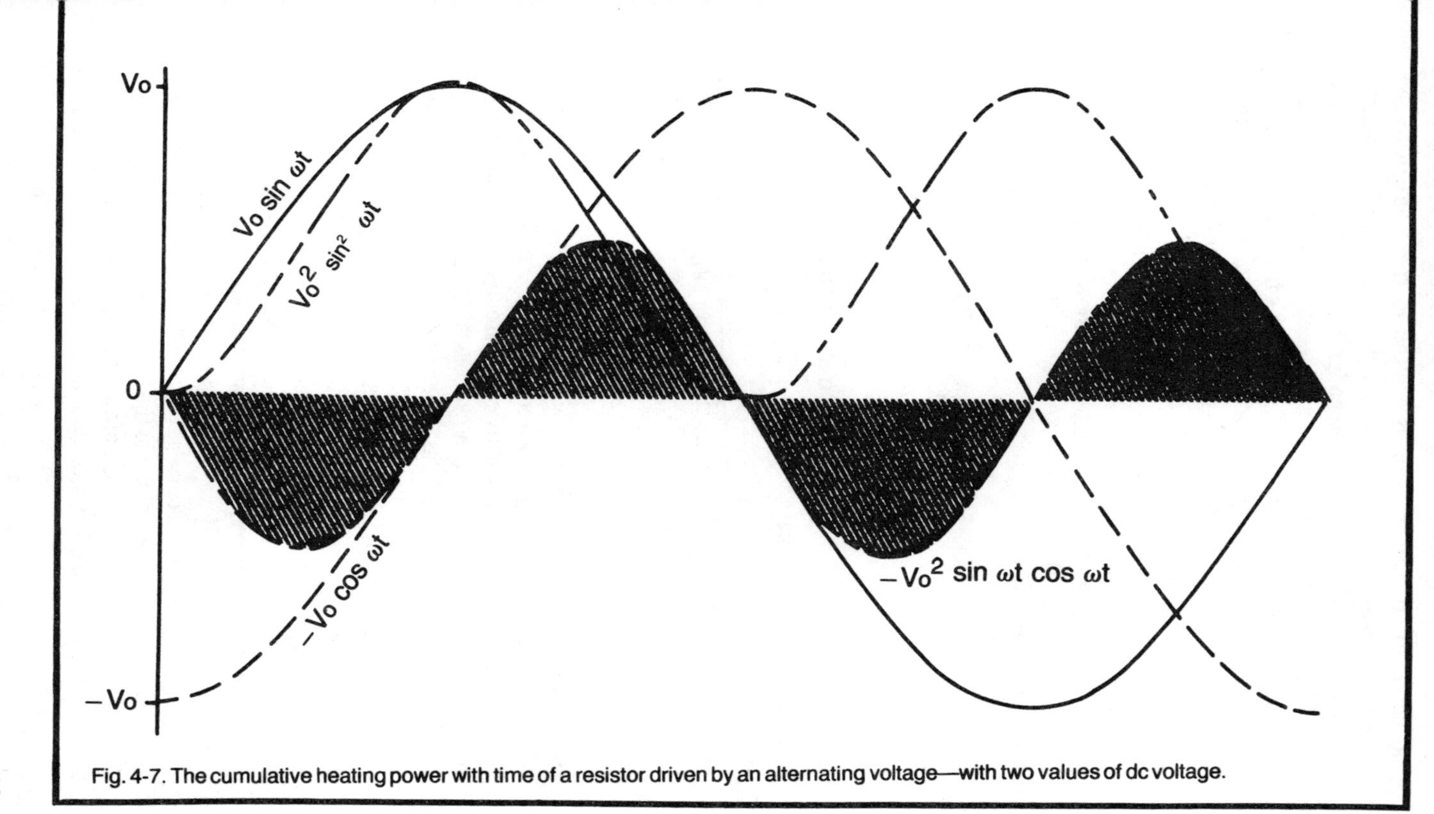

Fig. 4-7. The cumulative heating power with time of a resistor driven by an alternating voltage—with two values of dc voltage.

The curve for this expression is plotted in the figure with the assumption that $\omega L = 1$. You can see by comparing the hatched areas that there is just as much area under the line as there is above it when a full cycle or a half cycle is considered. This tells us that the inductor only stores the energy in the magnetic field and then gives it back on the subsequent quarter cycle. This nondissipated product, which is alternately stored and released, is sometimes called *reactive power* and is usually discussed in *reactive volt-amperes* (vars), or thousands of reactive volt-amperes (kilovars). The latter is usually abreviated kvar.

In the resistive case the matter is somewhat different. From Ohm's law we can obtain the current for a resistor excited by an alternating voltage:

$$i = V_o \frac{\sin \omega t}{R} \qquad (4.18)$$

where

R = the resistance in ohms.

From this we may obtain the expression for power:

$$\text{Power} = V_o \frac{\sin \omega t}{R} (V_o \sin \omega t) \qquad (4.19)$$

$$= V_o^2 \frac{\sin^2 \omega t}{R} \qquad (4.20$$

This curve is also plotted on the lower half of Fig. 4-7, for the case where $R = 1$ ohm. It may be seen that again, as in the inductive case, the power term varies at twice the frequency of the exciting voltage. Even more significant, however, is the fact that the squaring of the sin term renders all the values positive. This is *real* power, and it will make the resistor get hot!

For someone who wants to sell electric power for a living, alternating current presents somewhat of a problem. The customer could be using a great deal of your product and yet the long term average of the voltage and current you deliver him is zero. For this reason, alternating current is usually specified upon the basis of the long-term heat value. If we stepwise sum the products of the instantaneous voltage and current in the resistor (or integrate) over the period of a full cycle we obtain the curve shown in the top half of Fig. 4-7. Superimposed on this are curves for the power which is dissipated in the same resistor by two different voltages. From a visual inspection of the curve we see that the ac dissipates only one half as much power in the resistor as a dc voltage equal to the peak voltage. The equivalent ac voltage is called the rms voltage or root-mean-square voltage and is given by:

$$V_{rms} = \sqrt{0.5}\, V_o = 0.7071\, V_o \qquad (4.21)$$

It should be noted that *this relationship holds only for sine waves*. For other wave shapes, the relationship will be shown to be different in some later sections.

The rms voltage is the voltage that a dc source would have had to supply in order to provide the same heating in a resistor.

THE CAPACITOR

The capacitor, or condenser, has certain properties in common with the inductor but is significantly different. A capacitor is physically made with a pair of conductive plates spaced by an insulating material. It behaves very much like a small storage battery. When it is first connected across a constant voltage source (with a resistor to limit the current inrush) the current assumes a value equal to the source voltage divided by value of the limiting resistor. Very rapidly, the current will fall to zero and the capacitor will be found to have a voltage equal to the source voltage and will retain this for some time after it is disconnected from the source if no current is drawn. With a large computer grade capacitor and a small, high-resistance lamp bulb you can show that there is energy stored in the capacitor since it will light the bulb for an appreciable period. The energy stored in the capacitor is:

$$W = \frac{CV^2}{2} \text{ joules, or watt-seconds} \qquad (4.22)$$

where

C is capacitance in farads.

The charge in the capacitor is given in terms of the integrated current which flowed into the capacitor. The units for charge are ampere-seconds, or coulombs (q), The charge is related to the voltage by:

$$V = q/C \qquad (4.23)$$

where

q = coulombs $= \int i\, dt$ ampere-seconds

Now when an alternating potential is applied across a capacitor we obtain:

$$V_o \sin \omega t = q/C \qquad (4.24)$$

Differentiating the above equation with respect to time we obtain:

$$\frac{dq}{dt} = V_o \, \omega C \cos \omega t \tag{4.25}$$

Note the similarities and differences of this equation and equation 4.12. Whereas the current in the inductor is given by $-\cos \omega t$, this current is given by $+\cos \omega t$. Also, whereas the inductive reactance is given by $j\omega L$ the *capacitive reactance* is given by $-1/j\omega C$. By reference to the bottom curves of Fig. 4-6 we can see that the current *leads* the voltage in a capacitive reactance. For the steady-state condition we can write the relationship which is the dual of equation 4.15.

For a capacitor:

$$i = \frac{V_o}{-\frac{1}{j\omega C}} = \frac{V_o}{X_c} \tag{4.26}$$

A comparison of the ($\cos \omega t$) curve in Fig. 4-6 and the ($-\cos \omega t$) curve in Fig. 4-7 tells us that if, in a given circuit, $j\omega L = 1/j\omega C$ the effect of the two components will be to cancel one another.

PHASE ANGLE AND POWER FACTOR

Next, let us return to our example and the oscillograms of Fig. 2-5 regarding the drill press motor and see what sense we can make out of some of the conflicting readings. First of all we note that the ammeter told us that the motor was drawing 5 amperes and yet the oscillogram of Fig. 2-5 indicates a peak current of 7.2 amperes. When we look at the face of the ammeter, however, we note that it says "amperes rms." We check this out by the relation of equation 4.27 and we find that 5 amperes rms actually works out to 7.07 amperes peak. That seems to track reasonably well. The oscilloscope and current probe are calibrated in peak amperes and the ammeter in rms.

Actually the ammeter calibration is a little white lie. The mechanism of the ammeter actually reads the average current over a half wave. Now the average value over a sine wave half cycle is $0.6366\, i_o$ and the rms value is $0.7071\, i_o$ so the meter is calibrated to read high for dc currents by the ratio 0.7071/0.6366, or 1.1107. If we were only to measure sine waves the "fudge factor" would be of no consequence. However, in some of the controls to be studied later, the fudge factor will be grossly inaccurate and we will have to make corrections if we are to find out what is going on.

Now, referring to Fig. 4-8 we see the vector diagram resolving the 7.07 ampere motor current into its real and imaginary components. Because of the large phase angle, the imaginary component is considerably larger than the real component. The surprising thing is that the drill press motor is actually using only 55 watts and producing 0.074 horsepower while the reactive component works out to

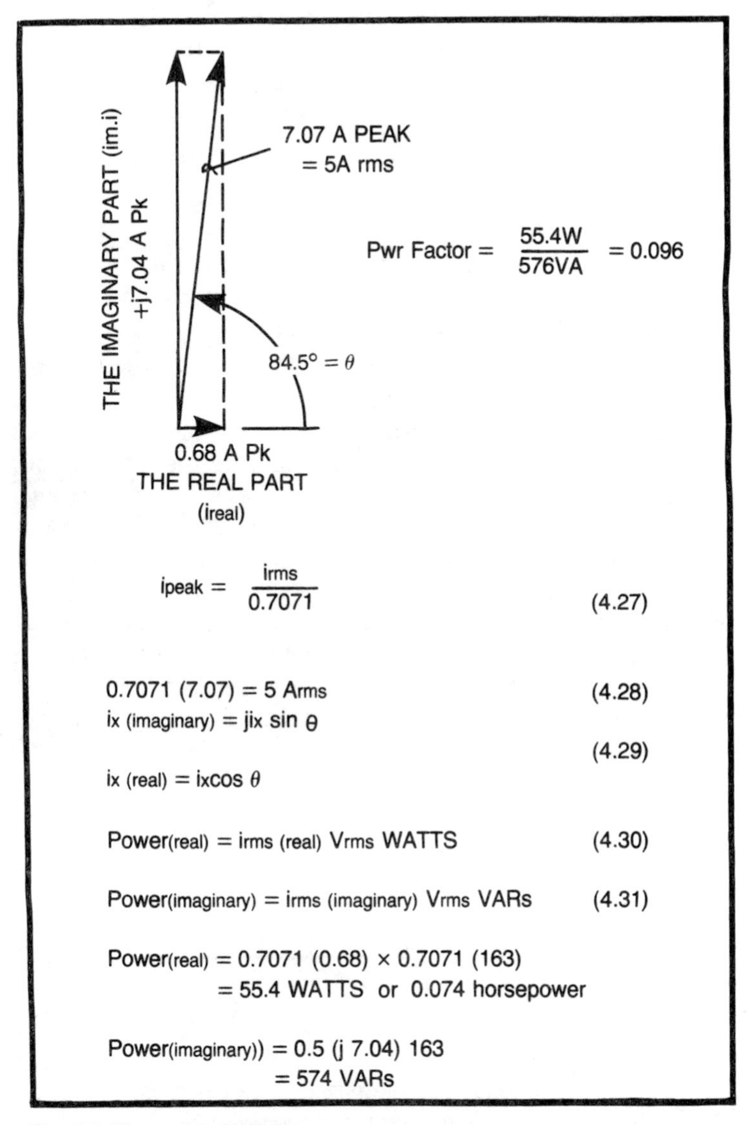

Fig. 4-8. The vector diagram.

574 vars. Now the motor is a venerable ball-bearing "COMPANION" purchased circa 1938 and the drill press has a tapered-roller bearing quill, so the whole assembly is a pretty free coasting outfit. Probably the main losses in the drive are in the V-belt. Thus it is perhaps not too surprising that the outfit is consuming very little power in the unloaded condition. As you start to actually drill, say through cast iron with a half inch drill, the peak current increases only slightly; however, the phase angle "walks" steadily toward the real axis. For the above conditions the peak current creeps up to something on the order of 7.8 amperes peak, and the phase angle falls to about 68°. This works out to a real power of:

$$\text{real } i_{\text{rms}} = 0.7071 \times 7.8\,A \times \cos 68^\circ$$
$$= 2.10 \text{ amperes rms}$$

which gives a power of:

$$\text{power} = 0.7071 \times 163\,V_{\text{peak}} \times 2.10\,A_{\text{rms}}$$
$$= 242 \text{ watts, or } 0.324 \text{ horsepower}$$

This value is more in keeping with the 1/3 HP stamped on the label.

It should be fairly obvious that the Rochester Gas and Electric Company (or Conn Edison, Consumers Power, or the Pacific Gas and Electric Company) has reason to be a mite unhappy about this sort of thing. Most of the time when the drill press is operating they have to supply me with 5 to 6 rms amperes of current. On the other hand, the wattmeter installed for the lab only bills me for 55 watts, the real power, when the drill press is unloaded, and only 242 watts when I am actually drilling with a heavy work load. Despite the j in front of them and the phase angle, there is nothing imaginary about the quadrature currents. They dissipate real power in the loss resistance of the lines and the pole transformer, and they will blow fuses just as effectively as if they had been in phase with the line voltage. If my lab were a large installation, I would be nicked for the power factor so that the power company would not have to bear the burden of carrying large currents in their system and not being able to bill me for the use of power.

The term *power factor* is defined:

$$\text{Power factor} = \cos\theta = \frac{\text{real power}}{\text{Total volt amperes}} \tag{4.32}$$

where

θ is defined as the phase angle between the voltage and the current drawn

Because of the difference in sign between the quadrature current drawn by a capacitor and an inductor, these components tend to cancel one another. In Fig. 4-9 we have drawn the various components of the motor as though they could be separated and have added a capacitor which will draw the same current at 60 Hz as the inductor. From the vector diagram at the left we can see that the imaginary components cancel one another. The result was one that surprised many of the early experimenters with ac. Meter A_1 does not display the sum of the currents of $A_2 + A_3 + A_4$; in fact, the reading is much smaller than the readings of either A_3 or A_4. The mathematical explanation that the currents in the inductive and capacitive branches are phase-opposed and simply circulate between the two units was developed largely by Charles Proteus Steinmetz during the late 1890s, working at the General Electric Laboratory in Schenectady, N. Y.

With this *power factor correction* the net current in the mains is reduced to a minimum; however, there are several points which should be noted.

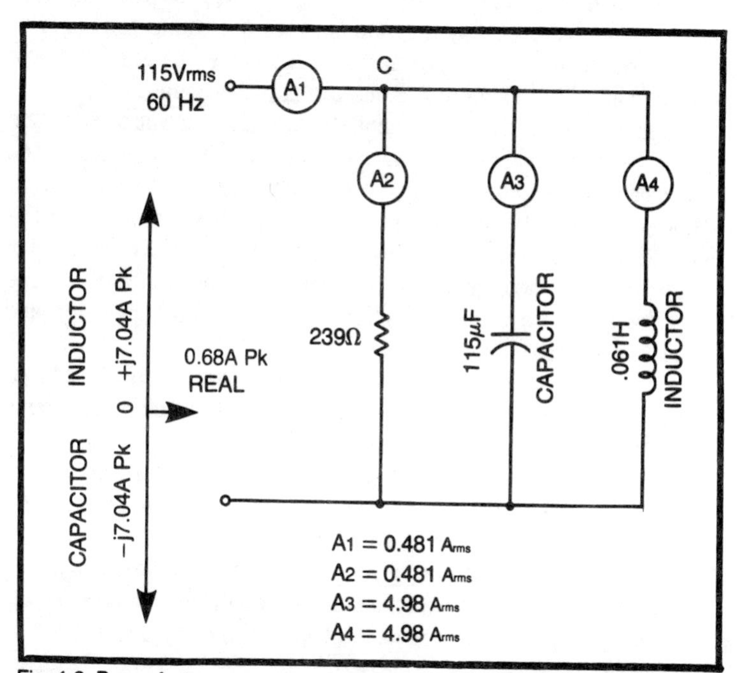

Fig. 4-9. Power factor restoration. A capacitor that draws a current equal to the inductor current is added to correct the system power factor. Note that the line current at point C is far smaller than the sums of the three branch-current readings.

1. Both the inductive reactance and the capacitive reactance are frequency dependent, and they vary in the opposite direction with frequency. For example, if the voltage were held constant and the frequency doubled to 120 Hz, the current in the inductor would be halved and the current in the capacitor would be doubled.
2. The power factor correction on a motor is load-sensitive. In the loaded case for our motor example, the current has risen to 7.8 amperes peak and the phase angle has fallen to 68°; therefore, the *im i* has raised slightly to:

$$\begin{aligned} im\ i &= 0.7071 \times 7.8\, A_{pk} \times \sin 68° \\ &= 5.11\, A \text{ rms} \end{aligned}$$

   In this case the error is relatively small, but it can become quite large in other motor types.

A load that has been *tuned out* in this way is said to be *resonant* and has a power factor of unity.

A second type of phenomenon that we shall touch upon is *resonant voltage multiplication*. Consider the case shown in Fig. 4-10 in which a resistor, capacitor, and inductor are connected in series across the ac power line. In this case the same current must flow through all of the elements. Since the inductive and the capacitive reactances exactly cancel one another, the input line will see only the resistive term. Using Ohm's law modified for ac from equation 4.16a we calculate the current to be 11.5 amperes in equation 4.33. However, we note that this same current must flow through the capacitor which has a reactance of $-j100$ ohms. As the calculation of 4.34 shows, the voltage across the capacitor is *much* higher than the line voltage. The calculations for the sizes of the inductor and capacitor are also shown.

The above effect is known as resonant voltage multiplication. This phenomenon will take place in nearly any ac circuit that contains both inductors and capacitors in series.

This is no mere mathematical curiosity! Any ac circuit containing inductors and capacitors in series should be expected to contain voltages far higher than the driving voltage.

Any ac circuit containing inductors and capacitors in parallel should be expected to contain currents far larger than the line current.

In more complex ac circuits both the voltage and current can be multiplied.

The last example was simplified because the inductance and the capacitance exactly canceled, leaving impedance Z with only a real term. For our final example, we shall modify the circuit of Fig. 4-10

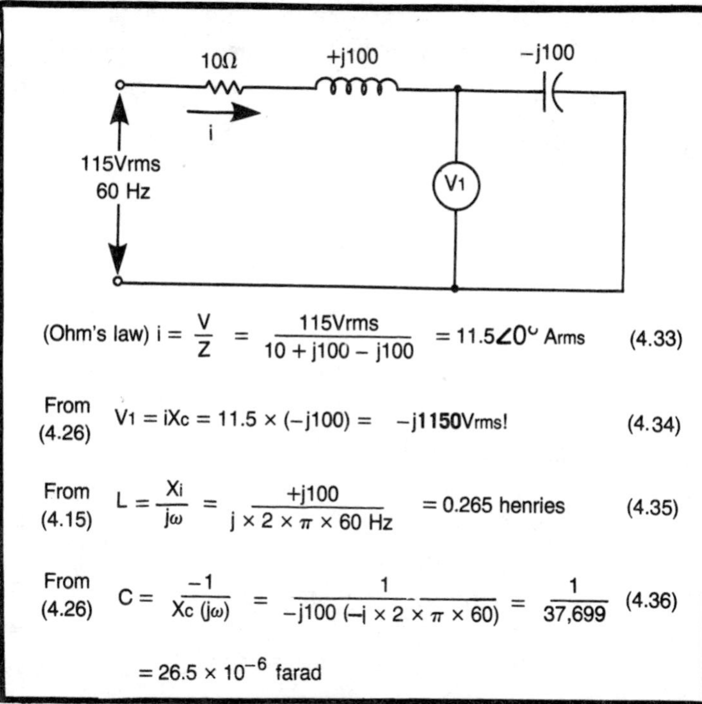

Fig. 4-10. Resonant voltage multiplication.

so that the impedance is not pure real. This is done by doubling the capacitance which, in turn, halves the capacitive reactance. Figure 4-11 shows this. In this case, we find that the denominator in the expression for current is not pure real and in fact contains a reactive term. Although vector division in rectangular, or Cartesian, notation is possible, it is somewhat easier to perform division and multiplication in polar or exponential notation. The rules for vector multiplication and division are shown in equations 4.39 and 4.41. The vector diagram at the bottom of the page illustrates equation 4.44 which notes that the sum of the voltage drops around a circuit is zero. This is just as true in the ac case as it is in the dc case PROVIDED THAT ACCOUNT OF THE PHASE ANGLE IS KEPT.

In general, the series resonant mechanism for restoring the power factor will provide the least circuit impedance, the largest line current, and the highest voltage, for a given size of circuit elements. Note that the current is considerably less in the case of Fig. 4-11 than it was in Fig. 4-10.

As a corollary, the line current is least in the parallel resonant case and the element currents are highest.

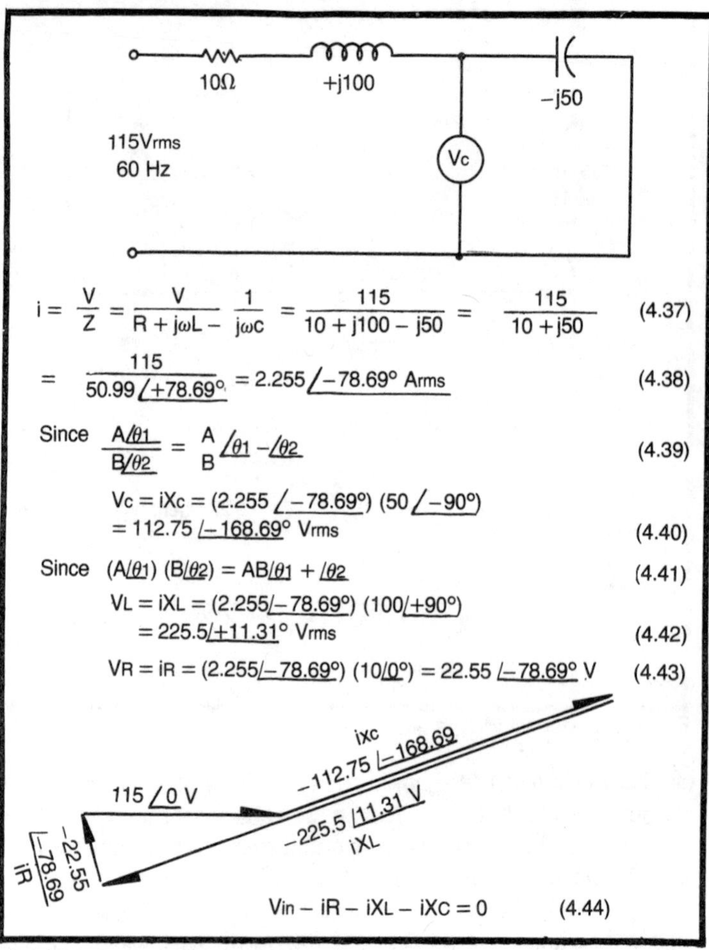

Fig. 4-11. An example of nonresonant circuit solution.

For further reading on the topic of circuit solutions and circuit impedance I would recommend:

Russell M. Kerchner and George F. Corcoran
Alternating Current Circuits
John Wiley and Sons, 1943
see Chapter 2 particularly

or

John A. Kuecken
Antennas and Transmission Lines
Howard W. Sams and Co., Inc.
Indianapolis, Ind.

5

Inductors and Transformers

The principal reason for the popularity and the common usage of ac electricity is the simplicity with which the operating voltage and current may be manipulated by means of transformers and inductors. The transformer serves to perform the same function for electricity that the gear train performs for mechanical rotary motion. The analogy is really rather direct.

TRANSFERRING FORCE

Consider for example a truck seated at the base of a hill which must be climbed. The thing that is needed is a large force at the periphery of the wheels that is much larger than available at the engine. To obtain this large force, the driver shifts into low gear which permits the engine to turn much faster than the wheels, and the truck will slowly climb. The driver has traded off a large rotational speed on the engine—which is analogous to a high current with limited voltage—for a low rotational speed with a large force, which is analogous to a high voltage, low current situation. In effect the driver has used the transmission to step up the source impedance presented to the load. Each of the gears, or speeds, in the truck transmission is comparable to a transformer tap allowing the selection of a different impedance transformation or "gear" ratio.

It is important to note that in neither case is the power altered. If asked, the truck driver would tell you that he downshifted to obtain more "power," but in fact he downshifted to obtain more force at the wheel periphery. Power is the product of force times velocity just as power is the product of voltage times current (or in the ac case, $Vi \cos \theta$). With both the transmission and the transformer, no new power is donated to the system; as a matter of fact, both tend to exact a "toll" for passage. The only thing to be done is to alter the force, or voltage, available at the output. The velocity, or current, is determined by the impedance at the output and the impedance characteristics of the source.

In both cases, the actual behavior is only an approximation of the simple force transformation; therefore, it is worthwhile for us to examine in some detail the behavior of real transformers with real loads since the departure from the simple, ideal behavior is very significant in many real-world designs.

THE INDUCTOR

The central issue in the understanding of the transformer is the inductor, so we will begin with a brief treatment of inductors. As in the earlier sections we shall be concerned primarily with the behavior of inductors and transformers. For this reason, we shall not delve more deeply than is required to obtain a picture of the behavior and some understanding of the measurements required to characterize a built-up device for the purpose of motor control. Those wishing to pursue the topic further are directed to the text by Stephen S. Attwood referenced in Chapter 4.

The basis of the inductor stems from the discovery by Oersted in 1820 that a current through a wire will deflect a compass needle, and that the converse is also true—that a magnet will deflect a wire with a current flowing through it. This work attracted the attention of Andre Marie Ampere, and within the year Ampere had demonstrated that there is an attraction between parallel wires carrying currents in the same direction and a repulsion between wires carrying currents in the opposite direction. Ampere then went on to demonstrate the fact that a helical coil of wire behaves exactly like a bar magnet.

These rules are so central to the operation of motors, inductors, and transformers that they are worth a little experimental effort to demonstrate and drive home. The apparatus shown in Fig. 5-1 is relatively easy to contruct in the most modest of facilities. A powerful magnet obtained from a burned out meter or from a toy store will suffice. The copper bail can be bent from a piece of house wiring with the insulation stripped off. It should be bent to hook through the screw eyes so as to swing freely. When the circuit is completed through a D size flashlight cell, a current on the order of

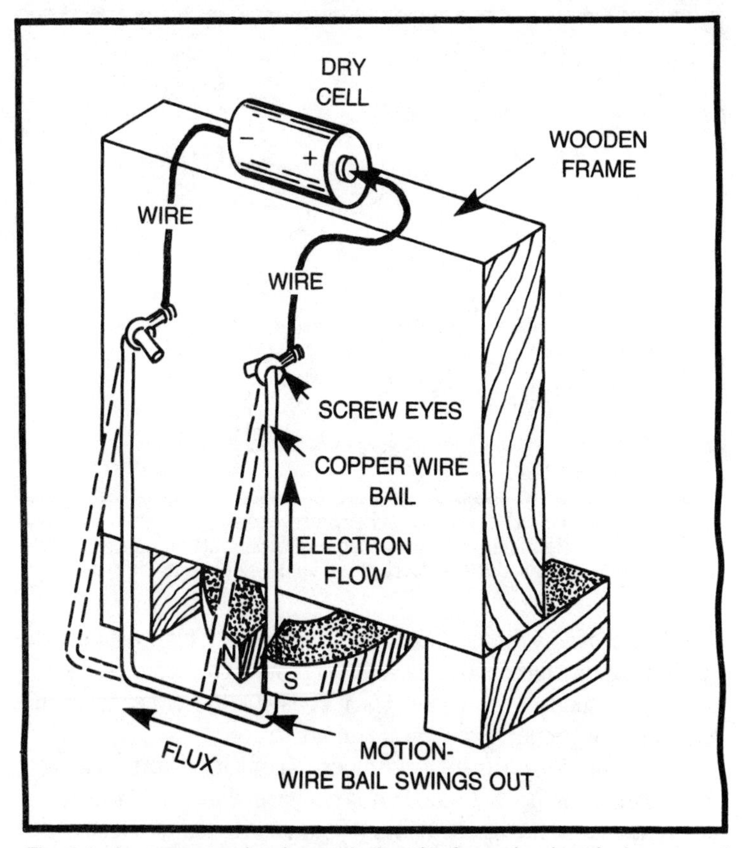

Fig. 5-1. An apparatus for demonstrating the force developed when current passes through a wire inserted in a magnetic field.

an ampere or more will flow through the bail, and it will kick out smartly from between the magnet poles. The north and south *seeking* poles of the magnet may be identified with a 50-cent pocket compass. Remember that the north pole of the magnet will *repel* the end of the compass needle marked *N* and the south pole will attract it. The action is very vigorous and unmistakable and it will help in the understanding of motor operation.

The diagram at the bottom of the figure illustrates the *right hand motor rule* describing the action. It should be noted that this is given with the "electronickers" concept that current flows in the direction that electrons travel; that is from the negative battery terminal to the positive terminal—external to the battery. For those who wish to maintain that current flows from the positive to the negative battery terminals (again external) the illustration should

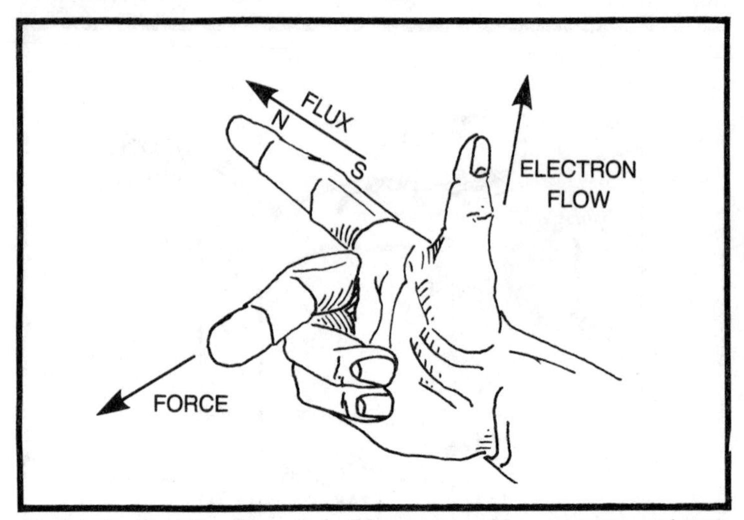

Fig. 5-2. The Right Hand Motor Rule. When any two of the parameters are known the third can be determined. The rule assumes that current flow passes from the negative battery terminal to the positive terminal—external to the battery.

show a left hand with the thumb pointing down. This would put the current flow in a direction consistent with the arrowheads on transistor, diode, and SCR symbols. Used consistently, either hand-rule will yield the proper direction for motor travel.

A similar demonstration of Ampere's law can be performed with the apparatus of Fig. 5-1 modified as shown in Fig. 5-3. The action is not quite as vigorous, but very distinctive nonetheless. It is noteworthy that this apparatus will kick the bail out regardless of the direction in which the battery is connected since reversing the battery terminals still leaves the currents in the parallel sections traveling in opposite directions. In order to show the attraction it is necessary to connect the device so that the current flows in the direction ABDC or DCAB rather than ABCD or DCBA.

The helical coil is illustrated in Figure 5-4. This is actually a rather common form of inductor, particularly when applied to very high frequency circuits. However, connected as shown, a coil with only a modest number of turns will be seen to have the same properties as a bar magnet, as can be shown with a compass. The magnetic behavior demonstrated when the current is flowing, despite the absence of any magnetic material was one of the very strong clues to the unity of electric and magnetic phenomena.

An interesting demonstration can be obtained with the *bounce motor* shown in Fig. 5-5. In this device a relatively limp coil spring such as a Slinky toy is arranged so that it hangs just barely in contact with a metal plate which is shown here connected to the negative terminal of a dry cell. When the contact is made a current of nearly an ampere will flow in the helical spring. As shown in the figure the current will flow in the same direction in each of the adjacent turns and the spring following Ampere's law, will tend to be drawn together. Thus it will lift its "foot" from the plate, break contact, and eliminate the compressive force. The spring will tend to oscillate up and down at its natural resonant frequency. This demonstration takes a bit of tinkering to get just right since the initial contact must be good enough to let a substantial current flow but still be light enough to permit the parallel current force to lift the spring and break the contact. Polishing the plate and the spring with steel wool helps. This motor has been used in whiskey ad displays where a cardboard figure of a girl serves as a weight for a hairspring and bounces slowly up and down powered by a single flashlight cell.

The compressive force on the solenoid is very considerable when the current rises to high levels. Under short-circuit loads, a

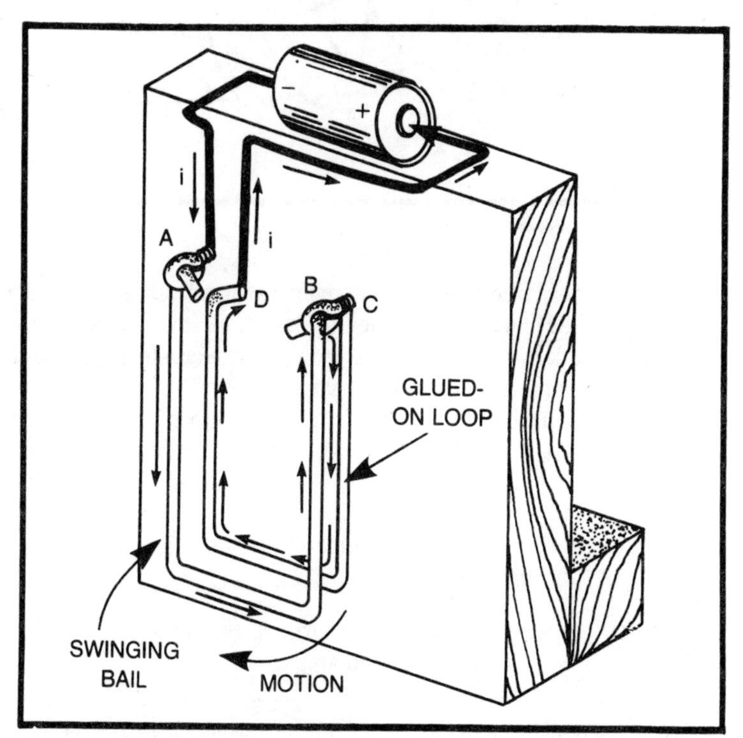

Fig. 5-3. Ampere's law demonstration.

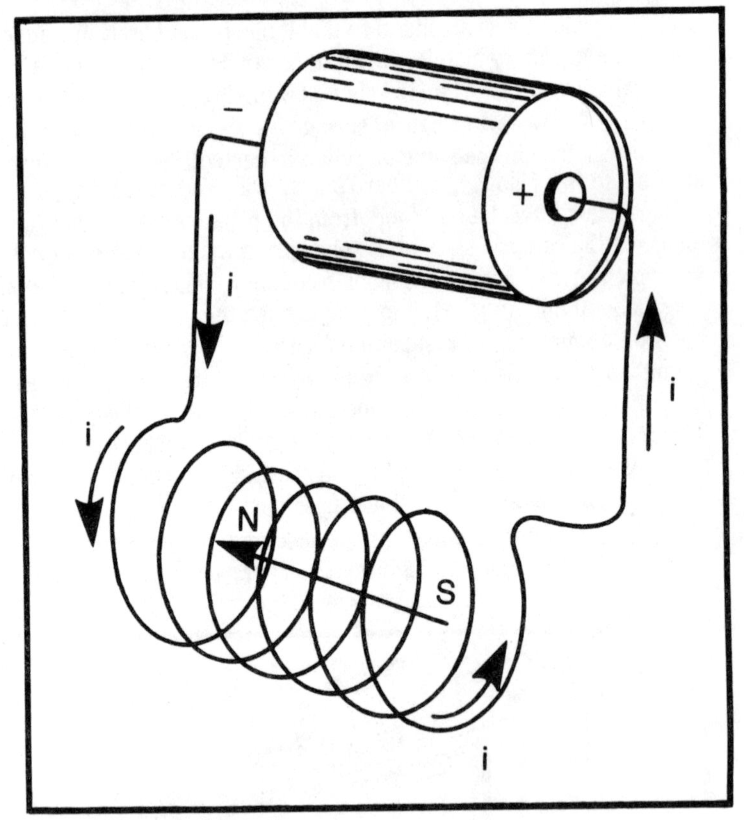

Fig. 5-4. The helical inductor.

transformer will frequently fail by having the windings tear loose and crush together thereby rupturing the insulation. It is this same force that is intended to produce the tremendous crushing pressures required to contain a sustained fusion reaction.

For a simple helical coil, such as the one shown in Fig. 5-4, it can be shown that the magnetomotive force at the centerline of the helix is given by:

$$\vec{H} = \frac{n\,i}{2\,r} \text{ amperes/meter} \tag{5.1}$$

where

n = the number of turns in the helix
i = the current in amperes
r = the radius of the coil in meters

The actual density of magnetic flux in the coil along the centerline is given by:

$$\vec{B} = \frac{\mu n i}{2r} = \mu \vec{H} \quad \text{webers/meter}^2 \tag{5.2}$$

where

μ is the permeability of the medium given in henrys per meter

The parallel to Ohm's law should be reasonably obvious for this relationship. The $\vec{H}$ is a forcing function like voltage, $\vec{B}$ is a response type function like current, and μ is an admittance type function similar to the reciprocal of resistance. Both $\vec{B}$ and $\vec{H}$ are vector quantities and have a direction associated with them. The quantity μ is not a vector property in general although in real ferromagnetic substances it can have a direction associated with it; that is, some irons are easier to magnetize in certain directions than in others. This is also true of certain crystalline substances such as spinels which will magnetize more easily in some directions with respect to the crystalline axis than in others.

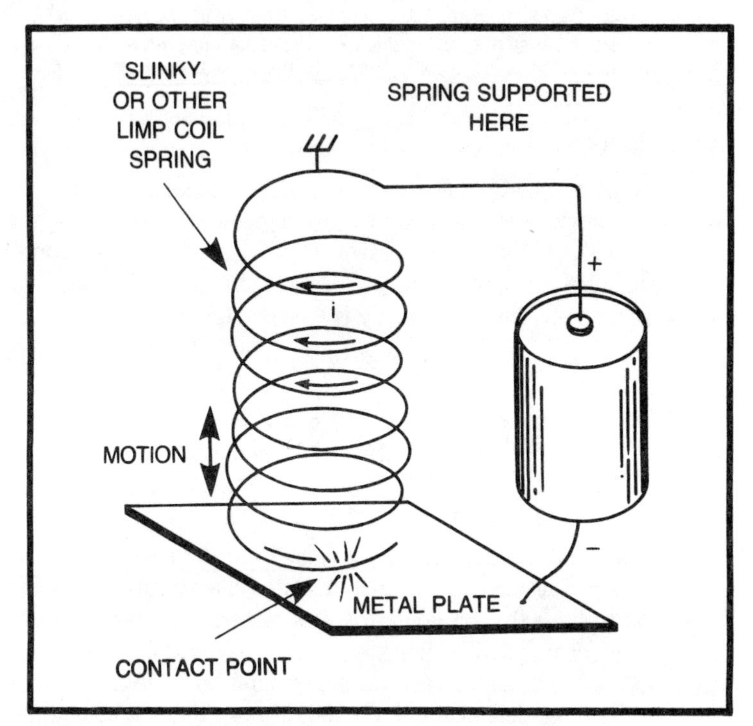

Fig. 5-5. The bounce motor.

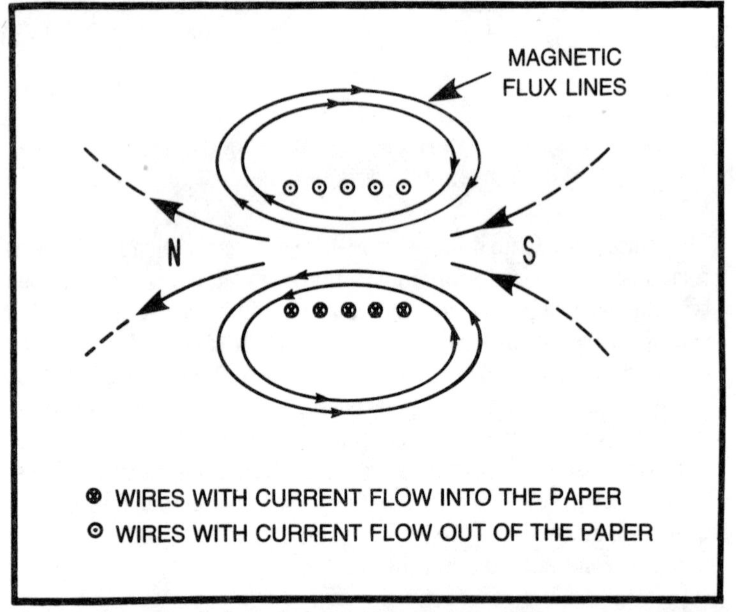

Fig. 5-6. The magnetic field of a helix. Although some of the magnetic lines are truncated here because of space restrictions, the actual lines are always closed loops. The field external to the coil is seen to be the equivalent of a bar magnet.

PERMEABILITY

The permeability of all materials is not the same; in fact it varies over rather wide ranges, and in ferromagnetic materials it is not even a constant, as will be shown later. It is common to quote permeability in terms of how magnetic a substance is with respect to air or a vacuum. The relationship is usually stated:

$$\mu = \mu_o \mu_r \tag{5.3}$$

where

μ_o = the permeability of air or a vacuum; this is equal to $4\pi \times 10^{-7}$ henry/meter

μ_r = the permeability of the material relative to the permeability of air or vacuum

Typical ferrites have permeability ranging from 10 to 5000 times the permeability of air, and irons tend to run into values from about 90 for cast iron to 55,000 for a good grade of transformer iron. Certain alloys used for magnetic shielding can attain permeabilities as great as 1,000,000. The fact that a magnet wound on an iron core can attain a magnetic field strength 50,000 or more times greater than could be attained with the same number of turns and current in air explains the fact that motors and relays are nearly always constructed with ferromagnetic cores.

In an inductor of the form shown in Fig. 5-4 the magnetic field is more or less of a dipole arrangement with the magnetic lines linking around through the windings in more or less compressed ovals as shown in Fig. 5-6. It is shown in the figure that the flux is far from uniform in a simple air core helix of this type. For this reason the calculation of inductance is somewhat more complicated for a helix than it is for a coil in which the magnetic flux is constrained to be uniform. In practice, in most power circuits, a relatively high permeability core constrains the flux to paths within the core.

Figure 5-7 shows a toroidal core with a set of windings on it. If the core material has a very high permeability, the magnetic field can be thought of as being entirely confined to the core material and the amount of magnetic flux which escapes the core can be neglected with negligible error.

The actual inductance of an inductor is caused by the energy stored within the magnetic field. This energy is given by:

$$W = \int ni \, d\phi \quad \text{joules} \tag{5.4}$$

where

ϕ = total number of flux lines
$\phi = \vec{B}A$
n = number of turns
A = the area of the flux field (meters^2)
i = current in amperes

Now if μ is constant and independent of field strength as in air, this integral may be solved:

$$W = \frac{ni\,\phi}{2} \quad \text{joules} \tag{5.5}$$

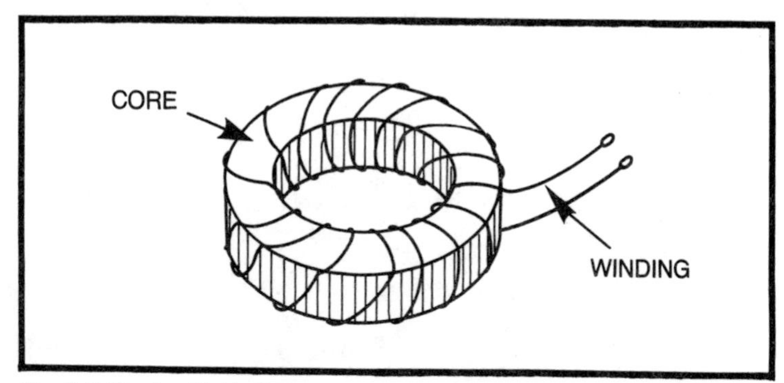

Fig. 5-7. The toroidal inductor.

The self-inductance of a coil is defined as the incremental number of flux linkages per incremental change in current or:

$$L = \frac{n\,d\phi}{di} = \frac{dW}{i\,di} \quad \text{webers/ampere (or henrys)} \qquad (5.6)$$

and

$$L = \frac{n\phi}{i}$$

This resolves to:

$$L = \frac{2W}{i^2} \text{ henrys} \qquad (5.7)$$

This is the same relation given in equation 2.5 which was arranged in the form

$$W = \frac{Li^2}{2} \text{ joules} \qquad (5.8)$$

This is the energy that appeared in the lively sparks jumping from the switch in the Henry magnet.

Now within the closed torodial magnetic path of the coil of Fig. 5-7 the magnetomotive force is:

$$H = \frac{n\,i}{s} \text{ amperes/meter} \qquad (5.9)$$

where

s = the average circumference of the toroidal core (meters)

The flux lines are all circles concentric with the axis of the toroid and if the toroid is relatively thin, that is if the inside diameter is close to the outside diameter, the flux is fairly uniform. Under these circumstances, and assuming μ is constant:

$$\phi = \vec{B}A = \mu HA \qquad (5.10)$$

$$= \frac{ni\mu A}{s} \text{ webers}$$

but

$$L = \frac{n\phi}{i} \text{ henrys} \qquad (5.11)$$

thus

$$L = \frac{n^2 \mu A}{s} \text{ henrys} \qquad (5.12)$$

It must be remembered that this relationship employs the absolute μ of the core material. The same relationship is sometimes given using the relative permeability. In this case is works out thus:

$$L = 4\pi \times 10^{-7}\ \mu_r\, n^2 A/s \text{ henrys} \qquad (5.13)$$

Power transformers and inductors are frequently wound on toroids or on split *E-I* or *U-I* cores, and for these the formula is directly applicable. For most commercial cores, catalog values are given for the average circumference.

THE TRANSFORMER: BACKGROUND

After its invention in 1831 by Faraday in England the transformer lay "fallow" for many years, preserved mainly as a laboratory curiosity. As we have seen, there was no source of altering current in common use save the interrupter, and the nature and workings of the transformer were but imperfectly appreciated. Neither the telegraph nor the telephone services felt the need of or appreciated the value of the transformer.

Some 47 years later, in 1878, Elihu Thompson invented the slipring alternator. The following year, at the Franklin Institute, Thompson demonstrated the use of an induction coil to step down current and increase voltage. Thompson was one of the founders of the Thompson-Huston Co. which was later to become The General Electric Co. Thompson had a rudimentary vision of the fundamental advantages of the ac system of lighting.

Edison contributed to the transformer in several curiously indirect ways. In 1877 he invented the carbon telephone transmitter which made telephony over appreciable distances practical and indeed, by 1886, J. J. Carty patented (U.S. Pat. 348,512) a "mutual inductance" hybrid transformer which permitted 2^{n-1} pairs of telephone conversations to flow over *N* pairs of wires.

In the early 1880s there was a widespread clamor for the installation of electric lighting systems. The rising success of the telephone systems springing up across the country had pre-sold the public on electricity. However, Edison's dc low voltage (100 to 150 volt) systems needed very large wires to carry modest amounts of power over very limited distances if i^2R losses were not to be troublesome. A school of thought sprang up which held that ac power generated at very high voltages could be shipped on small wires with little loss and then stepped-down to household levels at the destination.

Unfortunately, alternating-current phenomena were not clearly understood, and they can be a bit mystifying as shown in Chapter 4. Also, switching and safety equipment was nonexistent or home-brew in nature, and the high ac voltages were lethal. The decade was marked with debate and acrimony and Edison himself denounced ac as being without merit either scientifically or commercially.

However, the ac forces continued. In 1882 Lucien Gaulard of France and John Dixon Gibbs of England applied for a patent on an ac lighting and distribution system using "inductance coils" for voltage reduction. The U.S. patent rights were bought by George Westinghouse (the inventor of the railway air brake) and William Stanley was hired to perfect the system. Stanley found that the iron wire cores specified were difficult to make with small air gaps. He invented the core with the wires wound on the center bar of the "H." He stamped the cores from iron daguerretype (photographic) plates and insulated them with paper to prevent eddy-current losses. His transformers were the first with high enough coupling coefficients to permit primary-secondary isolation rather than simple inductive loading or autotransformer action.

Then, Nikola Tesla invented the rotating magnetic field concept and the induction motor in 1888. These were quickly followed by the synchronous and split-phase motors. The invention of good ac motors clearly made ac competitive for power as well as lighting.

Westinghouse installed the first ac lighting system in Great Barrington, Maine, in 1886. However, the real major blow to dc came in 1891 when Westinghouse installed the first large ac hydroelectric plant at Niagara Falls, N.Y. This plant incorporated many of Tesla's ideas and operated at 133 cycles.

In the final analysis, the ability of the transformer to raise or lower voltage cheaply, efficiently, and without rotating machinery, which required lubrication and attention, won the day.

However, the low voltage dc systems died hard. Portions of New York and Detroit were dc electrified until 1950. In the past decade there has been a trend to extra high voltage (1000 kV or more) dc for long-haul power shipping with solid state and gas tube inversion to 60 cycle ac at the destination (see J. Kuecken-U.S. Pat. 3,716,707). By the mid-1970s about 25 EHV-dc Systems were in operation with nearly 5000 miles of line. A 500 kV dc cable has been laid beneath the waters of the Skagerrak reaching depths of 1500 meters. The use of dc interties seems to be the path of the future. However ac local distribution remains unchallenged.

TRANSFORMER THEORY

In this section we shall be using conventional electrical engineering approaches to ac problems. This will include interchange between exponential (Ae^{je}) notation and complex ($a + jb$) notation. Time variation will be assumed implicity; that is, both $E = E_o \sin \omega t$ and $i = i_o \sin \omega t$ will be written simply as E and i unless otherwise specified. The reader is referred to:

Kerchner & Corcoran—"Alternating Current Circuits"
Ed II chapters I thru V published by John Wiley &
Sons, or some equivalent text for a fuller exposition

The Basic Transformer Circuit

To begin with, let us examine the basic transformer circuit as shown in Fig. 5-8. For the primary circuit:

$$E = i(R_1 + j\omega L_1) + j\omega M\, i_2 \qquad (5.14)$$

And for the secondary circuit:

$$O = j\omega M i_1 + i_2 \left[R_2 + j\omega L_2 + Z_1\right] \qquad (5.15)$$

These equations are generally known as *mesh equations* and are based upon the principle that the algebraic sum of the voltage drops around a loop is zero. The circuit of Fig. 5-8 is treated as if it were as shown in Fig. 5-9. The behavior of the magnetic coupling between the primary and secondary halves of the circuit (the paths of i_1 and i_2 respectively) has been resolved into an inductive reactance called the ***mutual inductance*** with the value $j\omega M_{12}$. This is somewhat of a mathematical artifice, however only slightly, since it is possible to physically measure mutual inductance. Similarly, the losses in the

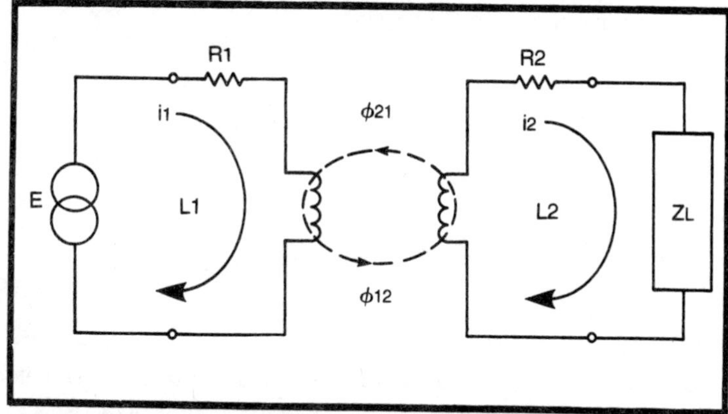

Fig. 5-8. The basic transformer circuit.

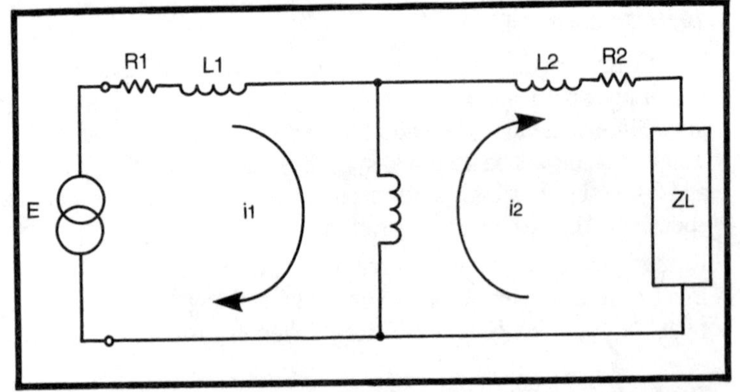

Fig. 5-9. The basic transformer circuit as it is drawn for use with mesh equations.

primary and secondary loops and the remaining inductance in the primary and secondary circuits have been resolved into R_1 and R_2, and L_1 and L_2, respectively. The output load is given as Z_L where:

$$Z_L = R_L + j\ (\omega L_L - \frac{1}{\omega C_L} \qquad (5.16)$$

This arrangement allows for an easy and flexible analysis of the situation without having to trace everything from primary to secondary through the flux linkages and their time-rate-of-exchange. Solving equation 5.15 for i_2:

$$i_2 = \frac{-j\omega M i_1}{R_2 + j\omega L_2 + Z_L} \qquad (5.17)$$

Substituting 5.17 into 5.14 yields:

$$\frac{E}{i_1} = Z_{IN} = R_1 + j\omega L_1 + \frac{\omega^2 M^2}{R_2 + j\omega L_2 + Z_L} \qquad (5.18)$$

In a high efficiency transformer:

$$R_1 << \omega L_L \text{ and } R_2 << \omega L_2$$

and

$$Z_{IN} \approx j\omega L_1 + \frac{\omega^2 M^2}{+j\omega L_2 + Z_L} \qquad (5.19)$$

Let us examine the makeup of some of these parameters before proceeding further. In equation 5.6 we defined inductance as:

$$L = \frac{nd\,\phi}{di} \text{ henrys} \tag{5.20}$$

by a similar token we note:

$$M_{21} = \frac{n\,d\phi_{21}}{di_1} \text{ henrys} \tag{5.21}$$

and

$$M_{12} = \frac{nd\phi_{12}}{di_2} \text{ henrys} \tag{5.22}$$

Thus M is concerned with the portion of the primary flux which links the secondary, and vice versa. Now, if we define a coefficient of coupling on the basis of this ratio we obtain:

$$K = \sqrt{\frac{\phi_{12}}{\phi_1} \; \frac{\phi_{21}}{\phi_2}} \tag{5.23}$$

$$K = \sqrt{\frac{M_{12}}{L_1} \; \frac{M_{21}}{L_2}} \tag{5.24}$$

(k is pure numeric)
Now, in a medium where μ is constant, the ratios in equation 5.22 have to be equal, thus $M_{12} = M_{21}$ and

$$K = \frac{M}{\sqrt{L_1 L_2}} \tag{5.25}$$

In air core transformers with widely separated windings K may take on values as small as 0.01 or less. This is called loose coupling and is generally used only in resonant circuits. Conversely, a well-designed power transformer may have $K = 0.99$ and is described as tightly coupled.

It is worthwhile to examine the physical significance and circuit effects of some of these parameters. Perhaps the easiest way to do this is with limiting-case examples.

In a power transformer, assume $K = 1$ and $R_1 = 0$. If we short circuit the output ($Z_L = 0$) equation 5.19 becomes:

$$Z_{IN} = j\omega L_1 + \frac{\omega^2 M^2}{j\omega L_2} \tag{5.26}$$

but from equation 5.25, for $k = 1$

$$M^2 = L_1 L_2 \tag{5.27}$$

therefore

$$Z_{IN} = j\omega L_1 + \frac{\omega L_1 \, \omega L_2}{j\omega L_2} \tag{5.28}$$

$$Z_{IN} = j\omega L_1 - j\omega L_1 = 0 \tag{5.29}$$

The input or primary circuit is short circuited also! This is an important result. It tells us that in this limiting case of a transformer with zero winding resistance and perfect coupling (i.e., every primary flux line links the secondary winding), a short circuit on the output would be indistinguishable from a short on the input.

In the converse case, if we open-circuit the secondary ($Z_L = \infty$). We see by inspection that only the primary self-impedance remains.

$$Z_{IN} \cong j\omega L_1 + \frac{\omega^2 M^2}{\infty} \tag{5.30}$$

This is also an important case because it describes the condition of the transformer in no-load operation. It can be shown that this case yields the highest peak flux within the core.

For a third case, representative of typical operation, $Z_L >> \omega L_2$ and:

$$Z_{IN} \cong j\omega L + \frac{\omega L_1 \, \omega L_2}{Z_L} \tag{5.31}$$

The voltage drop across the load is:

$$E_{\text{load}} = Z_L \, i_2 \tag{5.32}$$

and, from equation 5.17

$$E_{\text{load}} = \frac{-j\omega M i_1}{R_2 + j\omega L_2 + Z_L} \tag{5.33}$$

and from equation 5.18

$$E_{\text{load}} = \frac{-(j\omega M) Z_L E}{R_1 + j\omega L_1)(R_2 + j\omega L_2 + Z_L) + \omega^2 M^2} \tag{5.34}$$

For the lossless case $R_1 = R_2 = 0$ and for $K = 1$

$$E_{\text{load}} = \frac{-j\,\omega M Z_L E}{-\omega^2 L_1 L_2 + j\omega L_1 Z_L + \omega^2 M^2} \tag{5.35}$$

but from equation 5.25 for $K = 1$

$$L_1 L_2 = M^2 \tag{5.36}$$

therefore

$$E_{\text{load}} = \frac{-j\omega \sqrt{L_1 L_2}\, L_2 E}{j\omega L_1 Z_L} \tag{5.37}$$

$$E_{\text{load}} = \frac{-\sqrt{L_2}}{\sqrt{L_1}}\, E \tag{5.38}$$

but from equation 5.12, for a toroid

$$L = \frac{n^2 \mu A}{S} \tag{5.39}$$

therefore

$$E_{\text{load}} = \frac{-n_2}{n_1}\, E \tag{5.40}$$

The voltage is proportional to the turns ratio in lossless, tight-coupled ($k = 1$) transformers and is 180° out of phase, but pure real. Of course, anyone who has ever had anything to do with transformers knew that all along! Why go through all that long-winded math exercise for something so obvious?

The answer is that the departure from this behavior is very rapid with decreasing K. Suppose, for example, that only 10% of the primary flux does not link the secondary and vice versa.
then

$$K = 0.9 = \frac{M}{\sqrt{L_1 L_2}} \tag{5.41}$$

then

$$M^2 = 0.81\,(L_1 L_2) \tag{5.42}$$

and with $R_1 = R_2 = 0$ from equation 5.35

$$E_{load} = \frac{-.9 j \omega \sqrt{L_1 L_2} Z_L E}{-\omega^2 L_1 L_2 + j \omega L_1 Z_L + .81 \omega^2 L_1 L_2} \quad (5.43)$$

$$E_{load} = \frac{+0.9 j\omega \sqrt{L_1 L_2} Z_L E}{0.19 \omega^2 L_1 L_2 - j \omega L_1 Z_L} \quad (5.44)$$

only for the case: $\omega L_1 Z_L >> 0.19 \omega^2 L_1 L_2$ or $Z_L >> 0.19 \omega L_2$ does the load voltage approach

$$E_{load} \approx -0.9 \frac{n_2}{n_1} E \quad (5.45)$$

In other words, the output voltage has sagged by the ratio of the coupling coefficient with light loads or open circuit conditions.

For smaller values of Z_L the voltage on the load decreases rapidly, and the transformer is said to be poorly regulated. If the power line supply transformer did this, the lights would dim every time a motor started or an incandescent lamp was switched on. This can be very destructive of stability on a servo loop. While such behavior is poor in a power line transformer it may be actually desirable in such applications as neon sign transformers, arc welders, and ignition spark coils. In each of these cases the load is an ionizing affair which requires a higher voltage to strike than to sustain. Once the arc is struck, a constant current (or zero regulation) performance is desirable to limit the dissipation in the load to reasonable levels. This poor regulation is attainable directly in the transformer by either decreasing K or making one of the Rs appreciable, or it may be attained through addition of an external impedance. Reducing coupling has the advantage of limiting primary current without increasing heating.

With the secondary shorted but $K = 0.9$, equation 5.28 becomes:

$$Z_{IN} = \frac{E}{i_1} = j\omega L_1 + \frac{0.81 (\omega L_1 \omega L_2)}{j\omega L_2} \quad (5.46)$$

$$= j\omega L_1 - 0.81 j\omega L_1 = 0.19 j\omega L_1 \quad (5.47)$$

Thus, restricting K to 0.9 has lowered the open-secondary to shorted-secondary current ratio to about 1 : 5.

With small transformers in the range below about 200 VA, it is not unusual to find that the resistive terms are very significant and cannot be neglected. For example an Archer No. 273-1480 transformer rated at 24 V, 1.2 A output will actually deliver about 27 *V*rms no-load and will sag to about 21 *V*rms at the full 1.2 A rated output. The coupling coefficient is actually nearly 0.96, but the resistance of the primary winding is significant also, and the two combine to produce the sag. The equipment must be designed so that a soaring no-load voltage does not destroy components. On the other hand, some amount of sag may be desirable to limit currents in equipment which may be subject to occasional transient short circuits on the output.

Resonant Effects

Offhand, it would seem reasonable to expect the input current to even a poorly coupled transformer to be minimum with the secondary open circuited (i.e., $Z_L = 0$). However, this is not always the case! Recalling equation 5.18

$$Z_{in} = \frac{E}{i_1} = R_1 + j\omega L + \frac{\omega^2 M^2}{R_2 + j\omega L_2 + Z_L} \tag{5.48}$$

let us separate $Z_L = R_L + jX_L$ where X_L can be any combination of reactances.
Then

$$Z_{in} = R_1 + j\omega L_1 + \omega^2 M^2 \left(\frac{1}{(R_2 + R_L) + j(\omega L + X_L)}\right) \tag{5.49}$$

Separating reals and imaginarys and removing imaginarys from the denominator yields

$$Z_{in} = R_1 + j\omega L_1 + \frac{\omega^2 M^2 (R_2 + R_L)}{(R_2 + R_L)^2 + (\omega L_2 + X_L)^2} - j\frac{\omega^2 M^2 (\omega L_2 + X_L)}{(R_2 + R_L)^2 + (\omega L_2 + X_L)^2} \tag{5.50}$$

Now if

$$j\omega L_1 = \frac{+j\omega^2 M^2 (\omega L_2 + X_L)}{(R_2 + R_L)^2 + (\omega L_2 + X_L)^2} \tag{5.51}$$

then the reactive terms in equation 5.50 cancel and real terms remain; and:

$$Z_{res} = R_1 + \frac{\omega^2 M^2 (R_2 + R_L)}{(R_2 + R_L)^2 + (\omega L_2 + X_L)^2} \tag{5.52}$$

Now, for the special case where $(R_2 + R_L) < < (\omega L_2 + X_L)$, equation 5.51 reduces to:

special case

$$\omega L_1 \cong \frac{\omega^2 M^2}{\omega L_2 + X_L} \tag{5.53}$$

and from equation 5.24

$$\omega L_1 = \frac{\omega^2 K^2 L_1 L_2}{\omega L_2 + X_L} \tag{5.54}$$

Thus

$$\omega^2 L_1 L_2 + \omega L_1 X_L \cong \omega^2 k^4 L_1 L_2 \tag{5.55}$$

and

$$\omega L_1 X_L \cong \omega^2 L_1 L_2 (k^2 - 1) \tag{5.56}$$

therefore

$$X_L \cong \omega L_2 (k^2 - 1) \tag{5.57}$$

and for $K = 0.9$ as in the example of 5.41 through 5.47

$$X_L \cong 0.19 \omega L_2 \tag{5.58}$$

Substituting 5.57 into equation 5.52 gives

$$Z_{res} \cong R_1 \frac{\omega^2 K^2 L_1 L_2 (R_2 + R_L)}{[\omega L_2 + \omega L_2 (K^2 - 1)]^2} \tag{5.59}$$

$$\cong R_1 + \frac{\omega^2 K^2 L_1 L_2 (R_2 + R_L)}{\omega^2 L_2^2 (1 + K^2 - 1)^2} \tag{5.60}$$

$$\cong R_1 + \frac{\omega^2 K^2 L_1 L_2 (R_2 + R_L)}{\omega^2 K^4 L_2^2} \tag{5.61}$$

$$Z_{res} \cong R_1 + \frac{L_1}{L_2} \frac{(R_2 + R_L)}{K^2} \tag{5.62}$$

It may be readily seen that the second term of equation 5.62 can easily be much smaller than $\omega L\ (1-K^2)$ which would be the input impedance with the secondary short circuited (for negligible R_1).

This result is interesting in the fact that it tells us that a real (and imperfect) transformer can be made to draw more primary current with specific finite loads than with a short circuit on the output. A transformer which was designed to be self-protecting from short circuits can be blown up by a finite load!

Now a second interesting case takes place when the secondary alone is resonant (i.e., when $X_L = j\omega L_2$). Then equation 5.50 becomes:

special case

$$Z_{\text{res}} = R_1 + j\omega L_1 + \frac{\omega^2 M_2}{R_2 + R_L} \tag{5.63}$$

$$= R_1 + j\omega L_1 + \frac{\omega^2 K^2 L_1 L_2}{R_2 + R_L} \tag{5.64}$$

Note that the third term is pure real and that Z_{res} varies inversely with $(R_2 + R_L)$. Since this third term cannot go negative and be cancelled by R_1, this impedance is always higher than the open circuit impedance. Thus, we observe that in an imperfectly coupled transformer the input impedance can be lower than that obtained with the secondary shorted or higher than the input impedance with the secondary open.

To illustrate, let us consider a transformer with the following parameters:

$$R_1 = R_2 = 0$$
$$j\omega L_1 = j\omega L_2 = +j\omega/\omega_o\ 500$$

$$R_L = 50\ \Omega$$
$$X_L = j500\ \omega_o/\omega$$

Note that for $\omega/\omega_o = 1\ j\omega L_2 = -jX_L$; therefore, X_L is a capacitor. If we solve equation 5.51 for $\omega/\omega_o = 2$ we obtain a value of $K^2 = 0.75333\ldots$and at

$$\omega/\omega_o = 2 \qquad \text{(Eq. 5.52)}$$

$$Z_{\text{res}} = \frac{K^2\omega^2 L_1 L_2\ (R_L)}{R_L^2 + (\omega L_2 + X_L)} = 66.67\ \Omega$$

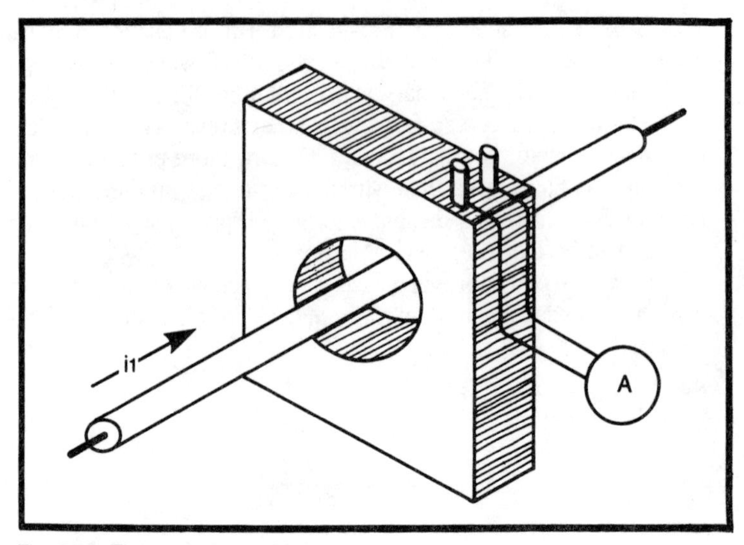

Fig. 5-10. The current transformer.

and at

$$\omega/\omega_o = 1 \qquad \text{(Eq. 5.64)}$$

$$Z_{\text{res}} = j\omega L_1 + \frac{\omega^2 K^2 L_1 L_2}{R_L} = 3765 + j500$$

We observe that the loose-coupled transformer can be deliberately designed to present a high input impedance at $\omega/\omega_o = 1$ and a relatively low input impedance at the second harmonic (i.e., $\omega/\omega_o = 2$) or some other frequency. This is frequently done in radio transmitters.

The Current Transformer

Another modification of the transformer finds wide use in metering circuits. Rewriting equation 5.17

$$i_2 = \frac{-j\omega M i_1}{R_2 + j\omega L_2 + Z_L} \qquad (5.65)$$

we note that for $j\omega L_2 >> R_2 + Z_L$ we obtain:

$$i_2 \cong \frac{-j\omega M i_1}{j\omega L_2} = \frac{-M i_1}{L_2} \qquad (5.66)$$

Note that ω drops out. This means that the induced voltage has no frequency sensitivity. This is the principle on which the current

Note that ω drops out. This means that the induced voltage has no frequency sensitivity. This is the principle on which the current transformer operates. The line to be metered is simply passed through the center of an iron toroid, unbroken as shown in Fig. 5-10. The ammeter is placed in the high inductance secondary.

Note that very high voltages can be developed with the secondary circuit opened. A current transformer should always be short-circuited when not in use.

In a clamp-on ammeter the core and winding is simply split to permit the action.

A current transformer was used to monitor the current waveform in the measurement set-up of Fig. 2-5. Since the oscilloscope is a high impedance device, a low resistance is used to load the secondary of the current transformer so that the condition of equation 5.66 is satisfied.

A current-transformer function can be obtained by adding a one or two turn winding to an ordinary small power transformer and *terminating all other windings in low resistance loads*. If a power transformer is used, the "no frequency sensitivity" statement must be modified by the nature of core losses at high frequency. A ferrite toroid of high permeability material can yield a current transformer that is flat from 40 Hz to several hundred megahertz.

Ferromagnetic Cores

The derivations of B, flux density, and H, magnetic field strength, in this chapter showed them to be related by μ-permeability as shown in equation 5.2

$$\vec{B} = \mu\vec{H} \tag{5.67}$$

In ferromagnetic materials μ is far from constant, and this factor must be accounted for in the design of inductors and transformers.

Modern magnetic theory holds that all magnetic phenomena are related to moving currents, even if only spinning electrons in atomic shells. This theory holds that the origin of ferromagnetism in materials such as iron, cobalt, and nickel is due to spinning electrons within the third, incomplete, shell of the atoms. These electrons create a magnetic moment. In neighboring atoms the magnetic moments are held parallel by quantum-mechanical forces. An aligned ensemble is known as a domain. In an unmagnetized material the domains are randomly oriented and cancel so that no gross magnetization exists. When an external field is applied the domains undergo a realignment, and their moments are added to the applied field, resulting in a great increase in flux.

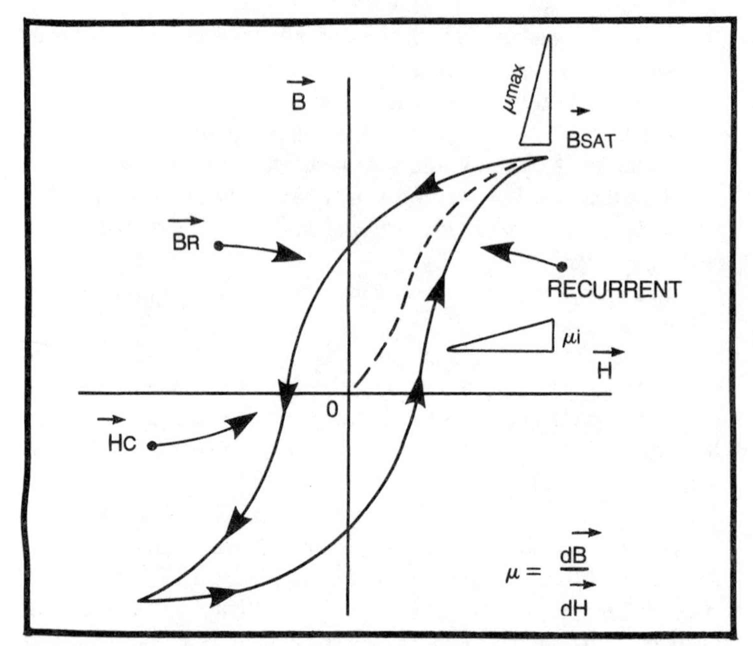

Fig. 5-11. The normal magnetization curve and hysteresis loop.

The theory serves to adequately explain experimental observations of μ which are required to obtain knowledge of any ferromagnetic material. The curve of Fig. 5-11 illustrates characteristic properties. On initial magnetization, the dotted curve is followed. The near-zero magnetization, the value (μi) is only about one-third to one-fifth as great as the maximum (μ_{max}). This corresponds to the existence of a domain threshold of energy required to "flop" the domains which are easiest to align. In the next region there is a near-linear, but offset, rise in $\vec{B}$ which corresponds to an increase to μ_{max}. In this region the domains are thought to be flopping from one "easy" condition to another "easy" state more nearly aligned with $\vec{H}$. In this near-linear, but offset, region the Barkhausen noise is very detectable.

The domain "flops" are quantized and a rushing noise will be heard as the magnet of Fig. 5-12 approaches the nail. The linear rise is actually composed of a series of tiny steps.

Following this, a further increase in $\vec{H}$ yields less and less increase in $\vec{B}$ until $\vec{B}_{MAX}$ is attained. After this, the material is fully aligned and $\vec{B}$ increases at the same rate as free space. On relaxation of $\vec{H}$ to zero the material follows the dotted path of Fig. 5-13 and winds up with a residual magnetization $\vec{B}_R$. Reversal of $\vec{H}$ will carry the value of $\vec{B}$ through the "S" shaped hysteresis loop shown in Fig.

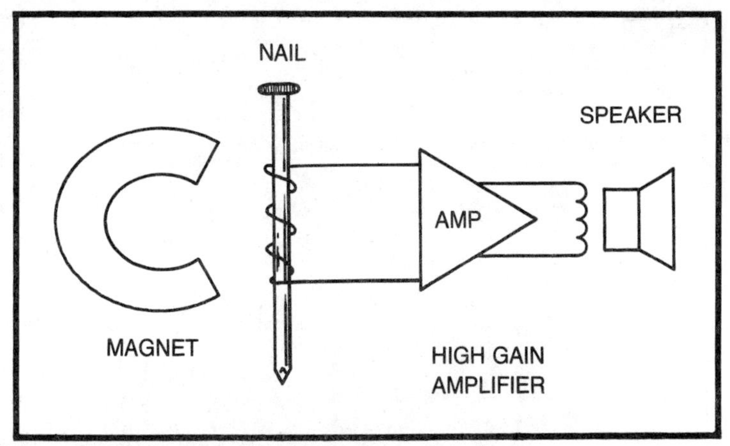

Fig. 5-12. Barkhausen noise demonstration.

5-11 with $\vec{B}$ going through zero at $\pm\vec{H}_c$ the coercive force of the material.

The actual size and shape of the hysteresis loop of an iron material varies not only with the chemical constitution but also with direction of rolling, condition of annealing, aging, and other factors.

In recent years a series of ceramic or composition type materials have been developed which exhibit ferromagnetic properties along with a high bulk resistivity which makes them attractive in many applications. These materials are ferrites, spinels, and

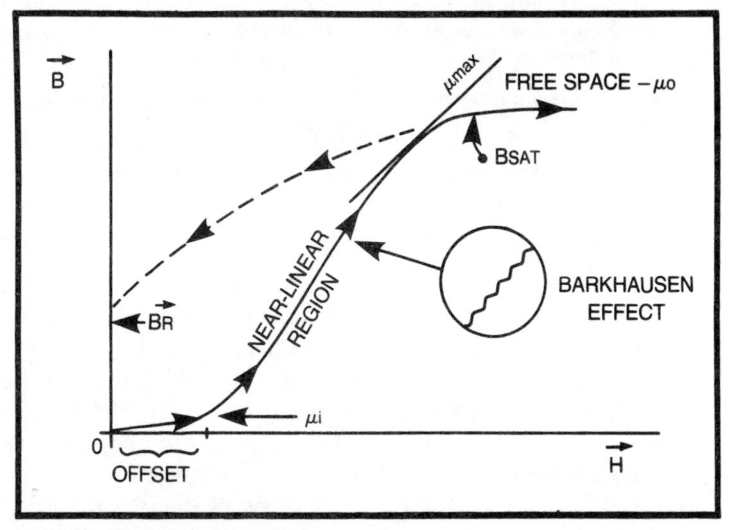

Fig. 5-13. The initial magnetization.

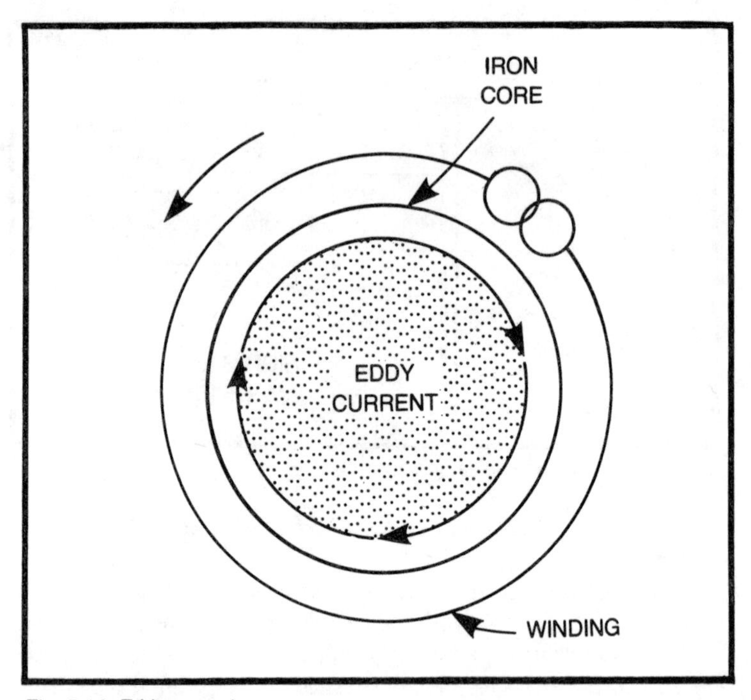

Fig. 5-14. Eddy currents.

carbonyl-iron or iron-powder compositions. They are widely used at higher frequencies.

Since our purpose here is not to delve deeply into the design of specific devices, we will not proceed much further into the design of iron core devices. The reader is referred to:

Stephen S. Attwood—Electric and Magnetic Fields, ch. 13—Dover Publications

Also, a very workable design procedure for small (a few KVA or less) transformers is included in:

Reference Data for Radio Engineers—ITT, pp 271-303.

We are, however, concerned with the principles of some of the sequelae of the facts and also with a little more detail on the ferrite devices since these are not so well represented in the literature.

In actual transformers we are confronted with the fact that a solid iron core excited with an alternating $\vec{B}$ (i.e., $\vec{B} = \vec{B}_1 \cos \omega t$) would experience an eddy-current induction; that is, eddy currents would be induced in the virtual "shorted-turn" on the skin of the core and would oppose $\vec{H}$ inside the core while contributing to losses (Fig. 5-14). This effect is countered by laminating the core from thin iron sheets, which are insulated from one another by varnish. A thickness of B & S 29 gauge is commonly used at 60 Hz (this is 0.014 inch or 0.0356 mm thick). For audio frequencies up to 10 kHz, thickness down to 0.001 inch is used. It is here that the ferrites, with their naturally high resistivity, show a clear advantage since eddy current losses are negligible. The bulk material is a reasonably good ceramic insulator. The powdered and carbonyl irons consist of tiny particles suspended in an insulating binder and thus also show a high resistivity in the bulk material. When measured by ac techniques, the eddy current and hysteresis losses are lumped together in the $\vec{B}/\vec{H}$ loop. The circuit of Fig. 5-15 can directly plot the hysteresis loop on the face of a cathode ray tube. The voltage in the primary winding must be sinusoidal to oppose the generator voltage, therefore:

$$\phi = \phi_M \cos \omega t = \vec{B}_m \quad A \cos \omega t \qquad \text{webers} \qquad (5.68)$$

and

$$V_2 = -n_2 \frac{d\phi}{dt} = n_2 \vec{B}_m \quad A\,w \sin wt \qquad \text{volts} \qquad (5.69)$$

$$\vec{H} = \frac{n_1 i_1}{S} \qquad \frac{\text{amperes}}{\text{meter}} \qquad (5.70)$$

where

A = core cross section area in meters
s = core magentic path length
$\vec{B}_m$ = max core flux

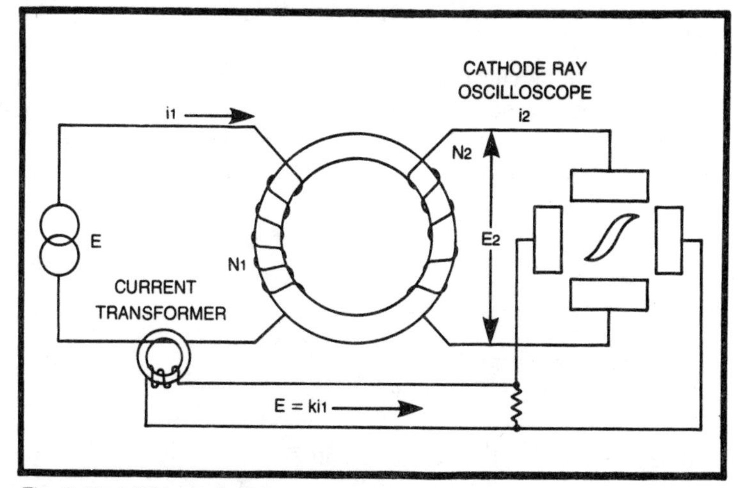

Fig. 5-15. A B/H curve plotter.

Note that if μ varies during the cycle, both $\vec{H}$ and i_1 must be nonsinusoidal. If a large resistor were placed in series with the primary winding, thus forcing i_1 to be sinusoidal, ϕ, $\vec{B}$ and V_2 would be nonsinusoidal. This waveform distortion is one of the sequelae of the hysteresis loop.

Note also the presence of the ω in equation 5.69. This tells us that higher frequencies require less flux in the core to yield a given voltage. The latter factor is very significant at high frequencies.

Equation 5.69 is frequently used in the design of transformers since it determines the number of volts per turn permissible on a given core. It is often given in the form:

$$\frac{V_2}{n_2} = \vec{B}_m A \omega \qquad (5.71)$$

since $\vec{B}_m$ is a property of the material and A is a property of the core construction this determines for a given core stack how many volts/turn are permissible without saturating the core. In small power transformers, it is always a fight to get enough turns of large enough wire through the window in the core. These units very frequently are designed to run well up on the $\vec{B}$ curve and will usually show a very sharply peaked current waveform in the no-load condition. This leads to increased hysteresis losses in the no-load condition which causes core heating. The no-load core heating is balanced against the wire resistance heating under loaded conditions. Fewer turns are required if the $\vec{B}_m$ limit is raised; therefore, larger and lower resistance wire can be used.

Hysteresis

Next, let us consider the area enclosed within the hysteresis loop. This represents the energy per cycle (or trip around the loop) per cubic meter of magnetized volume. This energy is dissipated in heat in physically aligning the domains and in eddy-current heating. Power loss is:

$$P_H = \text{volume} \times f \times \text{area of loop} \quad \frac{\text{joules}}{\text{second}} = \text{watts} \qquad (5.72)$$

where

$f = \text{frequency} = \dfrac{\omega}{2\pi}$

If we assume that the area of the loop is:

$$\text{area of loop} \approx 4\, \vec{H}_c\, \vec{B}_m \qquad (5.73)$$

we obtain a rough empirical estimate:

$$P_H \cong \text{volume} \times 4 f\vec{H}_c\vec{B}_m \qquad (5.74)$$

From the above it is apparent that with increasing frequencies materials with small values of $\vec{H}_c$ must be used if excessive losses and consequent core heating are to be avoided.

It should be noted that in Fig. 5-15 the secondary of the transformer is running essentially open-circuited; i.e., i_2 approaches zero.

Now a transformer operates because the flux from the secondary current tends to cancel the flux from the primary current. Thus:

$$\vec{H}_{tot} = \vec{H}_1 + \vec{H}_2 \qquad (5.75)$$

$$\vec{H}_{tot} = \frac{n_1 i_1}{S} + \frac{n_2 i_2}{S} \qquad (5.76)$$

and from equation 5.77

$$\vec{H}_{tot} = \frac{n_1 i_1}{S} + \frac{-n_2 j \omega M i_1}{S R_2 + j \omega L_2 + Z_L} \qquad (5.77)$$

and from equation 5.25

$$\vec{H}_{tot} = \frac{i_1}{S} \quad (n_1 - \frac{j n_2 \omega \sqrt{L_1 L_2 k})}{R_2 + j \omega L_2 + Z_L)} \qquad (5.78)$$

It may be seen that, with tightly coupled transformers, secondary current can considerably reduce B_{max} if i_1 remains constant, as in a tube or transistor plate or collector circuit which tends toward a constant-current action. In such installations a transformer will frequently burn up from core losses if the secondary circuit is accidentally opened. Of course, in a well-regulated or stiff, constant voltage, installation $\vec{B}_M$ is determined only by the input voltage, and i_1 rises accordingly, to cancel the reduction in flux from the secondary; thus heating on the core is essentially independent of the load current.

Table 5-1. Parameters of Ferromagnetic Materials.

Material	max (REL)	$\vec{B}_{sat}$	$\vec{B}_r$	$\vec{H}_c$	P_H/F	USE
Alnico XII			0.58	76,000	...	Perm. mag.
Carbon Steel			0.82	4,000	...	Perm. mag.
Supermalloy	1,050,000	0.8	.....	0.32	...	Mag. shield
Hypernik	70,000	1.65	0.73	3.2	3.2	Trans. (pwr)
Ferramic Q-1	400	0.33	0.18	.02640	0.0088	10 MHz Ferrite
Ferramic Q-3	42	0.26	0.14	.264	0.066	225 MHz Ferrite

The value of R used in the previous formulations is made up of a variety of factors including hysteresis and eddy current loss. It also contains the dc and skin effect resistances of the windings. It is common to assign the hysteresis and eddy current losses to the primary circuit in many calculations.

Table 5-1 is included to show the range of variation of parameters encountered in ferromagnetic materials.

At very high frequencies it is common to find that core heating dominates all other considerations in inductors or lightly loaded transformers. Note the Hypernik, used at 100 cycles and Ferramic Q_1 used at 10^7 cycles have a 10^5 difference in frequency but only about a 10^3 difference in P_H/F. Thus, whereas saturation is a major problem at power frequencies, it is seldom encountered at radio frequencies since the inductor would burn up from core losses first.

In passing, it is noteworthy that the catalogues supplied by the vendors of the materials are a fruitful source of design information. Much of the process of inductor design and transformer design is empirical in nature and one of the catalogues provides a good access to such design data.

EXPERIMENTAL EVALUATION

In the design of motor controls, it is frequently necessary to make measurements to determine some of the transformer parameters. These measurements can be readily made with rather modest laboratory equipment. Some of these tests will be outlined.

Primary Inductance and Voltage Rating. The primary inductive reactance can readily be determined with a voltmeter and ammeter and a measurement of the dc resistance of the primary. If the primary current is sinusoidal the primary impedance is given by:

$$Z_{\text{pri}} = \frac{E}{i} \approx R_{DC} + j\omega L_{\text{pri}} \tag{5.79}$$

and

$$j\omega L_{\text{pri}} \approx \sqrt{\frac{E^2}{i^2} - R^2_{DC}} \tag{5.80}$$

This measurement would, of course, be performed with all secondary windings open-circuited. The condition for sinusoidal current in the primary can be obtained by operating the primary at a fraction of its rated voltage. This reduced voltage can be obtained either from a Variac variable autotransformer or by use of a large valve resistor in series with the primary so that the voltage is held down. The relationships, of course, assume that the dc resistance and the dynamic ac resistance are identical. For most practical purposes this is true as long as the current is sinusoidal.

A somewhat better measurement can be obtained if a dual-trace oscilloscope is available or a scope capable of X-Y display on which the voltage/current Lissajous figure may be viewed. This differs slightly from the setup of Fig. 5-15 and is arranged to display the phase angle between the primary voltage and current. Figure 5-16 illustrates the connection employed. The discussion of the Lissajous figures is beyond the scope of this text, but a concise description is given in:

Frederick Emmons Terman
Radio Engineers Handbook
Edition 1 pages 947 and 948
McGraw Hill Book Co. 1943.

This reference is rather somewhat dated but well worthwhile. This measurement technique is simple and essentially foolproof in that it does not particularly require calibrated instruments. The dual-trace oscilloscope method is simpler to use; however, it is not foolproof in the fact that most dual-trace scopes are not truly dual-beam affairs but rather share the beam of the scope on alternate traces and can sometimes lie to you on the subject of phase difference between waves shown in the two traces.

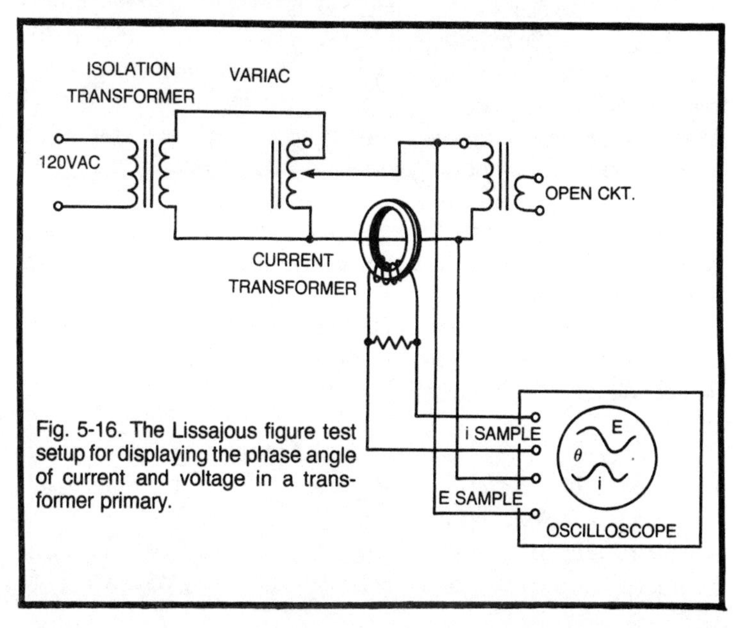

Fig. 5-16. The Lissajous figure test setup for displaying the phase angle of current and voltage in a transformer primary.

In any event, if the phase angle between the voltage and current waveforms θ can be found, the value of the resistive and reactive components can be calculated from:

$$j\omega L_{pri} = \frac{E}{i} \sin \theta \tag{5.81}$$

$$R_{pri} = \frac{E}{i} \cos \theta \tag{5.82}$$

This method gives the true ac resistance component for sinewaves in both voltage and current.

WARNING

The isolation transformer in Fig. 5-16 is not an ornament! The oscilloscope must be either isolated from the line or carefully connected so that its frame is attached to the grounded line terminal. Otherwise a blown fuse or a severe electric shock will result!

The voltage rating of the transformer winding can usually be found by carefully increasing the excitation until the current waveform ceases to be sinusoidal. As shown in Fig. 5-17 the effect of driving B_m too high is that the primary current will cease to be sinusoidal and will take on a definitely peaked characteristic. As noted earlier, this is accompanied by a definite and noticeable increase in core heating. The measurement arrangement of Fig. 5-15 is particularly revealing of this condition; however, the setup of 5-16 will usually detect the departure from the sine wave condition with sufficient accuracy. The best pragmatic test is simply to operate the transformer with no secondary load at the intended voltage and measure the temperature rise of the core with a thermometer. For a normal commercial-grade transformer any rise in excess of 40° C is excessive and the transformer is a fire hazard!

For a transformer that is expected to operate unloaded for any significant portion of the time, a core rise in excess of 25°C represents poor engineering practice.

If an oscilloscope is not available but a Variac and a voltmeter are, a carefully constructed plot of input voltage versus output voltage will reveal the core saturation by the presence of a departure from linearity.

This measurement can be performed on any of the windings of a transformer, but exercise caution since very large voltages can sometimes be developed on the unused windings.

Mutual Inductance. The mutual inductance of any winding to the primary on a transformer can be determined by connecting the

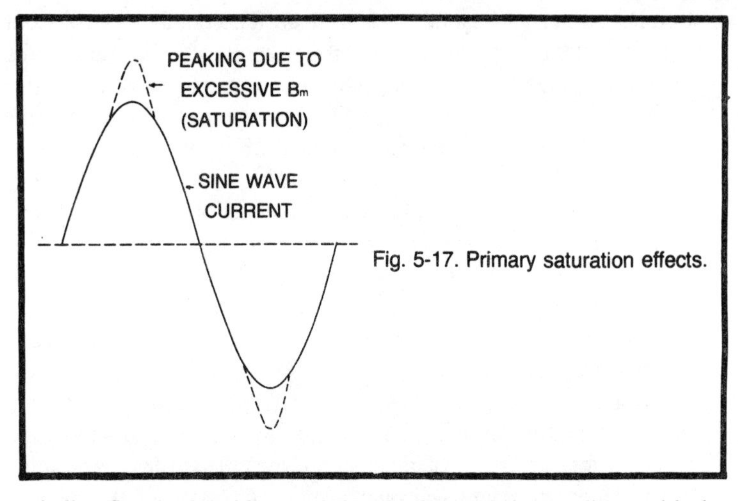

Fig. 5-17. Primary saturation effects.

winding first in the aiding and then in the bucking condition with the primary. The impedances can then be calculated using the techniques just described and the mutual impedance calculated using the relationship described below.

$$Z_{MAX} = j\omega L_1 + j\omega L + 2j\omega M_{12} \quad (5.83)$$

$$Z_{MIN} = j\omega L_1 + j\omega L_2 - 2j\omega M_{12} \quad (5.84)$$

The impedances calculated when the primary and the secondary are aiding the bucking are Z_{MAX} and Z_{MIN}, respectively. Since L_1 and L_2 can be determined separately, the value of M_{12} can be determined by solving the simultaneous equations. It should be noted that these equations ignore the effects of the resistances of the windings, but for most usable power transformers, they are sufficiently accurate to provide a reasonable measure of M_{12}. For a transformer that was designed for deliberately poor regulation, or with very small power transformers, this technique may not be sufficiently accurate to yield a realistic picture of the primary current for the shorted-secondary current. In this case, the best bet is to short the secondary and apply a greatly reduced primary voltage. Then measure the altered L_1 and the ratio of primary-to-secondary current. The method of equations 5.65 and 5.66 may then be applied, after R_2 and L_2 have been separately determined.

Volts Per Turn. In cases where it is desired to determine the number of volts per turn, it is frequently possible to slip a thin foil or wire winding of one or more turns between the coil binding and the transformer core around the outside of the existing winding. A gap is often left to permit assembly of the unit. The number of volts per turn is directly related to the core cross section and therefore to the power rating of the transformer. On small units of 10 VA rating, this may be as low as 75 mV/turn. A 100 VA unit would tend to run in the region of 0.5 V/turn and units in excess of 1 KVA will run upwards of 1 V/turn.

For testing the current handling capability of relay contacts and diodes, SCRs and transistors, it is handy to have a 100 or 200 VA transformer to which a secondary winding of five to fifteen turns has been added, using the largest wire that can be fitted or paralleled wires. Coupled to a Variac, this unit can provide a variable voltage source of high current capability. With five turns for example the unit can produce up to about 2.5 volts at currents up to about 80 amperes, for limited periods. The voltage is sufficient to supply the saturated forward drop of most good switching transistors and a series diode, etc., and will quickly permit testing of the devices without hazardous voltages or risk to the ultimate equipment.

Wire Temperature Rise. The traditional method of measuring the temperature rise of the wires within transformer windings is to measure the resistance at room temperature and then to measure the resistance after the transformer (or motor for that matter) has had enough operation time to permit the windings to stabilize. The temperature coefficient of copper wire is approximately 0.0039/°C. The temperature rise can be found from:

$$R_{HOT} = R_{COLD}\,(1 + 0.0039\,\Delta t) \qquad (5.85)$$

$$\Delta t = °C \text{ RISE}$$

Materials classified as AIEE class 0 which includes paper, cotton, silk, etc., without impregnation should not be exposed to hot-spot temperatures in excess of 90° C. If the motor or transformer is to be operated in ventilated room ambient of 30° C, a 40° or 50° C rise would be about the maximum safe rise. However, most electronic equipment will generate its own ambient well above room temperature and care should be taken to allow for the temperature rise within the equipment case.

WARNING

A large percentage of electrical fires are caused by overheated transformers. A commercial grade transformer can sometimes cause open flames before the insulation ruptures and blows the fuse.

Aside from the safety aspects, a transformer or motor will last a great deal longer if it is restricted to lower operating temperatures.

6
AC Motor Properties

The first alternator of any practical size was constructed by Elihu Thompson at the Franklin Institute in 1878. This was a bipolar machine which differed from the Pixii machine principally in the elimination of the commutator and its replacement with a pair of continuous slip rings. This machine developed a voltage and a torque closely approximating the relationships in equations 3.5 and 3-8. The machine would operate as a generator and under special circumstances as a motor. We will discuss the limitations shortly.

AC DEVELOPMENT

The real push behind the development of the ac machine came from Thomas Alva Edison, who felt that the use of alternating currents was "silly and dangerous." On December 31, 1879, in his 14,000th experiment, Edison started a test of an incandescent lamp which ultimately burned continuously for several months. In 1884, Edison's Pearl Street station was in operation in New York City and Edison hired a Yugoslavian immigrant by the name of Nikola Tesla. Tesla was to make the inventions that swung the infant industry over to alternating current.

During the incandescent lamp experiments, Edison found that the actual efficiency of the dynamo was very low. The principal source of the loss was the existence of eddy currents in the cast iron poles of the magnets. The induced currents in the pole and armature castings gave rise to an opposing torque as in the Arago disc and the Drag-Cup Speedometer. This wound up by wasting more than half of the shaft power pumped into the machine. This wasted power made the machine hot and limited the electrical capacity severely. By constructing the armature of laminated sheet iron with the sheets insulated from one another by an oxide or varnish layer, Edison was able to increase the overall efficiency of the dynamo to 90 percent. The reduction by a factor of five in the losses was the equivalent of increasing the output rating of a given size machine by a factor of five with the additional benefit that it required only about half as much coal to operate.

Edison, in his definitive investigation of the dynamo, also invented the use of fireproof mica insulation between the commutator bars.

However, 20 kilowatt Jumbo machines were still plagued with the problem of brush burning. The modern carbon brush will handle about 40 amperes per square inch in continuous duty (a starter motor runs at three times this level in intermittent duty). However, the Jumbo machines were required to handle about 200 amperes with brass brushes and without the benefit of commutation and equalizer windings which were not yet invented. As a result, the machines had to be shut down frequently to replace the brushes and to smooth the commutator surface.

Nikola Tesla was a flamboyant young man some nine years younger than Edison and they did not strike it off well. However Tesla stayed with Edison long enough to see the advantage of an electric motor which did not require the use of a commutator. In 1888 he patented the rotating-field concept for motor and alternator. The invention was sold to George Westinghouse and played a powerful role in the development of the electric industry.

To understand this invention, let us refer to Fig. 6-1. Here we see depicted a machine with two distinct sets of poles which may be energized independently. By comparison with the Pixii machine it may be seen that the machine is "inside-out" with the magnet rotating in the center, and the windings on the pole pieces. It is also not difficult to see that rotation of the magnet will induce a roughly sinusoidal voltage first in the windings of poles A-C and then 90° later in the windings of poles B-D. The machine will generate two sinusoidal voltages with a frequency ω and with a quadrature phase relationship. Referring to the curves at the bottom of Fig. 4-6 and assuming that the zero point matches the drawn position of the rotating magnet (north pole up), then the output of the A-C winding would be equivalent to the $A \cos \theta$ curve and the B-D winding output would be equivalent to the $A \sin \theta$ curve. A reversal of the rotation would reverse the voltage on the B-D winding so that it became $-A \sin \theta$.

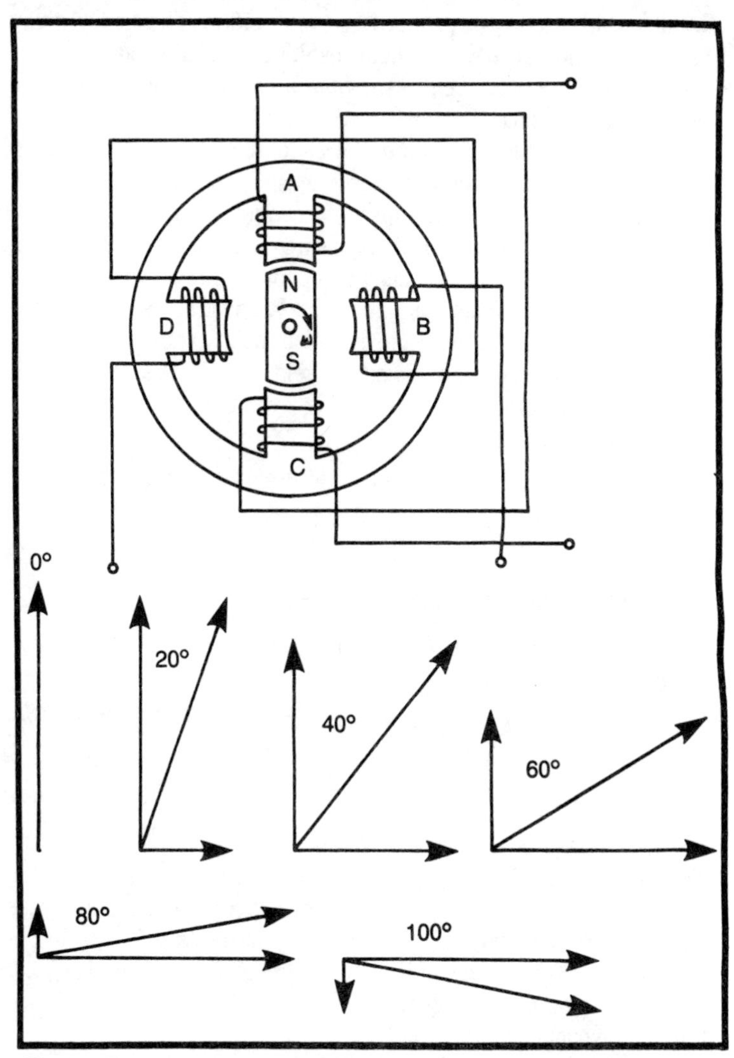

Fig. 6-1. The two-phase alternator. This machine is constructed with two distinct sets of poles which can be energized individually. The rotating element is a permanent magnet. The machine may be operated as either a motor or a generator. The associated vector diagrams illustrate the net magnetic field direction with the two windings excited by quadrature currents.

The principle of this dynamo is not difficult to see; however, the principle by which it will operate as a motor is not so apparent. The series of vector diagrams at the bottom of Fig. 6-1 illustrates the rotation of the magnetic field which would obtain if the A-C winding were energized with a current proportional to $A \cos \theta$ and the B-D winding with a current proportional to $A \sin \theta$. The heavy arrow represents the net magnetic field resulting from the addition of the two alternating fields. This is the *rotating field* principle which was developed by Tesla.

In order to obtain some understanding of the performance of this machine, let us consider the way in which the machine will generate its counter electromotive force and its torque. There are two rotational velocities to be considered: ω_m, the mechanical velocity of the armature in radians per second and ω_e, the electrical velocity of rotation of the magnetic field. We shall also have to define two different angles, ζ the angle by which the instantaneous position of the armature lags the magnetic field and ψ, the phase angle between the voltage and current in the winding.

The winding current:

$$i = i_o \cos (\omega_e t + \psi) \tag{6.1}$$

The armature position with respect to the pole:

$$\cos \alpha = \cos (\omega_M t + \zeta) \tag{6.2}$$

Substituting in equation (3.9) we obtain:

$$\text{Torque} = k \vec{\beta} i_o (\cos \omega_e t + \psi) \cos (\omega_m t + \zeta) \tag{6.3}$$

Using the trigonometric identity:

$$\cos A \cos B = 1/2 \cos (A-B) + 1/2 \cos (A+B) \tag{6.4}$$

we obtain:

$$\text{Torque} = \frac{k \vec{\beta} i_0}{2} \left[\cos ((\omega e - \omega m)t - \zeta) + \cos ((\omega_e + \omega_m) t + \psi + \zeta)\right] \tag{6.5}$$

The last equation, 6.5, is particularly interesting to us since the average value of the cosine terms is zero except in the special case where $\omega_m = \omega_e$. This tells us something interesting, *except for the special case where the motor is in precise synchronism with the line it develops no torque on the average and will not run.*

For the special case $\omega_m = \omega_e$

$$\text{Torque} = \frac{k \vec{\beta} i_o}{2} \left[\cos (\psi - \zeta) + \cos (2\omega t + \psi + \zeta)\right] \tag{6.6}$$

A similar argument can be applied to the double frequency term, $\cos 2\omega t + \psi + \zeta$. This expression also has a zero value when averaged over a complete cycle. The expression, therefore, reduces to:

$$\text{Torque} = \frac{k \vec{B} i_o}{2} \cos (\psi - \zeta) \tag{6.7}$$

This expression is maximum at $(\psi - \zeta) = O$ and reverses at $(\psi - \zeta) = \pm \pi/2$ reaching a negative maximum at $(\psi - \zeta) = \pi$, as shown in Fig. 6-2. The latter performance is interesting, in that it tells us a great deal about this machine. When the torque is positive,

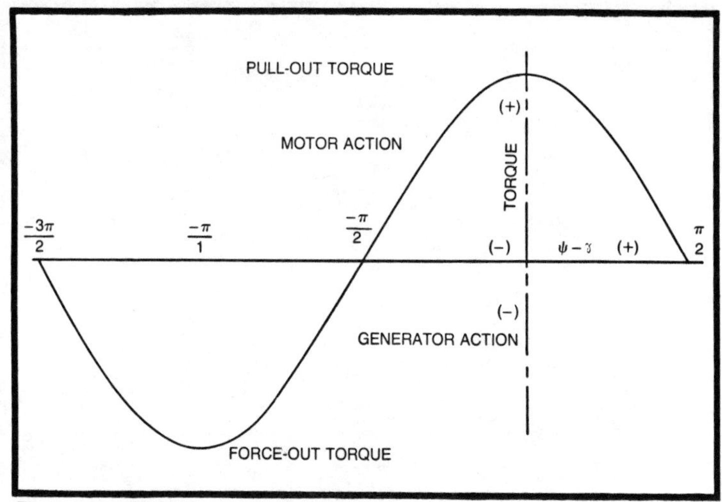

Fig. 6-2. The synchronous motor torque curve.

the device is running as a motor and delivering torque from the shaft. Conversely, when the torque is negative, the machine is absorbing torque from the shaft and delivering electric power to the line. Nearly all of the alternators that are used to supply utility power are machines of this general type and many of them must run on the electric line simultaneously. From the curve of Fig. 6-2 we see that the machine has a definite range of torque in which it will stay "locked-up" with the power line. If it tends to fall behind in phase, it will draw power from the line and speed up the turbine just enough to keep in step. Conversely it will absorb shaft power from the turbine and change only its phase angle up to the *force-out-torque* level without losing line synchronism. Running as a motor it will run in synchronism up to the *pull-out-torque* level without losing synchronism. Beyond this point the motor will pull out of synchronism and stall.

If the motor is lightly loaded the value of ζ falls to nearly zero and the value of ψ approaches $-90°$, and the motor behaves like a condenser, or capacitor. Use is made of this property for power-factor correction in electrical systems. A special variety of synchronous motor is built for this purpose which has no output shaft. In this condition it is referred to as a *synchronous condenser*.

THE AUTOMOTIVE ALTERNATOR

Since about 1960, the honors for sheer numerical quantity of annual production have fallen to the automotive alternator. This was

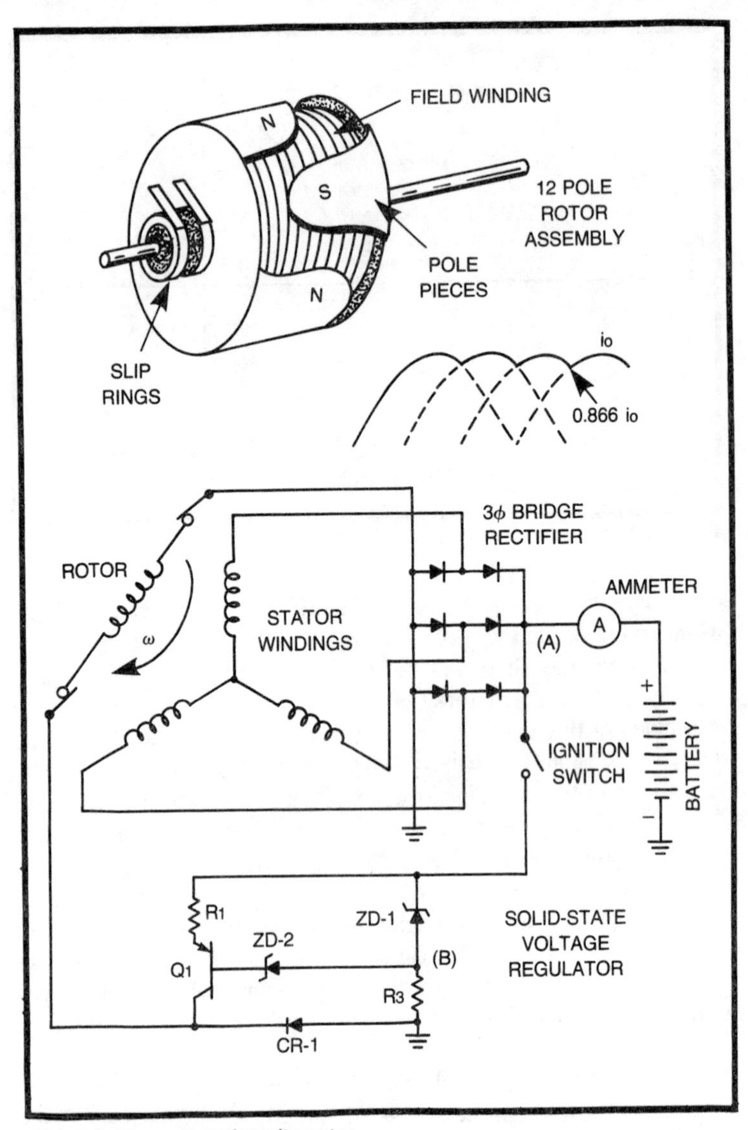

Fig. 6-3. The automotive alternator.

also the first machine in widespread production to be equipped with solid state components and controls. Figure 6-3 schematically depicts this machine.

In a typical automotive installation, the twelve pole alternator rotates at twice the crankshaft speed and thus turns up something like 1200 rpm while idling and 6500 rpm at 60 mph. It is driven to speeds like 10,000 rpm at maximum engine speed. The view of the rotor assembly shows the rather unconventional shape of the pole pieces, and the armature assembly. The field winding is a straight spool of wire which is easy and inexpensive to wind. The pole pieces themselves are fluted cup-shaped affairs which are press-fitted together and almost completely encapsulate the field winding. This design has the advantage of providing the very high centrifugal bursting strength required for such a high speed machine while at the same time lending itself to the extremely low cost, automated assembly required of automobile components.

The three-phase stator is wired through a three-phase bridge rectifier which serves two purposes:

1. It rectifies the ac output of the stator for battery charging.
2. It prevents the battery from discharging back through the field and stator. This latter point eliminates the use of a high-current relay which used to be required.

The use of the alternator became economically feasible when the cost of the six rectifier diodes came down to a suitable level.

The usual six cell lead-acid battery in the car develops about 12.2 volts at rest and should not be continuously charged at voltages much higher than 14.2 volts. If the battery internal resistance is 0.1 ohm, the battery charging current would be approaching 14.2 – 12.2 volts/0.1 ohm = 20 amperes at this level (actually the current will run a bit lower). However, since we have noted that the V_g is directly proportional to ω_m then in a speedup from 12.3 volts at idling to top speed, the alternator output would rise by a factor of 10,000/1200 = 8.33 to a level of 102.5 volts. Accordingly, it is necessary to provide some means of weakening the field by a factor of about eight in order to preserve the battery.

A simplified voltage regulator is shown in the schematic at the bottom of the figure. Diode ZD-1 is arranged to go into conduction at about 12.5 volts. With the motor stopped or running slowly, the only current through R3 is the base current of Q1. Diode ZD-2 serves to offset the voltage at the base of the transistor above point B and the transistor saturates with a current controlled by R1 and the field drop plug the V_{sat} of the transistor. This provides maximum field to the alternator. As the alternator turns, the voltage at point A will rise and ZD-1 will go into conduction, raising the voltage at B and reducing the forward bias of Q1. Now the current at A is actually a humpy three-phase rectified affair; however, the voltage is largely smoothed by the battery. Nearly the entire voltage increment will appear at point B, which will drastically lower the collector current of

Q1. This, in turn lowers the field current. Note that in this case Q1 is used in the linear mode and will be dissipating considerable power. Free-wheeling diode CR1 is there to protect the transistor from the inductive surge that would result if Q1 were suddenly cut off by a line transient.

THE INDUCTION MOTOR

While the synchronous motor is useful over a variety of applications, the *induction motor* has a far wider range of usefulness because of the self-starting capabilities. It was this motor that proved to be Tesla's first success. The induction motor has a far wider range of ability because it is capable of slowing down slightly to provide very large overload torques.

Imagine that you are riding on the rotor of a very large induction motor with a vertical shaft like a merry-go-round and that the rotor was turning at some speed ω_m. The *rotating magnetic field* would be rotating at the speed ω_e about you. If you had a compass in your hand, it would seem to you to be turning at a rate ω_s equal to the difference between the magnetic field speed and the mechanical speed. Someone on a stationary platform above you would see that the compass was actually turning at the magnetic field speed and staying in precise synchronism with it. Mathematically this would be written:

$$\omega_s = \omega_e - \omega_m \tag{6.8}$$

Now the field windings on your rotor would respond to this *slip frequency*, ω_s, with an induced voltage because the flux is changing.

$$V_i = n_f \frac{d\phi}{dt} \tag{6.9}$$

but

$$\frac{d\phi}{dt} = \omega_s \phi_o \cos \omega_s t \tag{6.10}$$

n_f = number of field turns

If the field winding is shorted through the slip rings, or directly on the stator, this would cause a current to flow in the field. This current would be controlled by the field resistance, the field inductance, and any external load resistance; thus:

$$i = \frac{Vi}{R_{ext} + R_{field} + j\omega_s L_f} = \frac{n_f \omega_s \phi_o}{\epsilon R + j\omega_s L_f} \tag{6.11}$$

(cos $\omega_s t$ is implicit—this is a steady-state equation)

The value of ϕ_o is not an independent variable but rather is related to the line current in the rotating field which is in turn determined by the CEMF and the winding reactance.

Another factor is of interest. In equation 6.11 we see that the field current and, therefore, the torque are proportional directly to the number of turns on the field winding and inversely proportional to the resistance and inductance of the winding. However, the inductance is proportional to the square of the number of turns on the winding. The presence of an n^2 term in the denominator tells us that a motor with a very small number of turns and with a very low resistance winding would have superior starting torque. In the *squirrel cage induction motor*, advantage is taken of this by casting the winding of aluminum or copper into the slots on the rotor. This provides a very low inductance and extremely low resistance. It also has the advantage that the cast winding shrinks and firmly binds the rotor lamination stack together. Frequently the ends of the casting will be fitted with cast-on fins to yield a centrifugal blower action. The physical appearance of a squirrel cage rotor is shown in Fig. 6-4 which illustrates the massive cast-on winding and the integral fan.

The performance of such a motor is illustrated in Fig 6-6, in the top curve. When the motor is first turned on ω_m is zero and $\omega_s = \omega_e$. With a polyphase motor and a true rotating field, a definite and substantial starting torque is developed. The starting torque is

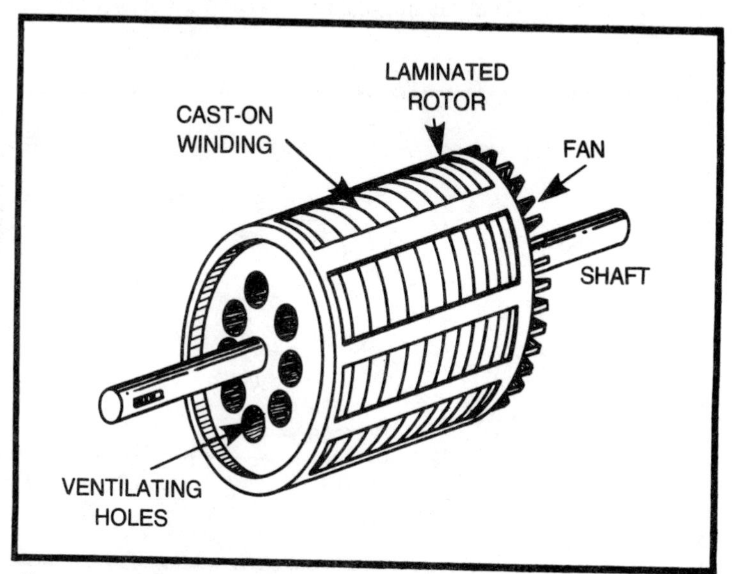

Fig. 6-4. The squirrel cage rotor.

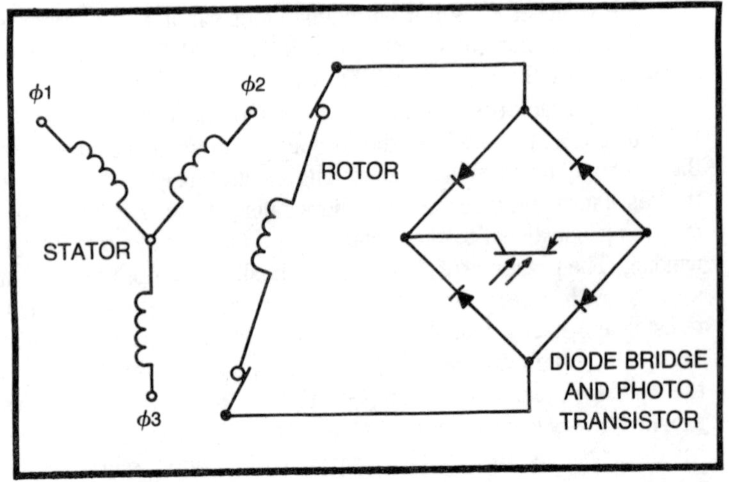

Fig. 6-5. The wound rotor induction motor armature resistance speed control.

limited to about 120 percent of the rated motor torque by the reactance of the rotor winding. As the motor picks up speed, the slip frequency falls and the rotor reactance becomes less and less significant. Above about 0.8 of the synchronous speed, the resistive component in the rotor winding begins to become effective and the torque curve descends rapidly. The torque curve actually goes negative at speeds faster than synchronous and the motor will act as a generator.

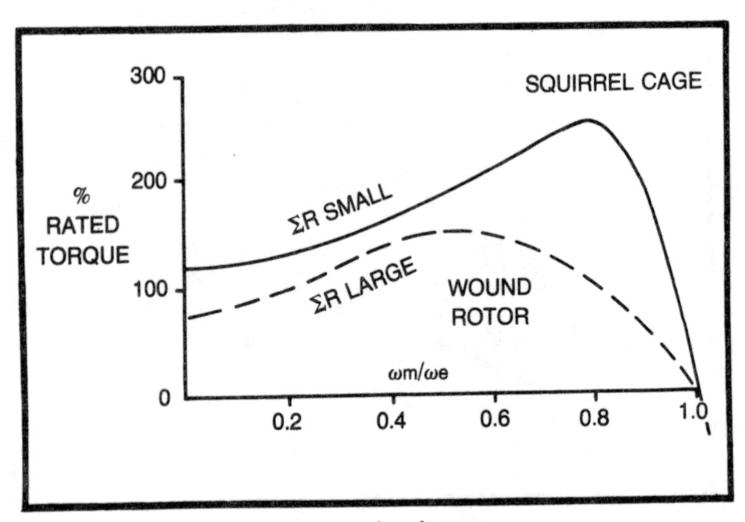

Fig. 6-6. Induction motor torque-speed performance.

For a *wound rotor* induction motor, the ratio of inductance to resistance is usually higher and the torque curve is flatter and peaks at lower fractions of synchronous speed. The wound rotor induction motor can be speed controlled over some limited range by means of varying the resistance of the rotor winding. This can be done with a rheostat or by a solid-state equivalent. The circuit of Fig. 6-5 illustrated a diode bridge rectifier and a photo transistor used for this function. It should be noted that the transistor is being used in the linear mode if it is acting as a variable resistor and would, therefore, be dissipating a considerable amount of power.

The curves at the bottom of Fig. 6-7 show the effect of varying the voltage on a relatively high resistance induction motor. It may be seen that a relatively small range of control is available. In this condition, if the torque load ever gets over the hump, a further decrease of speed will result which will slow the motor still further and the system will stall.

The chief advantage of the squirrel cage motor is its ability to absorb wide variations in torque loading with very little decrease in speed. It is a very "stiff" motor. The speed is determined almost entirely by the line frequency. This is the most common type of induction motor because of the low cost. It is seldom used in applications where the required control exceeds start-stop-reverse operations.

It is noteworthy that most synchronous motors have an induction winding added to permit them to start. At synchronous speed, the current in the induction winding falls to zero and the winding has

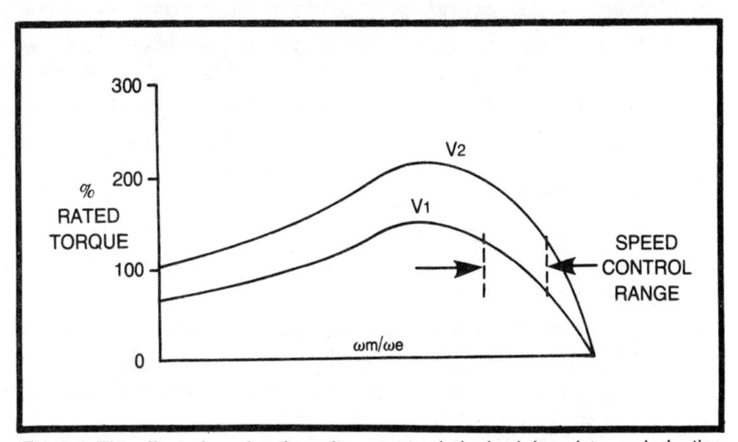

Fig. 6-7. The effect of varying the voltage on a relatively high resistance induction motor. Except on the wound rotor induction motor the speed range which may be controlled is very limited.

no effect. However, on starting, the winding will develop a substantial torque to get the motor up to *pull-in speed*. On large alternators, the induction winding is added to prevent the runaway of the machine in the event of field loss. This is particularly important in hydroelectric installations where the loss of the field would permit dangerous overspeeding before the gates could be shut down.

The principal element of interest to use in the area of motor control is the current which will be drawn by an induction motor. In the stalled, or locked-rotor, condition, the shorted turn effect tends to dominate the behavior and the motor will draw a large starting surge. The starting surge will be controlled largely by the effective stator resistance and the leakage inductance due to the imperfect coupling between the stator and the rotor. This may be measured on an existing motor by applying a voltage too small to permit the motor to start and measuring the phase and amplitude of the current.

Most squirrel cage motors tend to run at 1725 rpm on 60 Hz current. In this case $\omega_s = 0.04\ \omega_e$. Therefore the reactive effects of the rotor winding are small. The mechanical work done by the motor tends to add to the real component of the stator current. This does not have much effect on the total current drawn; however, it tends to increase the power factor. Systems which require load sensing usually work best by sensing the phase angle of the line current.

THE SPLIT-PHASE MOTOR

Up till now, we have been talking about *polyphase* motors. These are very common in industrial applications, but most household appliances are operated on single-phase current, since polyphase service is rarely found in homes, stores, and filling stations. If we refer back to Fig. 6-1 and imagine that only one of the windings was energized, we can see that the direction of rotation of the motor is ambiguous. It would shown no net starting torque and, once started, would run equally well in either direction. Accordingly Tesla invented the *split-phase motor* to permit single phase starts on an ac line. Once started, an induction motor can run on a single phase; however, much higher torque can be obtained with a capacitor-start/capacitor-run motor.

Figure 6-8 shows a two-phase induction motor, equipped for capacitor-start/capacitor-run operation. From the following equations it may be seen that the current in the first winding tends to lag the total voltage by nearly 90°.

$$i_1 = \frac{V - \text{CEMF}}{R_1 + j\omega L_1} \tag{6.12}$$

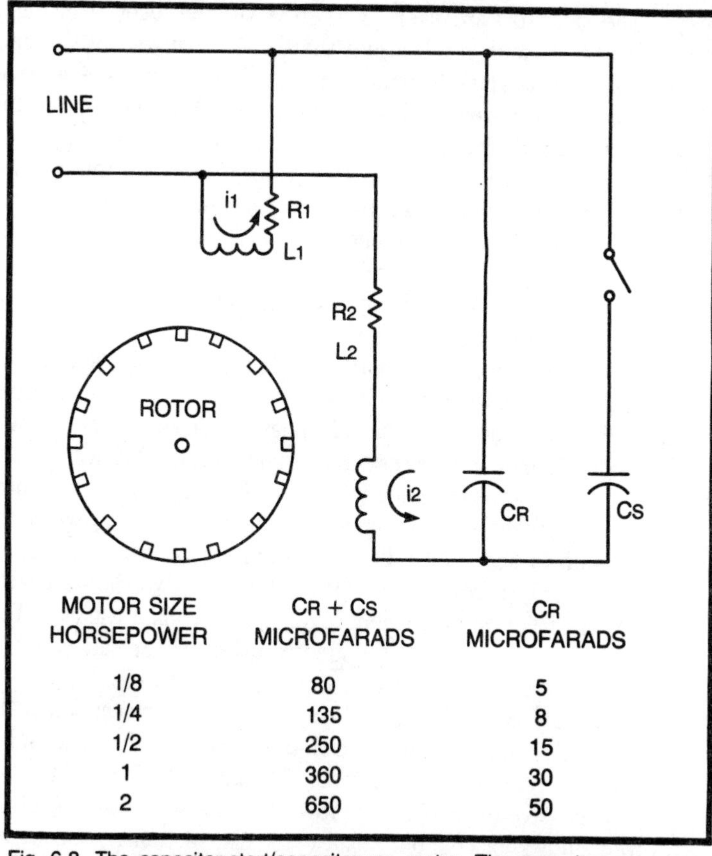

MOTOR SIZE HORSEPOWER	CR + CS MICROFARADS	CR MICROFARADS
1/8	80	5
1/4	135	8
1/2	250	15
1	360	30
2	650	50

Fig. 6-8. The capacitor-start/capacitor-run motor. The capacitors are motor grade ac electrolytics in the starting sizes and oil filled or paper in the run sizes.

and if $R_1 << \omega L_1$

$$i_1 \cong \frac{V - \text{CEMF}_1}{j\omega L_1} \tag{6.13}$$

$$i_2 \cong \frac{V - \text{CEMF}_2}{j\omega L_2 - j\ 1/\omega\ (C_s + C_R)} + R_2 \tag{6.14}$$

and if $\omega L_2 = \dfrac{1}{\omega(C_S + C_R)}$

$$i_2 = \frac{V - \text{CEMF}_2}{R_2} \tag{6.15}$$

The capacitors are inserted in series with the second winding to correct the phase angle. Because the CEMF changes both in magnitude and phase angle from starting to running conditions a single capacitor would be only a poor compromise. With only C_R in the circuit the motor will typically have a starting torque of only 30 to 60 percent of the rated torque. With both capacitors in, the unit will develop a starting torque as high as 450 percent of the rated torque; however, it will be over-compensated at higher speed and will not run up to as high a torque level. A centrifugal switch, therefore, is supplied to remove the starting capacitor from the circuit when the motor starts to pick up speed.

The actual split-phase motor developed by Tesla did not have the benefits of the large capacitors. It may be seen that by juggling the values of R_2 and L_2 a suitable phase shift can be developed. However, a winding with reduced L and increased R will tend to burn out if left across the line very long. A centrifugal switch must usually be included to disconnect the starting winding when the motor is up to speed. An alternative starter control operates by sensing the line current and removing the starter winding when the current has fallen to a sufficient level in the running winding. This form of starting control is frequently used on "sealed unit" refrigerators.

From a control standpoint there is a note of caution to be observed. A good starting control should have some mechanism such as a current sensor to protect the motor from failure to start due to excessive torque loads or low line voltage. A lightning strike will frequently knock out one phase of the distribution system in rural areas. The service voltage then drops to 55 volts or so. This is insufficient to get most appliance motors up to speed; therefore, they will not come off of the starting winding. It is not low enough to prevent the eventual burnout of the starting winding. A thermal current-sense breaker is probably the simplest and least expensive protection for the motor.

THE SHADED-POLE MOTOR

For phonographs and other low-power applications, a *shaded pole squirrel cage motor* (Fig. 6-9) is frequently used because of the low cost and single-phase self-starting characteristic. The flux in the portion of the pole face inside the shorted turn tends to lag the main flux. This yields a feeble to modest rotating field which causes these motors to develop a starting torque of 20 to 40 percent of the running torque. This is the simplest and cheapest form of ac motor. Because

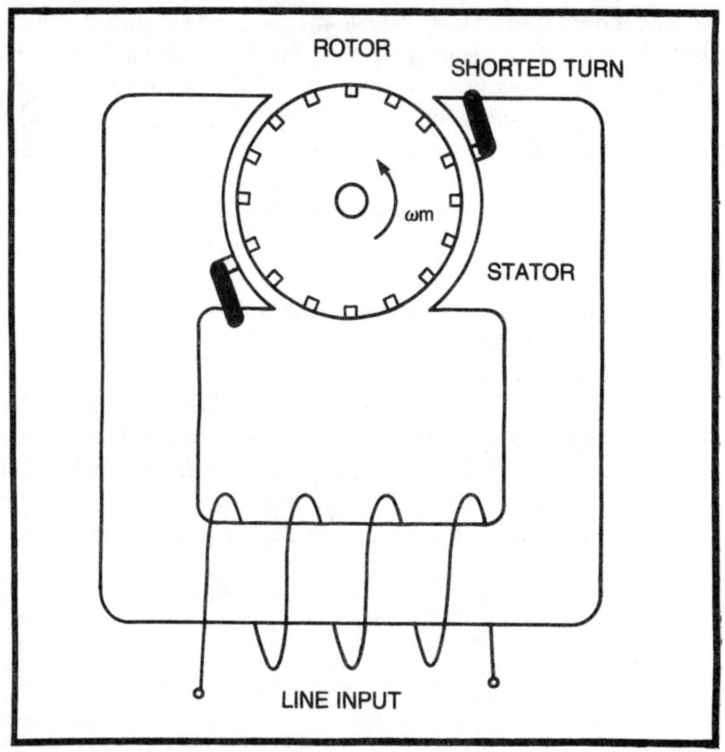

Fig. 6-9. The shaded-pole motor.

of the shorted turns it tends to be inefficient and to run hot. However, the low cost, silence, and simplicity tend to make it very popular for small fans, hair driers, rotisseries, and similar items in ratings from 1/50 to 1/8 horsepower. Above 1/8 hp the heating effects become too severe, and other types predominate.

HOW MANY PHASES?

We have examined two and three phase systems as well as single phase systems and it is perhaps worthwhile to consider the relative merits of the different phase numbers. If we hark back to equation 6.6 we see that a synchronous, single-phase alternator or motor will have two torque peaks per revolution—with the torque falling to a low level in between. With a two-phase system the torque peaks overlap and the minimum torque is about 87 percent of the maximum. In a three-phase system there are six torque peaks per revolution and the minimum torque is 91 percent of the maximum. Historically, we owe the use of three-phase power in commercial distribution to the fact that a double-acting steam engine also has two torque peaks per revolution and therefore a single-phase alternator could get out of phase and demand the maximum torque just when the engine could supply the least. The smoother torque load of the three-phase alternator made it much easier to control the steam engine.

In control and servo applications, however, the use of two-phase motors is very common. Two-phase servo motors are often designed to permit one of the windings to run permanently across the line without overheating, even in the stalled rotor condition. The second winding is then used for directional control and perhaps for torque or speed control.

By and large the induction motor is used in control applications only where the requirement is for start, stop, and reverse applications. For certain smaller applications, such as telescope drives, the line frequency may be derived from an inverter and the limited range of speed control required obtained by varying the inverter frequency.

7

Stepper Motors and Multipole Synchronous Motors

There is probably no type of motor more directly associated with electronic control than the stepper motor. The basic principle of this type of motor is very old and has been in use for a long time. It is only since the advent of electronic controls, however, that the stepper motor has achieved much commercial significance. The reasons for this rise in popularity stem principally from the remarkable degree of control which is attainable from these motors when a suitable electronic drive is applied. Without the electronic drive the stepper tends to lose many of its advantages. In this chapter, we shall explore a few of the properties of the electronically driven stepper motor and see how such units may be driven.

STEPPER MOTOR BASICS

Fundamentally, the stepper motor is nothing but a permanent magnet synchronous motor similar to the unit shown in Fig. 5-1; however, the actual mechanical details of design have been significantly developed by a number of manufacturers in order to render the actual unit more amenable to practical applications. Many of these innovations are protected by patents, and the motors offered by different manufacturers have a variety of different properties which tend to make certain types more suitable for one application than another. The range of properties is so wide that it is impossible to cover them here, so the reader is directed to the manufacturers' literature for an actual detailed assessment of properties. The treatment here shall touch upon only a few of the available types.

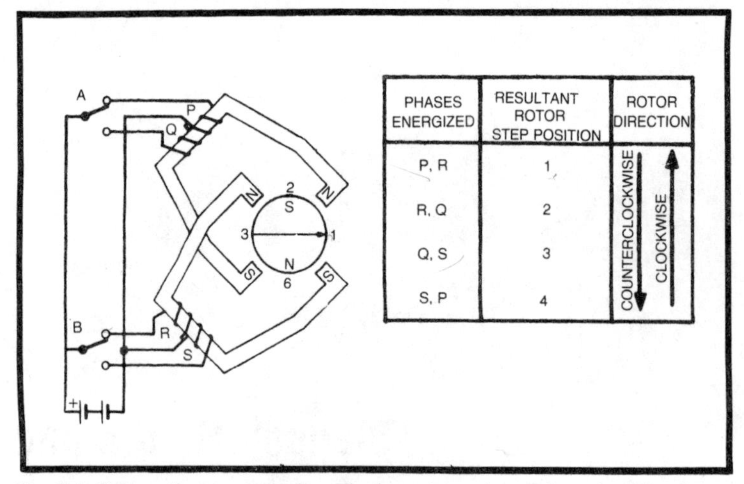

PHASES ENERGIZED	RESULTANT ROTOR STEP POSITION	ROTOR DIRECTION
P, R	1	COUNTERCLOCKWISE / CLOCKWISE
R, Q	2	
Q, S	3	
S, P	4	

Fig. 7-1. Schematic representation of a four-phase, two-pole motor. The direction of rotation can be reversed at any point.

Figure 7-1 schematically illustrates a four-phase two-pole motor. It may be seen that this unit differs from the permanent magnet synchronous motor of Fig. 6-1 only in the fact that the connection between winding A-C and B-D is tapped and a battery is used with a switching arrangement to replace the quadrature sine wave excitation. These changes do not alter the operation significantly, and the rotating magnetic field is still the fundamental principle of operation of the unit. If the sequence of energizing the poles in

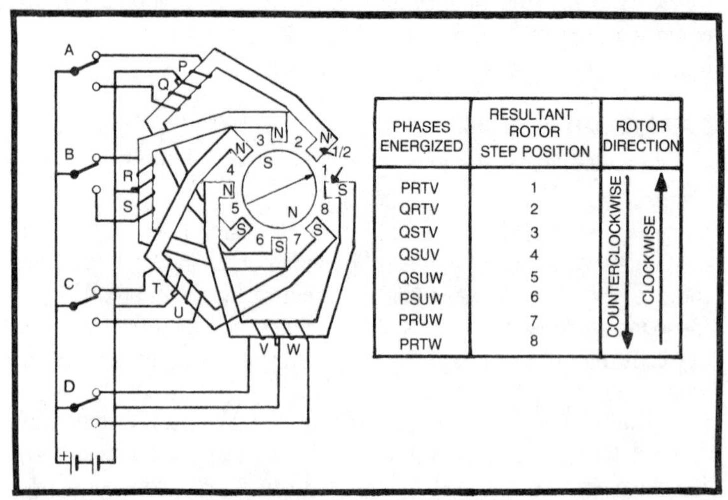

PHASES ENERGIZED	RESULTANT ROTOR STEP POSITION	ROTOR DIRECTION
PRTV	1	COUNTERCLOCKWISE / CLOCKWISE
QRTV	2	
QSTV	3	
QSUV	4	
QSUW	5	
PSUW	6	
PRUW	7	
PRTW	8	

Fig. 7-2. Schematic representation of an eight-phase, two-pole motor.

the figure is followed, the timing diagram at the bottom results, and it may be seen at the virtual stator-pole jumps around the circle in a series of 90° steps.

In the illustration of Fig. 7-2 we see the same principle expanded into an eight-phase two-pole motor. The timing diagram for this unit also is shown. Those familiar with logic design will recognize the timing diagrams as being identical with the outputs of a *Johnson*, or *ring counter*, as illustrated in Fig. 7-3. If we consider that the

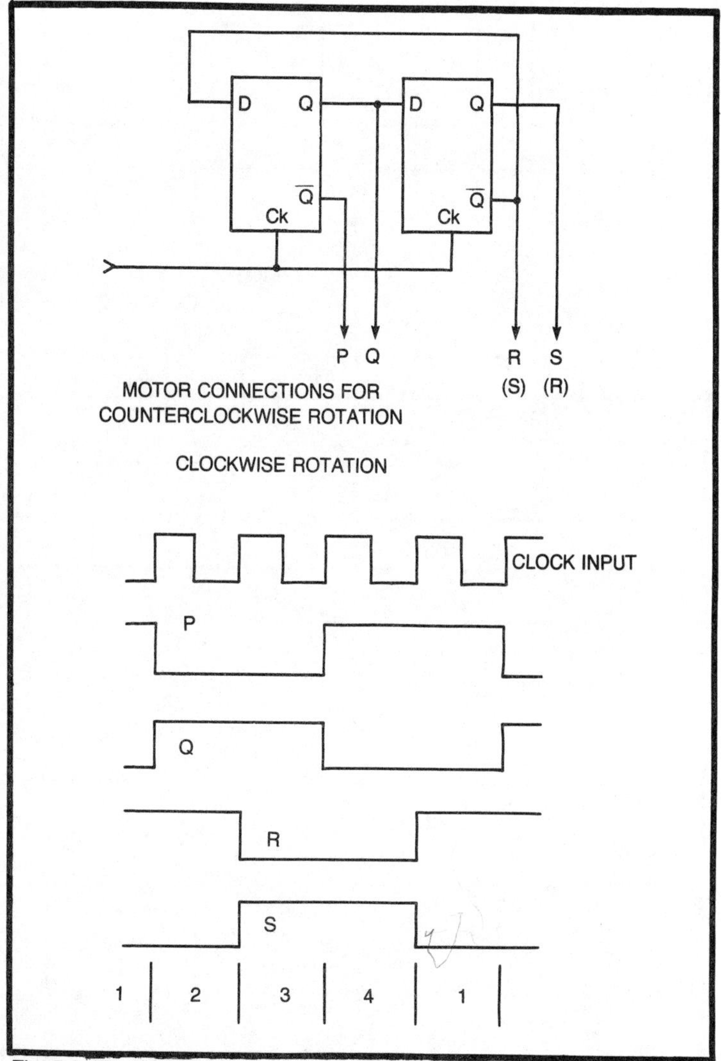

Fig. 7-3. The ring counter.

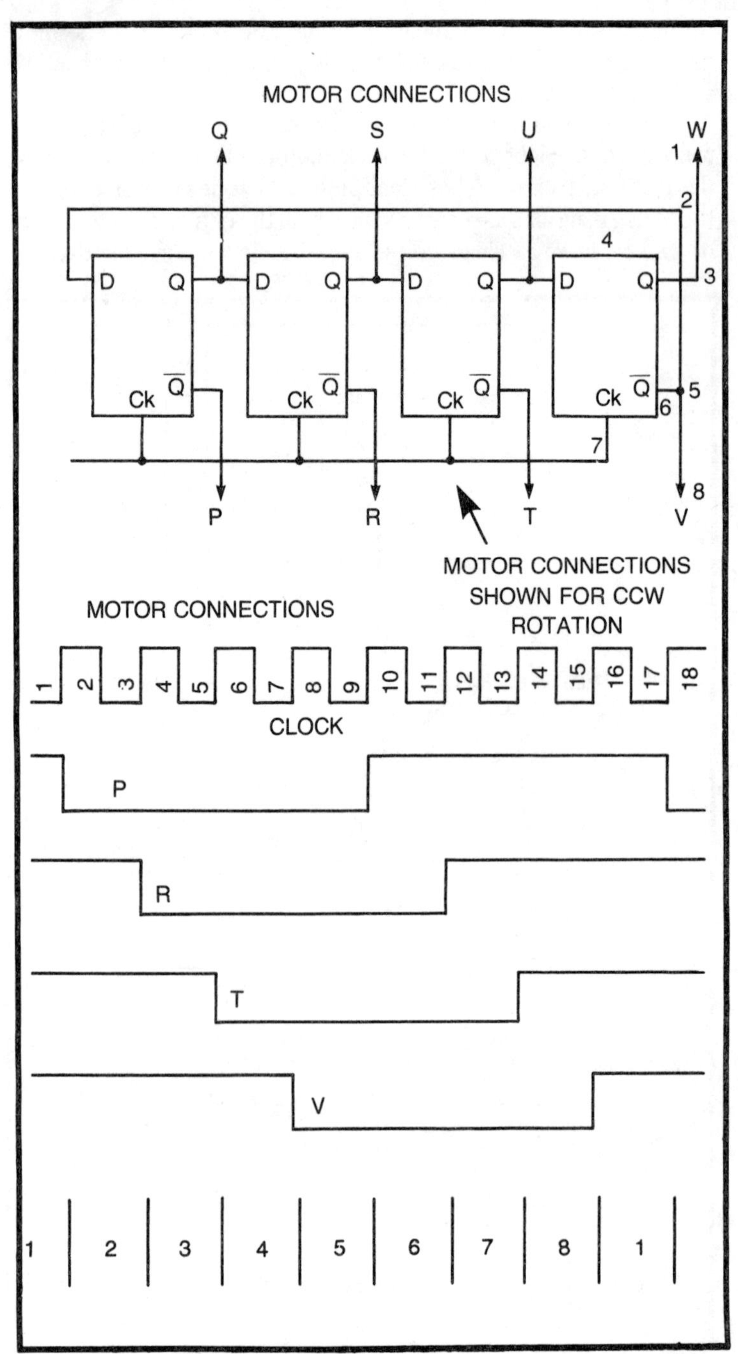

Fig. 7-4. The eight-phase ring counter.

counter is initially cleared, then the $\overline{Q}$ outputs will be high and the Q outputs will be low. The first rising edge on the clock will shift a 1 into the Q of the first flip-flop. The second rising clock edge will shift a 1 into the second flip-flop. This in turn makes the $\overline{Q}$ output of the second unit low which in turn forms a low data input on the first. The resulting timing diagram is shown with the schematic. Note that interchanging the R and S leads will serve to reverse the phase rotation and the direction of the motor. However, it should be noted that this change of rotation direction would be accompanied by a movement of one step. For instance, suppose that the motor were in position 1 with P and R energized. Then if we swap R and S the rotor will assume position 4 which calls for P and S energized. A reverse command would have to take account of this extra step if the position accuracy were to be maintained.

Figure 7-3 shows the ring counter expanded to provide the eight outputs for the eight-phase unit. Note that here the reversal is not as simple. The sequence obviously may be reversed by simply reversing the entire order, i.e., taking the V and W outputs from the left hand flip-flop and the P and Q outputs from the right hand, etc. This is not very satisfactory, however, since the unit would jump on reversal at any setting where P, R, T, and V were not all ones or all zeros. A more satisfactory arrangement may be obtained by making use of a left-right shift register as shown in Fig. 7-5. With this arrangement, the motor may be operated in either direction without jumping at changeover or reversal.

The advantages of being able to control not only the motor speed and direction of rotation but also the precise angle through which the shaft steps are relatively obvious. Circuits that will count out a known number of pulses are very easily built using digital integrated circuits (ICs). Furthermore, the logic can readily keep track of the motor position so that an open-loop system can be used. The later is a great cost saving and removes a substantial source of instability in the system. The ability to give the motor a digital command and then to know that it did just that and at the precise rate is one of the reasons that this form of motor is so widely used in digital control applications.

Harking back to our discussion of the PM synchronous motor, we recall that equation 6.5 showed that it had no average starting torque and that it was necessary to add an induction winding or some other mechanism to get the motor up to synchronous speed. It seems fair to ask just what exempts the stepper motor from this characteristic. The answer comes from a couple of places. First of all, we noted that the output torque of the synchronous motor was a

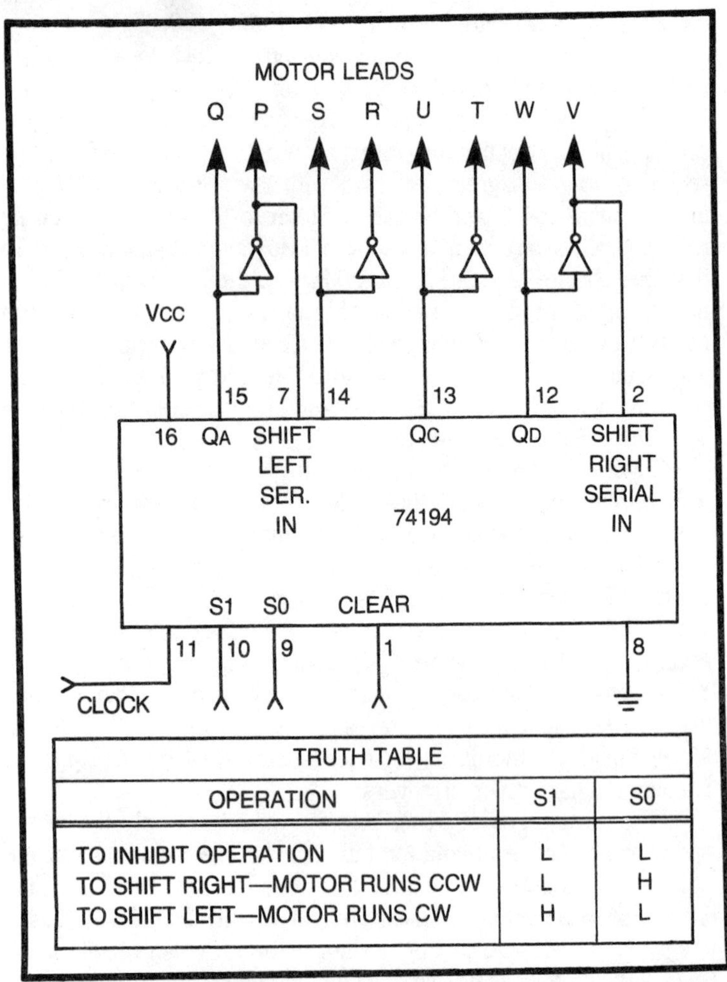

TRUTH TABLE		
OPERATION	S1	S0
TO INHIBIT OPERATION	L	L
TO SHIFT RIGHT—MOTOR RUNS CCW	L	H
TO SHIFT LEFT—MOTOR RUNS CW	H	L

Fig. 7-5. The bidirectional shift register ring. Inputs S1 and S0 should be changed only when the clock is high. A low on the clear line will return the motor to position 1.

function of the angle between the field of the rotor and the rotating magnetic field of the stator. The torque was capable of going both positive and negative so that if the rotating field was turning rapidly with respect to the rotor, the *long term average torque* was zero, and the motor would not start. On the other hand, if the rotor is low enough in inertia to manage to accelerate to synchronous speed in the first half cycle, the motor will start and lock-up in synchronism. In general the principal difference lies in the act that stepper motors are small and therefore have relatively little inertia. In addition they have a great many poles, so a full electrical cycle represents only a fraction of a revolution and synchronous operation does not require a great deal of acceleration. *There is always some speed at which a PM synchronous motor will start in stepper fashion, provided that the frictional or static load is not too great.* Accomplishment of such a start may require very sophisticated control of both current and frequency to prevent motor burnout.

For a motor designed for stepper operation, the design is tailored to maximize the torque and minimize the intertia. Figure 7-6 shows a smaller stepper built by the North American Philips Controls Corporation. Their smaller motors have 12 poles for 15° and 7.5° step angles and 24 poles for 7.5° and 3.25° step angles. In the illustration a cutaway view of the motor is shown. The pole pieces for the stator are actually strips pressed inward from the cover, which also serves as part of the pole structure. The donut coils shown top and bottom create an axial magnetic field which is concentrated by the pole pieces. The arrangement is more or less an inside-out version of the pole structure of the automotive alternator rotor shown in Fig. 6-3. These coils are wound bifilar, that is with parallel insulated wires which provide the equivalent of the *P* and *Q* and *R* and *S* windings. The rotor is actually two rotors which are displaced by a half step from one another and mechanically tied together. Each donut-shaped stator winding represents two of the phases. This is a four-phase unit. For an eight-phase unit, there would be four bifilar donut windings and four stators. This is a relatively inexpensive type construction and is most effective in smaller motor sizes.

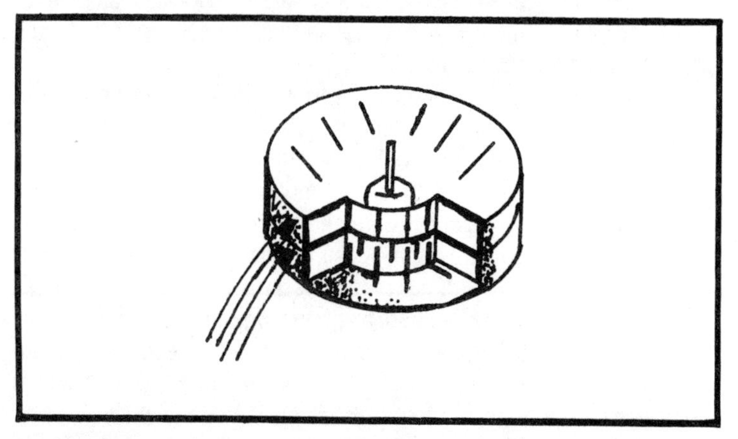

Fig. 7-6. The North American Philips stepper motor. This shows a cutaway view of a typical four-phase permanent magnet logic stepper version.

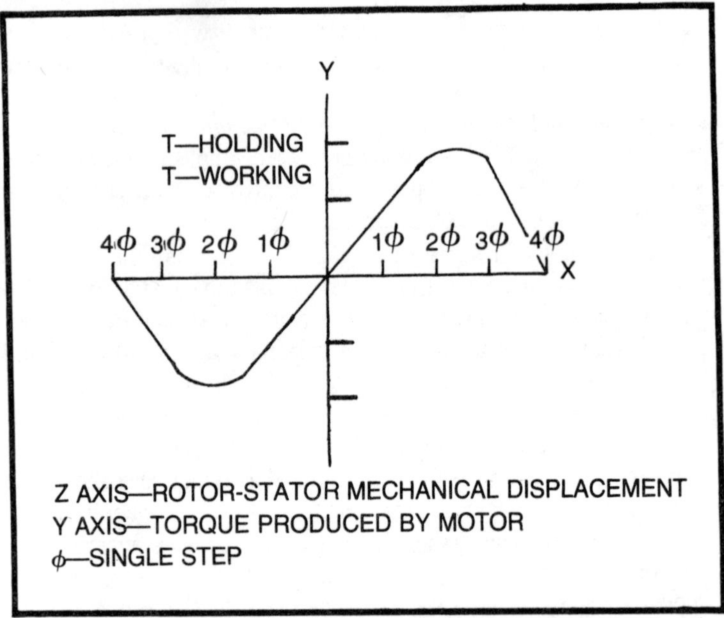

Fig. 7-7. Maximum rotational force for an eight-phase stepper occurs when the rotor leads or lags the stator by two steps.

From a review of the timing diagrams, it may be seen that the eight-phase motor has the advantage that only one of the four energized windings switches during any step while three remain constant. This means that the eight-phase unit will have more torque for starting in a given size than a four-phase unit which switches half of its field on any step. There are, of course, twice as many parts in the eight-phase unit and in its drive so that the economic advantage goes to the four-phase unit.

As with any synchronous motor, the stepper delivers no net torque with zero angle between the field and the rotor as shown in Fig. 6-2. The curve of Fig. 7-7 shows this property translated into steps for the two- and four-phase motors. It may be seen that maximum rotational torque is obtained when the motor is as much as two steps behind for the eight-phase motor and one step behind for the four-phase motor. Of course, the starting torque for the eight-phase motor must be defined from the one-step point since it is always presumed that the motor is never more than a single step behind. In a case where there may be an external load applied temporarily, the eight-step motor will spring as much as two steps away and then return to near zero when the load is removed. With any more torque, the motor will break away and eventually come to rest in some different position when the load is removed. In the case of the four-phase motor the breakaway point comes at the single-step level.

There are several different torque ratings which must be considered for these motors. The curves of Fig. 7-8 define several other items. The *pull-in torque* is a plot of the torque versus frequency for the allowable friction load that the stepper motor can start without missing any steps. The *pull-out torque* is a plot of torque versus frequency for the allowable *friction* load which the stepper can move at its maximum stepping rate *after* it has been slewed up to speed. The area between these curves is the *slew range*. Operation in this range must be kept close to either the pull-in curve or the pull-out curve. The area in between is termed an *unstable range*, and the motor may jump a step or more if operated in this range continuously. The unstable range arises because of the torsional springiness implied by the curve of Fig. 7-7 and the inertia of the motor and torque load of the system. This tends to constitute a *mass-spring system* in rotation and the area between the curves is a function of the natural oscillatory period of the mass-spring system.

The difference between the pull-in and pull-out rates is due to the inertia of the motor and the system. In a completely inertialess motor with only a friction load, the pull-in and pull-out curves would be identical.

From Fig. 7-7 we see that the eight-phase motor is developing maximum torque, and therefore maximum acceleration, when the rotor lags or leads the stator by two full steps. In a design for maximum acceleration (and deceleration) the drive should be designed to run the step rate up (or down) until the two-step lag (lead) is obtained. A photochopper on the motor output shaft will detect the shaft position and provide an output pulse for every complete step.

This form of operation is referred to as *ramping*. As shown in Fig. 7-9 operation in a ramping mode requires much the same kind of control as exercised when driving an automobile. In cases where the distance to be covered is adequate, the system starts at the maximum pull-in rate, accelerates to the maximum slew rate, and anticipates the point where maximum deceleration must be applied in order to come to a halt at the destination without overshoot. However, it should be noted that not all destinations are sufficiently distant to permit attainment of the maximum slew rate. This is shown on the figure in case 2. Here, the system must shift from the lag to the lead condition near the halfway mark.

This form of operation is more common in plotters, densitometers, and such instruments, where a relatively fixed load is to

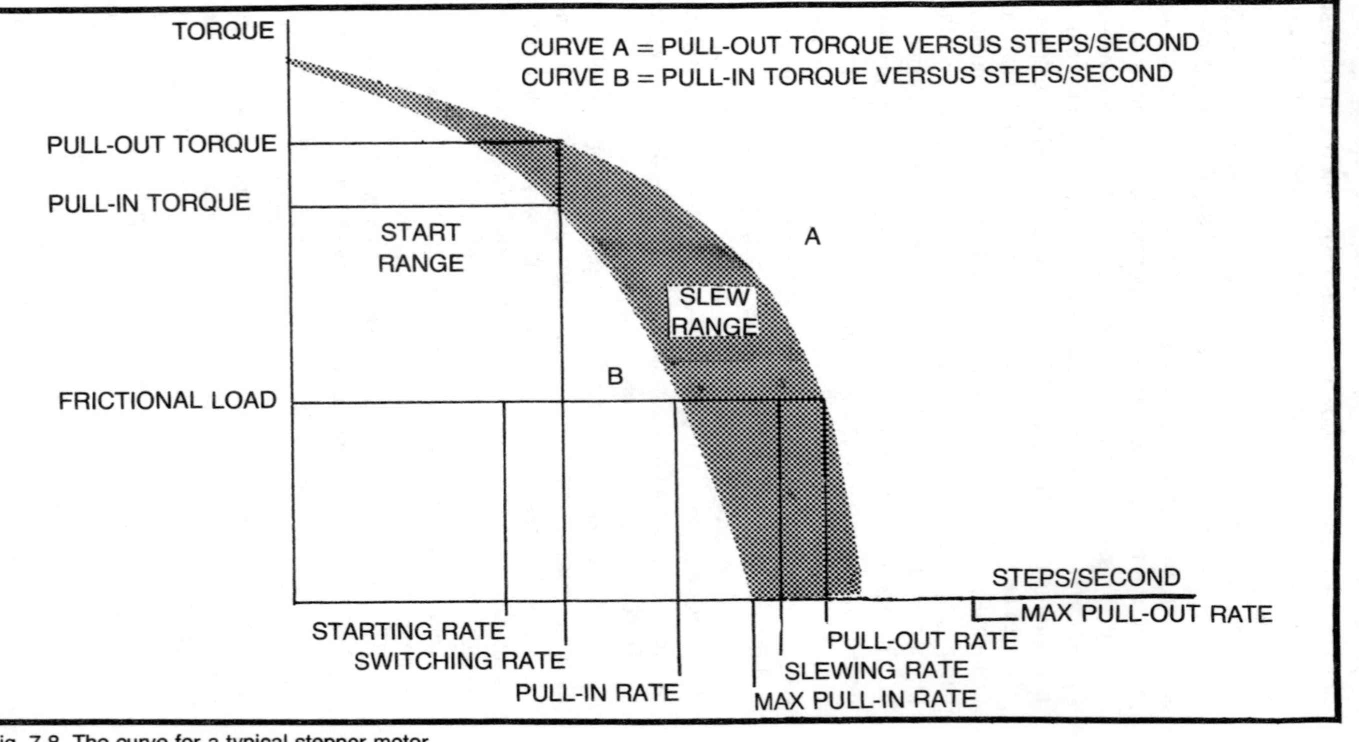

Fig. 7-8. The curve for a typical stepper motor.

be moved. In machine tools and other applications, it is quite common that the load will be variable. A heavy casting on the table will considerably increase the inertia of the load. The ramping rate and the maximum pull-in rate are very significantly affected by the inertia of the load; therefore, the development of a ramping control must be considerably more sophisticated since the unit has to sense on the upward ramp what the system inertia was and then determine the maximum allowable slew rate and the maximum ramp down rate while the system is in transit. The degree of "smarts" required of a *load inertia sensing ramping system* is such that a microprocessor or minicomputer would be almost mandatory.

In practice, the actual rates required on machine tool motion while actually cutting are in the single-step range in the form, "advance 0.001," "wait," "advance 0.001," etc. It is only when going from one hole location to the next that ramping would provide any

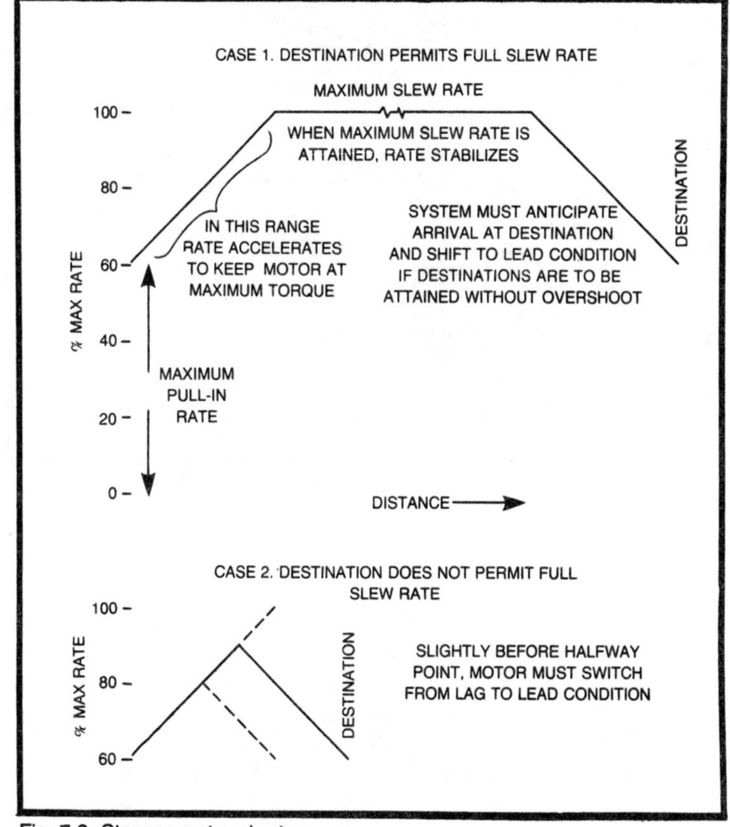

Fig. 7-9. Stepper motor slewing.

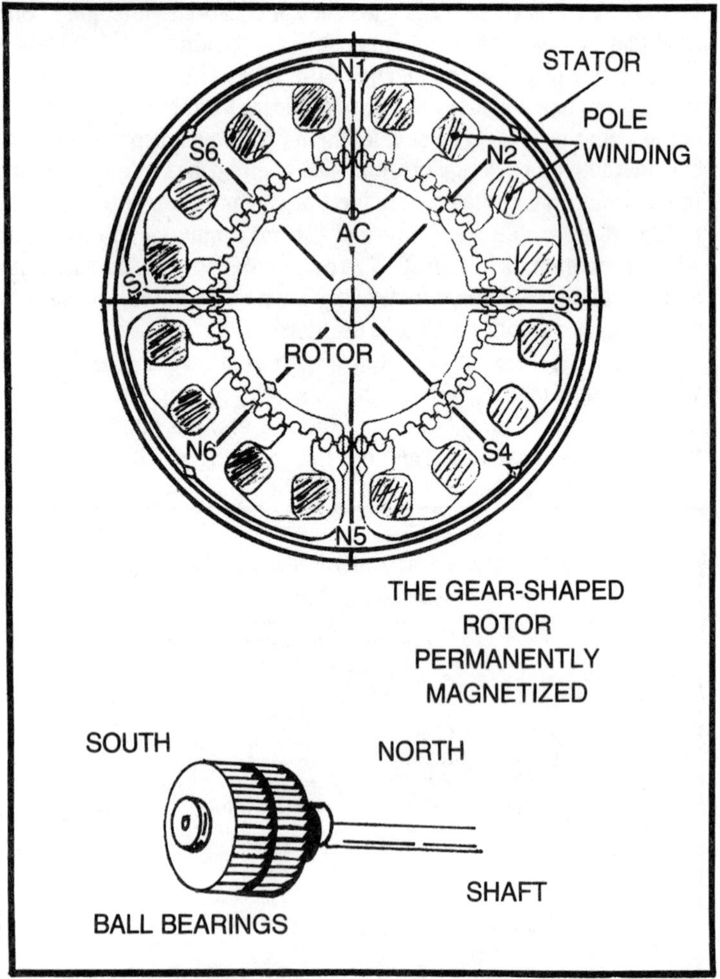

Fig. 7-10. The Slo-Syn stepper motor sold by Superior Electric Co.

improvement in performance. The provision for maximum speed ramping is, therefore, not common in these applications.

Stepper motors of a somewhat different configuration are sold by the Superior Electric Co. Figure 7-10 shows the cross section of a four-pole (pair) Slo-Syn stepper. Note that a number of "teeth" have been machined into the surface of the rotor and the stator poles, and that the pitch of the teeth is different for the rotor and stator. The unit shown in the illustration is designed for a 1.8°/step rotation. Note that the teeth are lined up on poles 1 and 5 as shown in the illustration and misaligned to various degrees on the remaining poles. After a 1.8° clockwise rotation, the teeth will be aligned on stator poles 2 and 6. A second pulse would align the teeth on 3 and 7 and a third on 4 and 8. The fourth pulse would restore the alignment to 1 and 5 and the cycle would repeat.

In all, it may be seen that it would take four steps for the motor to advance one tooth width. Since the rotor has 50 teeth, this means that 4 × 50 = 200 steps are required for a full revolution of the rotor, or stated differently, the rotor advances 360°/200 steps = 1.8°/step.

With this motor, if a 20 thread-per-inch lead screw would advance the tool or table 0.00025 inch per step, then four steps would be required for a 0.001 inch advance. Superior Electric offers these motors with step angles of 0.72°, 1.8°, 2.0°, 5.0°, 7.5° and 15°.

Figure 7-11 shows the performance and step sequence for a motor of this type equipped with both "standard" and bifilar windings. The performance of the bifilar unit is obviously somewhat better, but this comes at the expense of some added complexity.

The detailed selection of a stepper motor for a given application hinges heavily upon the nature of the load. The balance between friction and inertia is particularly critical. A subsequent chapter shall deal with the topic of inertia in motor systems since this is of interest in nearly all types of motor systems, not just stepper motors. However, the inertia loading and system coasting is particularly important in stepper systems operated in the open-loop mode.

THE MULTIPOLE SYNCHRONOUS MOTOR

In Fig. 6-1, the machine shown had just one pole-pair on the rotor, and it would complete a full mechanical revolution per electrical cycle. It is not hard to imagine that if the rotor magnet were transformed into a cross-shaped structure with two pole-pairs, and the stator structure similarly subdivided that it would take two electrical cycles to complete a mechanical revolution. Multipole induction motors are widely offered in the integral horsepower sizes by motor manufacturers, and in motor sizes larger than 50 horsepower nearly all induction motors are of multipole design and offer shaft speeds in fractions of the 1750 rpm common to two-pole designs.

The Superior Electric Slo-Syn design has the feature that it can operate either as a stepper motor or as a *permanent magnet multipole induction motor*. The motor shown in Fig. 7-10 shows that the motor has eight salient (protruding inward) poles. These are ener-

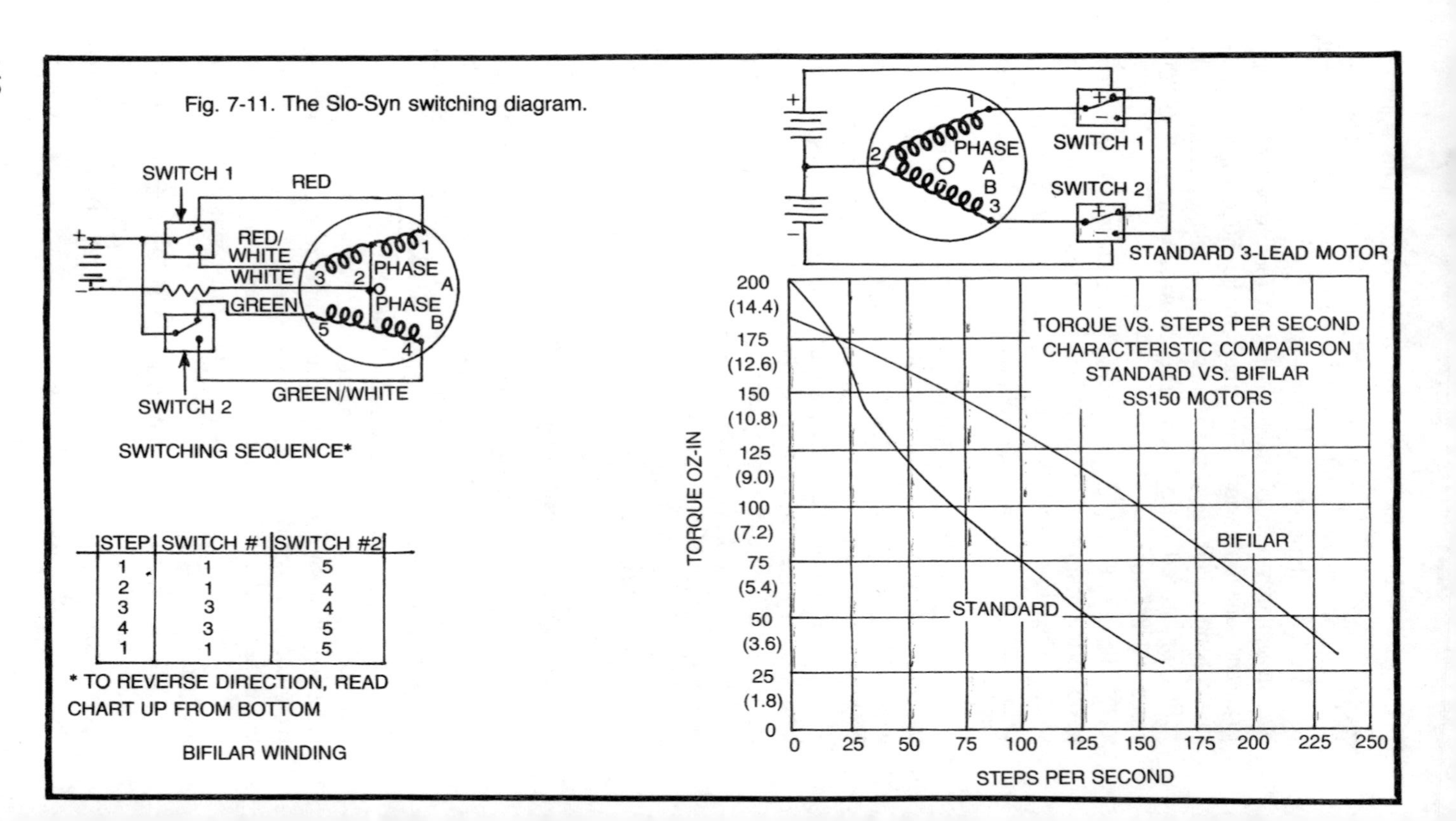

STEP	SWITCH #1	SWITCH #2
1	1	5
2	1	4
3	3	4
4	3	5
1	1	5

Fig. 7-11. The Slo-Syn switching diagram.

gized with a two-phase four-pole winding. Poles N1, S3, N5 and S7 are energized with one phase and poles N2, S4, N6, and S8 are controlled by the opposite phase. The stator teeth are cut on a pitch of 48 per circle although only 40 are present because one tooth has been dropped for interpole spacing. The rotor that is illustrated at the bottom of the figure has a pitch of 50 teeth, or two more than the rotor. It has the form of a pair of spur gears which have been indexed by a half-tooth spacing. The rotor is actually an axially magnetized permanent magnet. One "gear" is a south pole and the other is a north pole.

Because of the difference in pitch between the rotor and the stator only two rotor-stator tooth alignments can take place for any position of the rotor. This results in the vernier-like action described earlier.

Because of the misalignment of the teeth on the stator an amount of residual torque is provided by the rotor, even for zero displacement. The manufacturer claims that this gives the motor the ability to start and stop instantaneously.

Figure 7-12 shows the connections used to drive the motor from a single-phase ac line. After a little consideration, it can be seen that a full cycle on the line is equivalent to four steps in the sequence from the tabulation included with Fig. 7-11. With the line frequency at 60 Hz, therefore, the motor will run at the equivalent rate of $60 \times 4 = 240$ steps per second $\times$ 1.8°/step = 432°/sec = 72 rpm.

When this motor is run from the power line, the instantaneous phase of the input is not under the control of the designer. The direction in which the motor instantaneously starts is a function of two things: the position of the rotor at the instant of the start and the

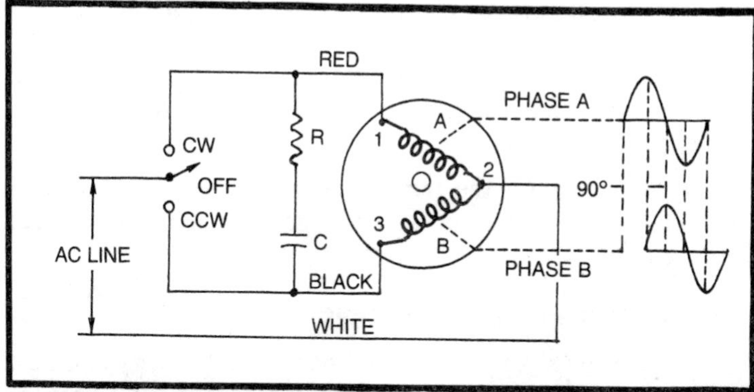

Fig. 7-12. Slo-Syn motor operation.

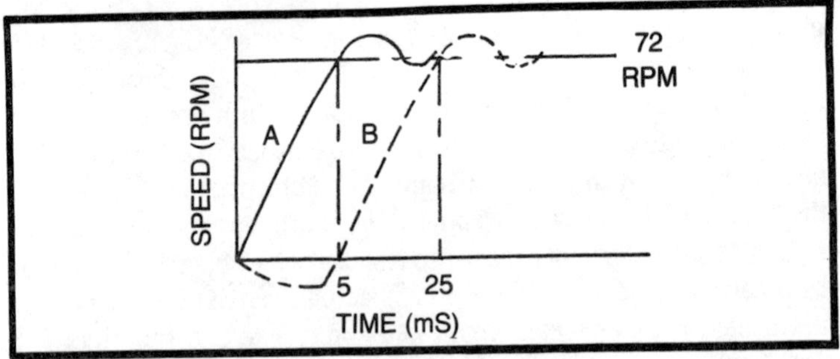

Fig. 7-13. Starting performance of the Slo-Syn motor.

phase of the excitation at the instant of application. The motor is prone to start instantaneously in the wrong direction and then reverse and rotate in the proper direction one half cycle later (16.7 milliseconds at 60 Hz). Once rotation commences in the proper direction, it takes the motor only 5 milliseconds to come up to speed. This performance is illustrated in Fig. 7-13 which also shows that there is some overshoot in the rotor speed. In certain cases, such as on machine tools, the instantaneous reverse might be intolerable since it would jam the tool into the work rather than withdrawing it, and could result in tool breakage. This can be prevented by the use of external circuitry which permits the circuit to close only upon the zero crossing of the line voltage. There are also other cases, however, where the possible 16 millisecond delay might be undesirable. In such cases, a synthesized sine wave drive which can be reset to zero phase in a few nanoseconds could be employed.

Generally speaking these motors will start up within 1 1/2 cycles of the applied frequency and will stop within 5° of the mechanical position where the power was removed without any braking circuitry or mechanical braking being used. The magnetic "detenting," or cogging action, of the rotor teeth is sufficient for this performance. It should be noted that the presence of large external loads can very easily negate this performance. The catalog rating for the maximum moment of inertia cannot be exceeded if proper operation is to be obtained.

On these motors the starting and running currents are nearly identical, so special precautions for surge protection are not generally required.

HYSTERESIS SYNCHRONOUS MOTORS

One of the most common and popular of the very small size synchronous motors is the *hysteresis synchronous motor*. This type is most commonly employed in very small ratings such as in clocks and timers. These motors employ a multipole construction and frequently derive the rotating field from a shaded-pole construction. The rotor is usually stamped from a steel that is magnetizable but relatively easily demagnetized. On starting the rotating field demagnetizes the rotor, and the induced eddy currents cause the motor to start as a squirrel-cage, shaded-pole induction motor. As the rotor approaches synchronous speed, the magnetism "freezes" in the rotor, and the unit accelerates up to synchronism. These are relatively low-torque, low-performance units, but they are cheap and perform quite adequately in the lightduty clock and timer applications.

A somewhat higher performance and higher power type employs a capacitor-start and capacitor-run mechanism to derive the two-phase excitation. These motors will frequently have a cast-on low resistance squirrel-cage winding for superior starting torque. This type is frequently supplied with a built-on gear reduction unit and very substantial low speed torque can be obtained. This type of motor is frequently employed in such applications as telescope drives and recorder drives where very good speed regulation is required along with substantial torque on both starting and running.

For both of these types the nearly instantaneous start and stop type of operation is not readily obtained.

THE ROTARY SOLENOID

An important subclass of the stepper motor is the rotary solenoid shown in Fig. 7-14. In this device a solenoid has been modified by pressing a set of helical inclined ball races into the face of the armature of the solenoid so that the closure of the solenoid is accompanied by an angular twist of perhaps 36°. The device is often equipped with an interruptor cam, which breaks the coil circuit at the end of the rotation, and a toggle mechanism to hold the interruptor open until the armature springs all the way back out. In electrical switching applications, the shaft will often drive a selector switch disc through a ratchet clutch. This is an open-seeking servo. For example, if the power were supplied through terminal C the device would cycle around in steps until the notch in the selector switch was rotated beneath contact C, whereupon the system would halt and remain in that position until another terminal was energized. A variety of wafer switches can be attached to the selector switch shaft. This type of stepper motor is frequently used for remote switch operation with multiple-switch decks, in items as diverse as radios, electronic instruments, and pinball machines.

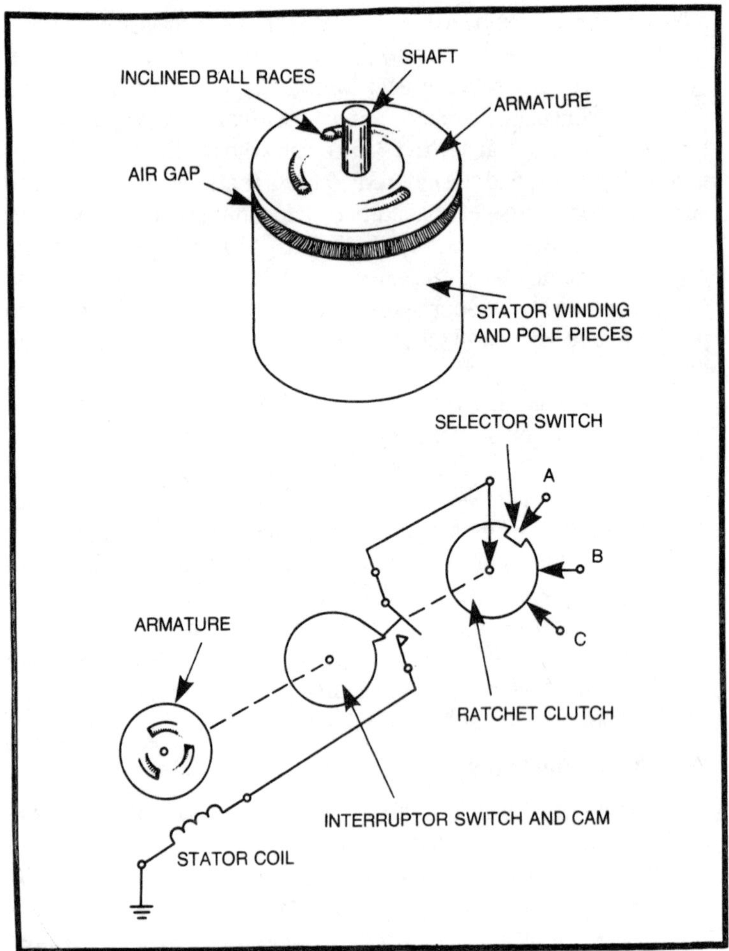

Fig. 7-14. The rotary solenoid.

Typically the solenoid will draw rather substantial "slugs" of current from the supply during the brief switching interval. Numbers like 5 amperes at 28 volts for 15 milliseconds are typical. Because the inductance of the solenoid changes during the closing and opening of the solenoid, the waveform of the current is somewhat different from a straight RL charging cycle. There are very large spikes developed unless a free wheeling diode is included across the coil. Inclusion of some RFI filtering is also generally required if any sensitive electronic equipment is to be operated nearby or from the same supply line. This is not an elegant servo but it is simple and effective.

8
The Universal (AC/DC) Motor

A number of years ago, before the dc illumination plants originally installed in larger cities had been replaced, it was necessary that the motors used on fans, adding machines, vacuum cleaners, and similar appliances be capable of operating on either ac or dc mains. Since the early 1950s, this reason has largely dissappeared; however, the motor designs remain with us for other reasons. By and large, you will find that most of your electric hand tools (except the variable speed types) are perfectly capable of operating on dc. Mixers, razors, and vacuum cleaners also tend to belong to this group. The development of high voltage transistors eventually will probably change or eliminate this situation, but at this writing, it remains.

THE ADVANTAGES

One of the principal reasons for the persistance of these designs has been the power-to-weight ratios attainable with brush commutated motors. It was noted earlier that the torque attainable from a brush type motor is given by:

$$T = K_1 \mathrm{B}\, ilnr_a\, (f_{(\theta)}) \tag{8.1}$$

for the simple Pixii type machine model where:

K_1 = a construction constant
B = flux
l = length of armature
r_a = radius of armature
n = number of armature turns
$f_{(\theta)}$ = flux distribution constant

It was also noted that if the signs of B and i are *both* reversed, the motor rotation and the algebraic sign of T is unaltered.

To reverse the direction of rotation of a shunt or series type motor, it is necessary to reverse either B or i, but not both.

This tells us that a series or shunt type dc motor is fundamentally capable of operation on ac well as dc since the field flux, B, reverses with the current, i.

Now, if we assume that our motor were to do work by wrapping a cord around a drum and lifting a weight, the work done would be equal to the weight lifted times the velocity.

$$\text{Work} = \text{torque} \times 2\pi r \times S \qquad (8.2)$$

Where (in convenient English units)

Torque = motor torque (lb/ft)
$2\pi r$ = circumference of drum (feet)
S = rotational speed in revolutions/sec

The important point here is that the output work is directly proportional to speed. Now, at ordinary line frequencies, 50 Hz in Europe and 60 Hz in the United States, we see that synchronous speed for a two pole motor is only 3000 or 3600 rpm, and we have noted that induction motors generally tend to operate at speeds only about 49 percent as great, at 1460 or 1750 rpm, respectively. This is not a very high speed for a small piece of machinery. Obviously, a considerable increase in output power would be available if the motor could rotate faster, without sacrificing torque.

In general, the weight of an electric motor is determined by the armature and field iron, and the copper windings. These are about equal in an induction motor or a brush-commutator motor of the same physical size. The latter course must carry the penalty of the weight of brushes and commutator. However, owing to the fact that the speed of a brush-commutator motor is not tied to the 50 or 60 Hz line frequency, much higher speeds can be developed, and a considerable net advantage in horsepower/weight ratio goes to the brush-commutator type.

A typical one-speed electric hand drill will turn the drill bit at something on the order of 1100 rpm. The chuck is geared down from the armature by a ratio of seven or nine and the armature will turn 8,000 to 10,000 rpm. This motor will develop three to five times the power of an induction motor of equivalent size, just considering the speed advantage.

In actual practice, the induction motor will tend to be current limited by its self-inductance. Because of the high no-load speed of the drill motor, the counter EMF is very high at no-load and drops drastically as the motor is loaded. In a series-wound drill, the current and field will increase sharply. This in turn raises the torque. A typical 1/4 inch capacity drill will develop about 1/4 hp at full load with a total weight of 1 1/2 pounds (for limited periods). This can be compared with the 15 or more pounds of a continuous duty induction motor with a 1/4 hp rating. Because of the high rotor speed, the cooling fan of the drill motor is considerably more efficient, therefore enhancing the ability of the smaller motor to dissipate heat.

Hand-held power tools like a builder's (circular) saw, a hand drill, sander, or grinder must operate over a very wide range of torque loading. These devices take advantage of the exceptional ability of the series-wound motor to increase its output torque with reduced speed. As the rotor speed is reduced, the CEMF falls and the current rises. The CEMF is given by:

$$\text{CEMF} = K_m SBi \qquad (8.3)$$

where

K_m = a motor constant
S = rotor speed (rev/sec)
B = field flux density
i = field and armature current

However, in a series motor, up to the point of field saturation, B is a function of i. Therefore:

$$\text{CEMF} = K_{m2} Si \qquad (8.4)$$

where

K_{m2} = the motor constant including the proportionality between B and i

The value of i is given by:

$$V_{in} = i\,(R_t + j\omega L_t) + K_{m2}\,Si^2 \qquad (8.5)$$

where

V_{in} = input voltage to motor
R_t = total resistance
$j\omega L_t$ = motor inductance

Rather than belabor the mathematics, we will consider only the dc case for motor torque. Simplifying equation 8.1 by lumping fixed motor parameters, we obtain:

$$\text{Torque} = K_{m3}\,K_{m2}\,i^2 \qquad (8.6)$$

where

$$K_{m3} = K_1 lnr_a f(\theta)$$

Now, if in equation 8.5, the term $i(R)$ were negligible, we would obtain:

$$\text{Torque} = K_{m3} K_{m2} \times \frac{V_{in}}{K_{m2} S} \tag{8.7}$$

$$= \frac{K_{m3} V_{in}}{S} \tag{8.8}$$

For a given torque load, the speed is directly proportional to V_{in}. Also for a fixed V_{in}, speed is inversely proportional to torque. It should be noted that even in this case, the speed would be kept finite in the "no-load" condition by the commutator drag, cooling fan load and the work required to churn the grease in the gearbox.

In a practical hand-tool motor, the resistive term is not negligible compared to the CEMF term. As a matter of fact, for many years, sewing machine motors were controlled in speed by a foot pedal rheostat in series with the motor. With this arrangement, the speed regulation and starting torque are relatively poor. In order to get enough torque to break the motor loose and start the machine without aid, it was usually necessary to depress the pedal fully. The machine then would accelerate to high speed before one could back off on the pedal. For slow operation, it was usually necessary to depress the pedal for the desired speed and give the machine flywheel a push by hand. While Bertha, the sewing machine girl, could manage this, it took a fair amount of practice.

In general, when one wants to control the speed of a power tool, it is because the speed itself is critical, and not the torque loading. A good example of this is the sabre saw. When cutting wood, the speed can be high in softer woods but when cutting maple the speed must be reduced to avoid overheating the blade. Progressively slower speeds are required for plastics, aluminum, iron, and steel. However, the torque required of the motor does not fall, but actually increases. Ideally, the speed control should control speed independent of the torque load.

Figure 8-1 ranks three open-loop speed controls in ascending order of regulation. The variable voltage control at (b) has better regulation than the series resistance because R_t is held to a minimum. The phase-control unit at (c) operates by turning the supply voltage on for only a fraction of the time. It tends to have an even better regulation than (b) because at any given speed setting, when the current is on, it is higher than the current at (b). This favors the CEMF term which is proportional to i^2 compared to the

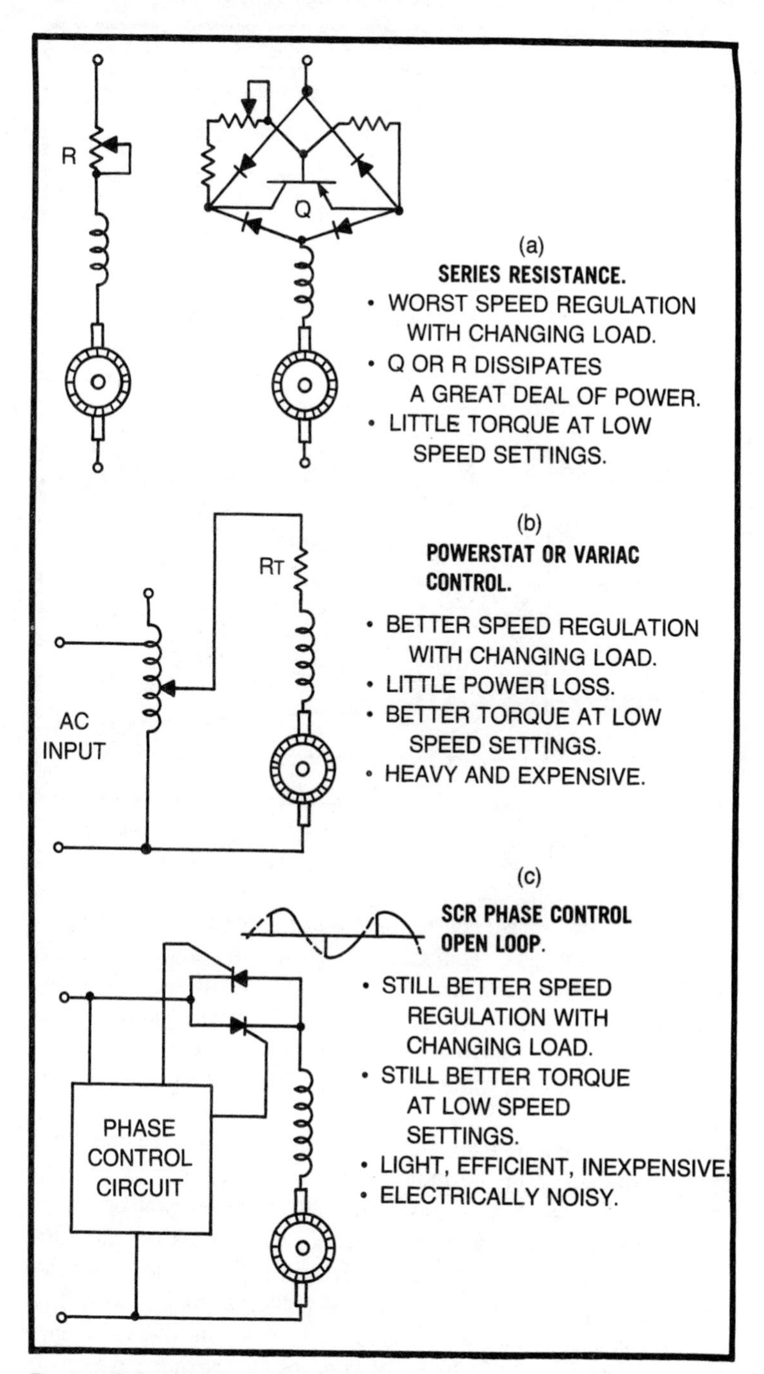

Fig. 8-1. Three open-loop speed controls.

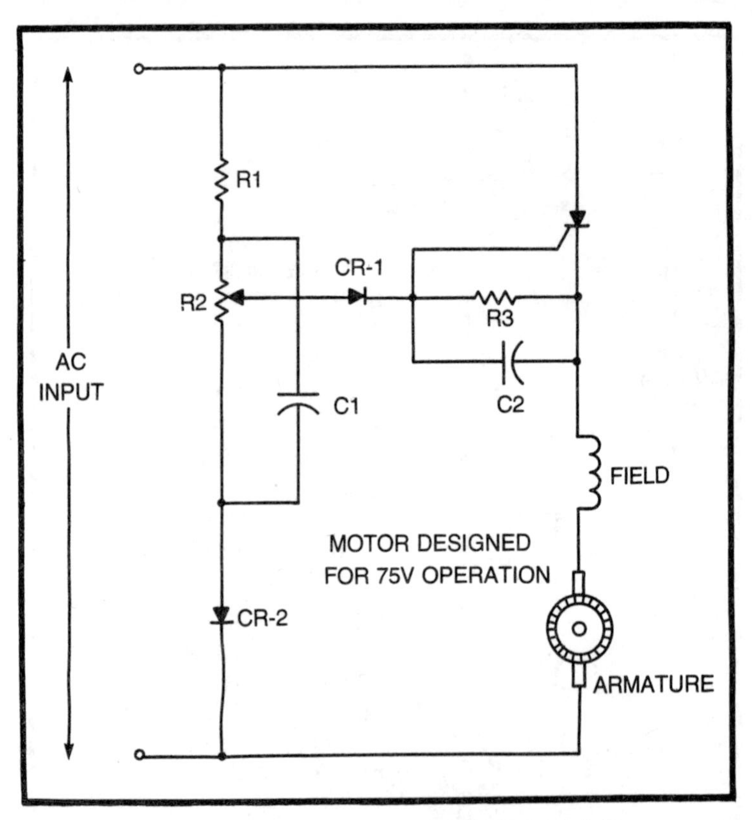

Fig. 8-2. The ac-dc series motor control with CEMF feedback (half-wave operation).

$i(R + j\omega L)$ term. Phase control tends to give the motor a "growly" or "rattly" sound and requires very good line filtering to avoid excessive *electromagnetic interference* (EMI). However, it has become relatively inexpensive, is small, light, and need not dissipate much heat since the control devices operate in the switching mode. A subsequent chapter will give some details of phase-control systems.

By far the best regulated form of speed control is the servo type in which the speed of the motor is measured and compared to a reference. The speed of the motor may be measured by an optical or electro-mechanical one-shot mechanically coupled to the shaft. Unfortunately, these devices tend to add considerably to the cost of the overall product and the series-wound universal motor is usually employed in cost-sensitive applications. A similar and even stronger argument applies to the inclusion of a permanent magnet and a sense winding; therefore, conventional tachometer type speed sensors are usually not found built into these motors.

Figure 8-2 illustrates a form of speed control in which the CEMF is sensed.

THE INVERTER TYPE UNIVERSAL MOTOR

For aircraft applications, where high power-to-weight ratios are at a premium, the use of 400 Hz systems has been standardized since the beginning of World War II. At 400 Hz a two-phase induction motor will operate at speeds between 11,700 and 12,000 rpm. In the past, the cost of deriving the necessary two-phase high frequency current in both money and weight has prohibited the use of such systems for commercial variable-speed tools. However, the rapidly falling cost of integrated circuits and high voltage transistors makes this alternative increasingly attractive.

Figure 8-3 illustrates a simple type of variable high frequency speed control. In this unit the ac line voltage is rectified into positive and negative dc, and the unit as shown is not strictly ac/dc. However, it could be made so with some additional complexity.

The control signal causes the VCO to oscillate at eight times the desired frequency. Flip-flop acts as a divide-by-two to ensure the squareness, or duty cycle, of the output wave. Flip-flops FF-2 and FF-3 are arranged to provide a two-phase output signal. This logic was shown with timing diagrams in Fig. 7-3. When the Q output of FF-1 goes high, $Q5$ cuts off, thus cutting off $Q4$. At the same time, $Q6$ saturates, thereby saturating $Q3$. Winding ϕ 1 is now connected to $+HV$. When the Q output of FF-1 goes low $Q5$ saturates, thereby saturating $Q4$. Transistor $Q6$ cuts off, thereby cutting off $Q3$. Winding $\phi 1$ is now connected to $-HV$. The operation of $\phi 2$ is identical except for the fact that reversing switch $S1$ has been provided.

Over the range of F_o = 900 cycles a motor with properties of a typical small aircraft type 400 Hz (nominal) motor will operate successfully with the torque decreasing with increasing speed, due to the inductive reactance of the windings limiting the current. Below 300 Hz it may be necessary to install a switching type regulator to limit the current. This, in turn, will cause the torque to fall.

Since the motor is operating on square waves, there is a substantial harmonic content in the field excitation; this can cause heating in some motors. A subsequent chapter will describe means by which the harmonic content can be significantly reduced.

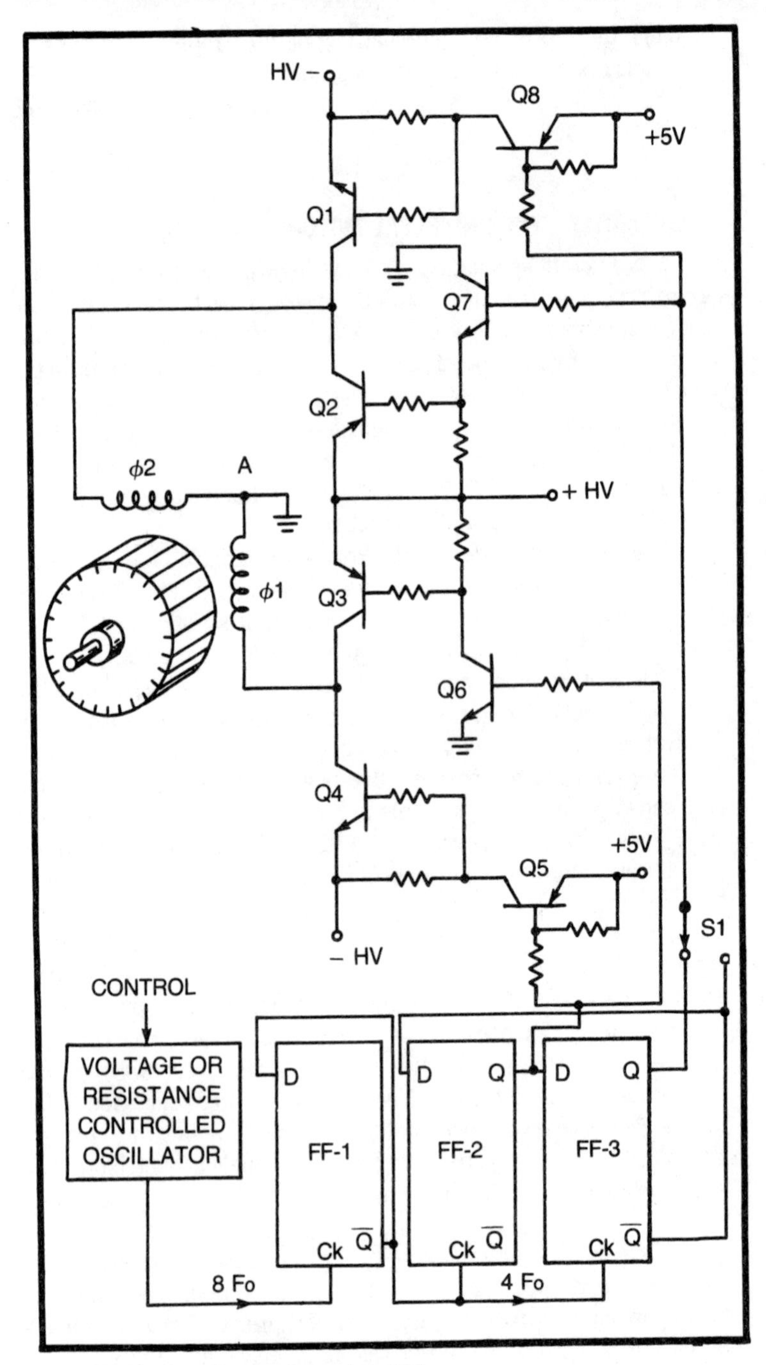

Fig. 8-3. The inverting type speed control.

CAUTION

The control technique of Fig. 8-3 is capable of driving some motors up to speeds where bursting of the motor or of an external load—such as a squirrel cage blower—can occur. Use care. Check the maximum speed specifications.

Low cost, high performance hand-tool motors designed for variable frequency operation are not currently marketed, but it seems likely that they will not be long in the offing. The advantages of a motor with the torque stiffness and starting torque of a two-phase induction motor and the advantages of brushless operation seem too great to remain ignored. The simplicity of the motor proper will greatly offset the cost of the drive control.

9

Inertia and Friction

This chapter is intended as a brief review of the subject matter to indicate its reflection upon motor selection and control design. For those interested in a more complete and rigorous discourse I would recommend:

Fred B. Seely and Newton E. Ensign
Analytical Mechanics for Engineers
John Wiley & Sons Pub. New York, NY

INERTIA

In all but the very slowest and least efficient positioning systems, the effects of inertia must be considered and, in many cases, a unidirectional potential energy may also be involved. Consider the simple example of Fig. 9-1. The car must accelerate from zero velocity, climb the hill, then brake to a halt at the bottom. On the way up, the motion is opposed by friction, inertia, and the force of gravity. On the way down, inertia and the force of gravity oppose friction. Friction may have to be increased by applying the brakes in order to come to a halt at the bottom.

Some of the motor control systems previously described are not particularly concerned with these effects. For example, a speed control on a fan or on an electric hand drill may often be designed without too much concern for inertial effects. In positioning systems, however, this is seldom the case. The control and drive for a milling machine table, an elevator, or the head drive on a floppy disc

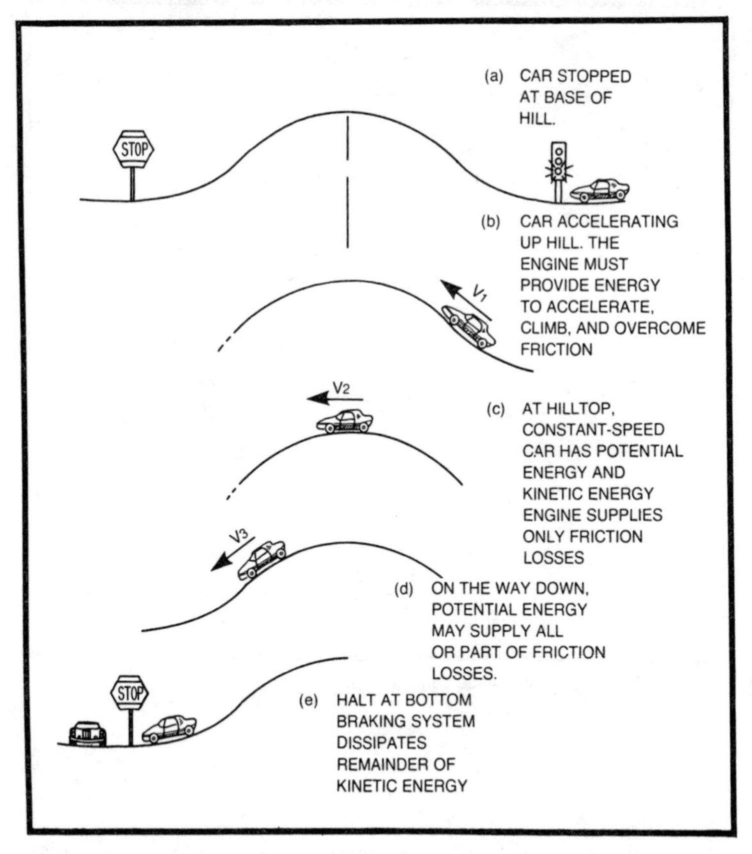

Fig. 9-1. Resolution of the force of gravity. The force of gravity is always vertically downward. This must be opposed by two forces, the force on the car suspension and the force required to climb. The suspension force is always normal (at a right angle to the roadway). The force required to climb is applied parallel to the roadway.

memory must all be concerned with the problem of starting a physical mass, moving it to the desired location, and stopping it without crashing into something.

POTENTIAL ENERGY

Figure 9-2 shows that the force to overcome gravity is not constant over our curved hill but is instead the component of gravity projected normal to the roadway. From the illustration we can see that the force and the work required *just to climb* can vary in a rather complicated way. The work to climb (ignoring acceleration) is:

$$U_P \int_0^s F\,ds \text{ foot-pound} \tag{9.1}$$

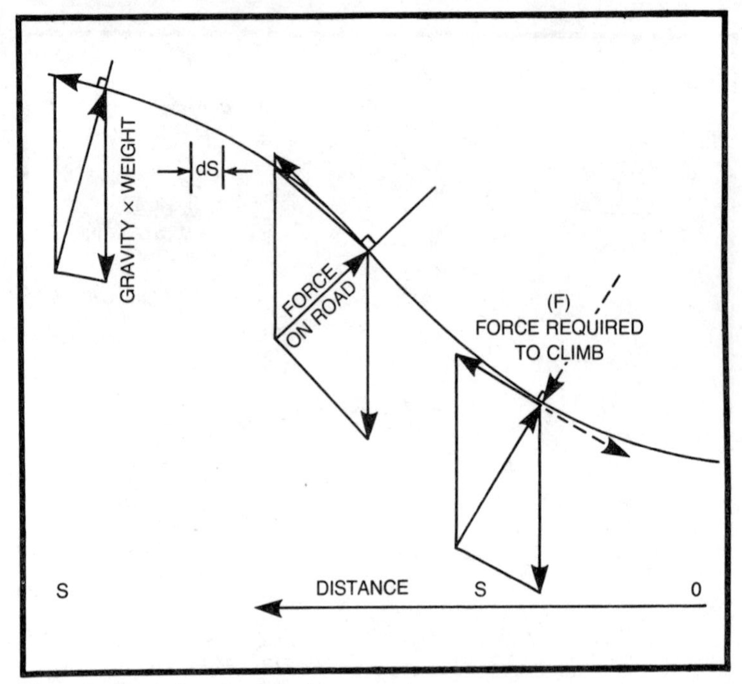

Fig. 9-2. Resolution of gravity force.

with

F in pound
s in feet

The power at constant speed is:

$$hp = F \frac{ds}{dt} \times \frac{1}{550} \text{ horsepower} \qquad (9.2)$$

with

dt in seconds

The term work is simply equal to the product of force times distance. In Fig. 9-2, it represents the *potential energy* of the car when it is at the top of the hill. This energy is the same as if the car had been lifted straight up by a crane and placed atop the hill. In other words:

$$W\,h = U_P = \int_0^s F ds \text{ foot-pound}$$

This simply says that the potential energy acquired by the car in reaching the hilltop is the same regardless of the path. On descending, this energy must be given up into kinetic energy or friction (heat).

KINETIC ENERGY

The concept of kinetic energy, or energy of motion, stems from Newton's laws of motion. In order to change the velocity of a physical body, a force must be applied. In *mass units*:

$$F = ma = \frac{w}{g}\, a \tag{9.3}$$

where

F = force in pounds
a = acceleration in ft/sec^2
w = weight in pounds
g = acceleration due to gravity, at sea level = 32.2 ft/sec^2
m = mass in slugs

The choice of the system of units is a rather complicated subject. As stated in this equation it is in "English engineering," or mass (slug) units. The unit of mass in the equation is a derived unit which is arranged so that force and acceleration can be in common units. If the unit of mass had been selected as the pound, these would be gravitational units and then force would be a derived unit given in *poundals*, where 32.2 pounds of force equal 1 poundal. This usage is not too common in English system calculations. However, the following is frequently used in metric calculations:

$$F = ma$$
$$\text{dynes} = \text{grams} \times \text{centimeters/sec}^2$$

Here the dyne is equal to force in grams × 981 cm/sec^2. More recently, the MKS or meter-kilogram-second system has been gaining acceptance. In this system

$$F = ma$$
$$\text{Newtons} = \text{kilograms} \times \text{meters/sec}^2$$
$$1 \text{ Newton} = 9.81\ kg \text{ force}$$

It is not unusual to find mixed units such as torque in in.-oz (mass units) *along with moment of inertia in lb-in.*^{2} (gravitational units) *in a single calculation in catalog examples. Always check the units carefully.*

Equation 9.3 may be integrated to yield the velocity after some time.

$$V_t = V_o + at \text{ feet/sec} \tag{9.4}$$

where

V_o = Velocity in ft/sec at the beginning of time period t

t = time period of the uniform acceleration

The distance traveled is given by another integration.

$$S_t = S_o + V_o t + \frac{at^2}{2} \tag{9.5}$$

where

S_o = position at the beginning of t

The *kinetic* energy supplied to a body is equal to the integral of force times distance. This can be shown to be equal to:

$$U_k = \frac{mv^2}{2} = \frac{\omega}{2g} v^2 \tag{9.6}$$

$$= \frac{lb \times ft^2/\text{sec}^2}{6.64 \text{ ft/sec}^2} = \text{lb-ft}$$

The term power is a constant times work per unit time:

$$\text{Power} = \frac{U_k + U_p + FVt}{550t} = \text{horsepower} \tag{9.7}$$

where

U_k = change in potential energy in foot-pounds
U_p = potential energy
F = force of friction in pounds
V = velocity in feet per second
t = the period in seconds during which U_p and U_k took place

Energy calculations can sometimes yield a simple approach to an otherwise rather difficult problem. For example, let us consider a simpler case than Fig. 9-1. A car weighing 3220 pounds accelerates from a halt up an inclined plane at a constant rate of acceleration, such that 20 seconds later at the top of the plane it is going 80 feet per second and it has climbed 100 feet.

The potential energy gained is simply:

$$U_p = 100 \text{ ft} \times 3220 \text{ lb} \tag{9.8}$$
$$= 3.22 \times 10^5 \text{ ft-lb}$$

The kinetic energy is given by the velocity:

$$U_k = \frac{mv^2}{2} = \frac{\omega}{9} \times \frac{v^2}{2} \tag{9.9}$$
$$= \frac{3220 \text{ lb}}{32.2 \text{ ft} \ \ /\text{sec}^2} \times \frac{(80)^2 \text{ ft}^2/ \text{ sec}}{2}$$
$$= 3.2 \times 10^5 \text{ ft-lb}$$

If we neglect friction, then the power required of the engine is:

$$\text{Power} = \frac{U_k + U_p}{550 \times 20 \text{ sec}} \tag{9.10}$$
$$= \frac{(3.2 + 3.2) \times 10^5 \text{ ft-lb}}{550 \times 20 \text{ sec}}$$
$$= 58.18 \text{ hp}$$

Since the acceleration was constant, therefore minimum, and friction was neglected, this is the least possible power required to produce the changes in U_p and U_k. If the acceleration had not been uniform, then at the instant it was greatest, the greatest power would have been required.

Friction represents somewhat of a different matter. In general, friction on a slow-moving sliding body is given by a coefficient multiplied by the weight or force loading.

$$\text{Sliding friction force} = k_s W \text{ pounds} \tag{9.11}$$

where k_s is a pure numeric with no units. Actually, this is seldom a good approximation even for slow moving bodies since it depends upon the condition of the two surfaces, the presence of oil or moisture, and other factors. In addition, there is always a *static* friction component which must be overcome in order to "break loose" and start the body in motion. This is usually greater than the sliding coefficient by a significant factor.

In a real machine, as opposed to a theoretical machine argument, there is usually a friction component which is proportional to velocity:

$$\text{Dynamic friction} = K_d VW \text{ pounds} \quad (9.12)$$

And most often, there is a viscous damping force

$$\text{Viscous damping} = F_{VF} = K_v V^2 W \text{ Pounds} \quad (9.13)$$

Note: In the above equations, to provide harmony in the units K_d is in units of $1/v$ and K_v is in units of $1/v^2$.

Viscous damping was mentioned earlier. The V^2 dependence is typical of the sort of thing one gets when stirring something like oil or maple syrup. It can also arise from churning grease in a gearbox, or from air resistance of a car or airplane or fan, or water resistance of a boat.

There is a variety of ways of predicting viscous damping in machines, but they are well beyond the scope of this book. On the other hand the coefficient of viscous damping is of great interest in the design of a control system. If the machine to be controlled is available or if a similar machine can be obtained, the coefficient of viscous damping can usually be measured by a variety of means.

1. Determine the input power required to maintain a constant speed, for a variety of speeds. This is fairly easily done with a dc motor drive by measuring the input power and accounting for the motor losses as described earlier. If the speed and power are both known, the coefficients for equations 9.12 and 9.13 can be calculated with reasonable accuracy.
2. Shut down the power with the machine at full speed and measure the time required for the system to coast between V_{max} and any other velocity. If one assumes that the damping is entirely viscous the method and data of Table 1-1 may be used.
3. Shut down the power with the machine at full speed and measure the distance the machine will coast to a halt. Use the Table 1-1 method.

On a closed-loop servo

4. If it is safe, with power off and at the widest deadband setting:
   a. Reduce the deadband **slightly**.
   b. Apply a step input error.
   c. Power up the system and note response.
   d. Return to step a if system is overdamped.
   e. When a slighly underdamped condition is obtained, back off slightly.
   f. Measure the deadband.

g. Calculate viscous damping coefficient from the Table 1-1 method. Assume that the deadband corresponds to a velocity decay of 0.5.

The example which follows shortly is designed to give you some further insight into the use of Table 1-1. These techniques are not too elegant from the viewpoint of a mathemetician, but they are often the best that one can manage on a real machine. On something like a jet fighter plane, calculation of the damping of the aircraft itself and the damping of the various control surfaces can occupy the services of a large computer and an office full of engineers for a year or more. Furthermore, in the final analysis, the computation relies heavily upon empirical data obtained from the previous aircraft and from wind tunnel tests.

Getting back to our little car problem, let us see how we would calculate the damping on the car from a simple measurement. We shall assume that the damping is viscous in the format given in Table 1-1.

Since the power expended in friction is a function of force times distance, it is maximum at maximum speed, except in the rare case where the friction force falls with speed. Let us suppose that the road leveled off at the place where we were going 80 feet per second. Next, suppose we kick the car into neutral and measure the time required for it to decelerate to half speed, or 40 feet per second. If we look at Table 1-1 for the entry for $V/V_{max} = 0.5$ we find:

$$V/V_{max} = 0.5015$$
$$kt = 0.7000$$
$$\frac{K\Sigma S}{V_{max}} = 0.5059$$

We could interpolate but the 3 percent error is not too great so we will accept the values as close enough. Now, suppose we find that it took the car 32.56 seconds to slow to half speed. Then from the table, we see that:

$$kt = 0.700 \tag{9.14}$$

but

$$t = 32.56 \text{ sec}$$

thus

$$k = 0.7/32.56 = 0.0215 \qquad 1/\text{sec}$$

Now if we assume that the velocity expression is

$$V = V_{max}\,(e^{-kt}) \tag{9.15}$$

The first derivative of this function is:

$$dV/dt = kV_{max}\,(e^{-kt})\ \text{ft/sec}^2 \qquad (9.16)$$

now, at the instant we go out of gear, $t = 0$ therefore

$$dV/dt\ (\text{max}) = kV_{max}\ \text{ft/sec}^2 \qquad (9.17)$$

and in this case, our maximum acceleration would be

$$dV/dt_{(max)} = 0.0215 \times 80\ \text{ft/sec}$$
$$= 1.72\ \text{ft/sec}^2$$

Note incidentally that the units for k are the reciprocal of seconds. Now returning to equation 9.3 we can calculate the force of friction.

$$F = \frac{Wa}{g}\ \text{Pounds} \qquad (9.18)$$

$$= \frac{3220\ \text{lb}}{32.2\ \text{ft/sec}^2} \times 1.72\ \text{ft/sec}^2$$

$$= 172\ \text{Pounds}$$

and the power being expended by the motor is

$$hp = \frac{\text{Force} \times V}{550} \qquad (9.19)$$

$$= \frac{172\ \text{lb} \times 80\ \text{ft/sec}}{550}$$

$$= 25.0\ \text{horsepower}$$

During the deceleration, the car would have traveled a distance given by the third table entry;

$$\frac{k\Sigma S}{V_{max}} = 0.5059$$

$$\Sigma S = \frac{0.5059 \times 80\ \text{ft/sec}}{0.0215\ 1/\text{sec}}$$

$$= 1882\ \text{feet}$$

We can also see that the total distance travelled by the car when it finally came to a rest would be very close to V_{max}/k or 3721 feet, about 7/10 of a mile. If we were to assume that the same damping applied during the climb, then at the peak of the hill the motor would be having to donate the 58.18 horsepower for the increase in kinetic and potential energy plus the 25 horsepower (at 80 feet per second) required to overcome the damping.

TORQUE AND ROTARY INERTIA

Since the overwhelming majority of electric motors are rotary devices, we are generally more interested in torque than in linear force. The designer of a motor control system is, therefore, most often concerned with inertia in rotational units. With conventional ac or dc motors the matter of concern is the length of time it may take the motor to start or to stop the load. This involves the determination of the resulting overload conditions and the placement of limit switches with respect to mechanical stops, and so forth. In the special case of stepper motors and slow speed synchronous motors the matter of inertia is even more critical since motors of this type have a very sharply defined torque limit for any given excitation voltage. The motor must be capable of not only supplying sufficient torque to overcome static friction, viscous damping potential energy increment, and the inertia of any load plus the inertia of the rotor itself, but also of accelerating this load to full speed in one or, at most, two cycles. Otherwise the system will simply stall and fail to start or it will overshoot on stop. For this reason, stepper and slow speed synchronous motors are always carefully rated by the manufacturer in terms of starting torque, rotor inertia, and other properties of interest.

The term torque is simply defined as the force exerted on or by a shaft at the end of a lever:

$$\text{Torque} = \text{force} \times \text{radius}$$

In general English engineering units this is given in pounds of force and feet of radius, and for integral horsepower motors the calculations are usually performed in such units. However, for fractional horsepower electric motors, torque is generally quoted in ounce-inches. This may be converted by:

$$\text{Torque} = \frac{\text{force (oz)}}{16\ \text{oz/lb}} \times \frac{\text{radius (in)}}{12\ \text{in/ft}} = \text{oz-in.}/192$$

Or conversely:

$$\text{Torque} = \text{ft-lb} \times 192 = \text{oz-in.}$$

If we assume that a motor is doing work at a constant torque load, for example by lifting a weight by winding a cord around a drum, the distance traveled by the weight in one revolution of the drum is two times the radius of the drum. Figure 9-3 shows this situation along with the calculation for potential energy and horsepower.

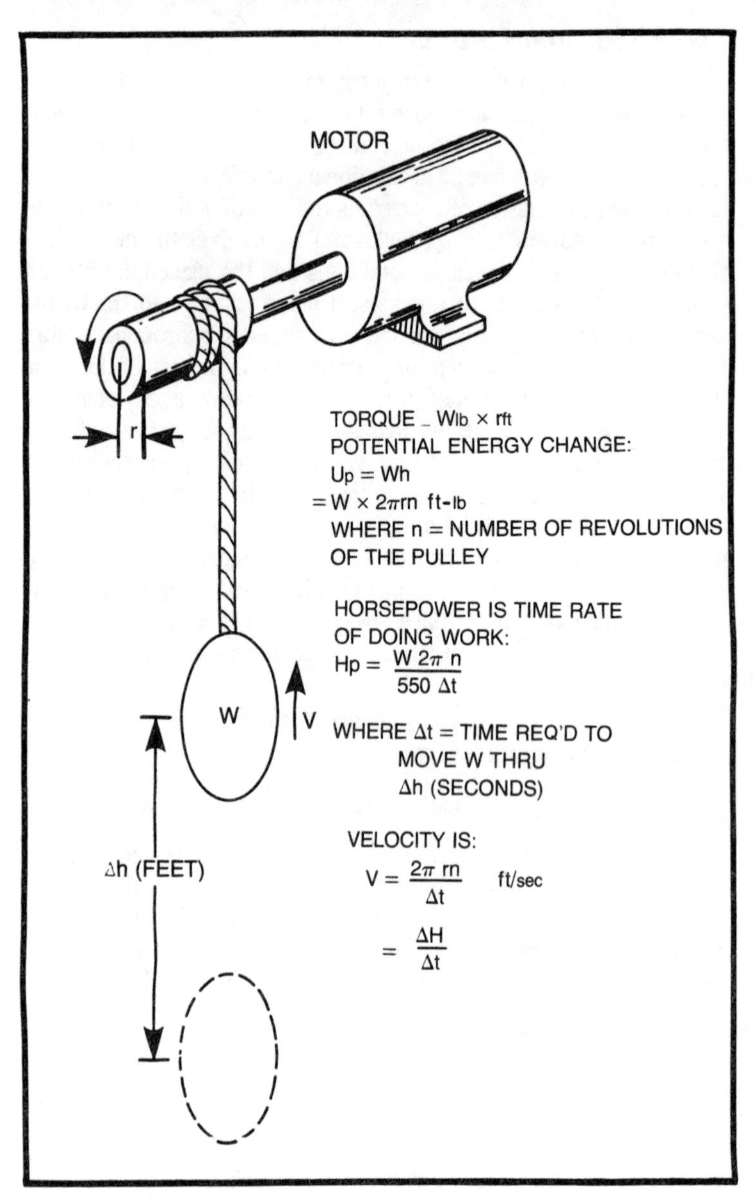

Fig. 9-3. A motor lifting a constant torque load.

From the figure we can derive the expression for the kinetic energy which is the dual of equation 9.6. If we assume that the motor has no mass or inertia, then the kinetic energy of the system is given by:

$$U_k = \frac{Mv^2}{2} = \frac{W\,v^2}{g\,2} \text{ Foot-Pound}$$

and substituting for v we obtain:

$$Uk = \frac{W\,(2\,\pi r n)^2}{g\,2\,(t)^2} \text{ foot-pound}$$

and the units are

W in pounds
r in feet
t in seconds
g in ft/sec^2

From this we see that the units do indeed cancel out to yield energy in foot-pound.

ANGULAR ACCELERATION

The rotary equivalents of Newton's laws of motion, as shown in equations 9.3, 9.4 and 9.5, are as follows:

$$\text{Torque} = I\,\alpha \quad \text{foot-pound}$$

where

I = moment of inertia
α = acceleration in radians/sec^2

The acceleration term is related to linear acceleration:

$$a = r\alpha \qquad \text{ft/sec}^2$$

The dimensions of r are naturally feet/radian. From the above it is obvious that the moment of inertia, I, must contain the dimension feet. The most common shapes to be turned are a cylinder and the hollow cylinder. It can be shown for the cylinder that the moment of inertia about the axis is:

Cylinder

$$I = \frac{M\,r^2}{2} = \frac{W\,r^2}{g\,2} \text{ slug-feet}^2$$

To calculate the moment of inertia of a hollow cylinder one may simply calculate the moment as if the cylinder were solid and then subtract the moment of the missing part in the center. This works out to:

Hollow cylinder

$$I = \frac{W}{2g} (r_1^2 + r_2^2) \text{ slug-ft}^2$$

(Note that the + in the above equation is not an error, it stems from the factoring of an $r_1^4 - r_2^4$ term and the subtraction of the missing mass.)
The term α is the first derivative of angular velocity ω:

$$\frac{d\omega}{dt} = \alpha \qquad \text{radians/sec}^2$$

By integrating equation we obtain the dual of equation 9.4:

$$\omega = \omega_0 + \alpha t \qquad \text{radians/sec}$$

A second integration yields the total position θ in the dual of equation 9.5

$$\theta = \theta_0 + \omega_0 t + \frac{\alpha t^2}{2}$$

The zero subscripts refer to the initial position and the initial velocity at the start of time interval t.

As an example, let us consider the starting reaction on a laundry tub filled with water, as in a spin drier. If the tub has a radius of 1 foot and a water depth of 1 foot, the volume of water is:

$$\text{Volume} = \pi r^2 \times \text{depth} = 3.14 \text{ ft}^3$$

With a weight of water taken as 62.5 lb/ft^3 the weight is 196 lb. The moment of inertia is

$$I = \frac{W r^2}{2g} = \frac{196 \text{ lb} \times 1 \text{ ft}^2}{2 \times 32.2 \text{ ft/sec}^2}$$

$$= 3.05 \text{ slug-ft}^2$$

Now, suppose we assume that we would like to have the unit up to 90 percent of terminal speed 600 rpm = 10 rev/sec = $10 \times 2 \times \pi = 6.28$ rad/sec within 20 seconds of the start. If we assume that a viscous damping takes place such that motor and tub speed follow the relation:

$$\omega = \omega_{max} (1 - e^{-kt}) \qquad \text{radians/sec}$$

let us calculate the torque required of the motor initially.
The derivative of the last expression is:

$$\alpha = \frac{d\omega}{dt} = \omega_{max}{}^{k}\,(e^{-kt}) \qquad \text{radians/sec}^2$$

now, at t = 20 sec we require ω to be 90% ω_{max}.
therefore

$$e^{-kt} = 0.1$$
$$kt = 2.303$$
$$k = 0.1151$$

thus

$$\alpha = 6.28 \text{ rad/sec} \times 0.1151$$
$$= 7.23 \text{ rad/sec}^2$$
(k has the dimension 1/sec)

then the torque is:

$$\text{Torque} = I\alpha = 3.05 \times 7.23$$
$$= 22 \text{ lb-ft}$$

This is a pretty husky torque. If we assume that we intend to use a condenser-start, condenser-run motor with a starting torque about 450 percent of running torque we obtain a running torque rating of 4.89 lb-ft. The motor power rating is then:

$$\text{Horsepower} = \frac{4.89 \times 2 \times \pi \times 10 \text{ rev/sec}}{550} = 0.56$$

The start and run currents (full load) for a typical capacitor-start, capacitor-run motor of this size will approach 25 amperes and 8 amperes, respectively, on a 120 volt unit. This is a pretty good fraction of the capability of a circuit fused at 15 amperes. A starting winding type induction motor of this size simply would not start up fast enough to stay within the heat impulse envelope of a 15 ampere fuse with a load of this size.

HOLLOW ROTOR MOTORS

For certain types of servo mechanisms it is necessary to have the very fastest possible response, i.e., to have the motor up to operating speed in the least possible time from a rest, or to have the motor stop from full speed in the minimum distance. We have seen with the stepper motors that this response can be as rapid as one line half cycle, approximately eight milliseconds. However, these motors tend to accelerate up to speed mainly because they are inherently rather slow and they do not have all that far to go to get up

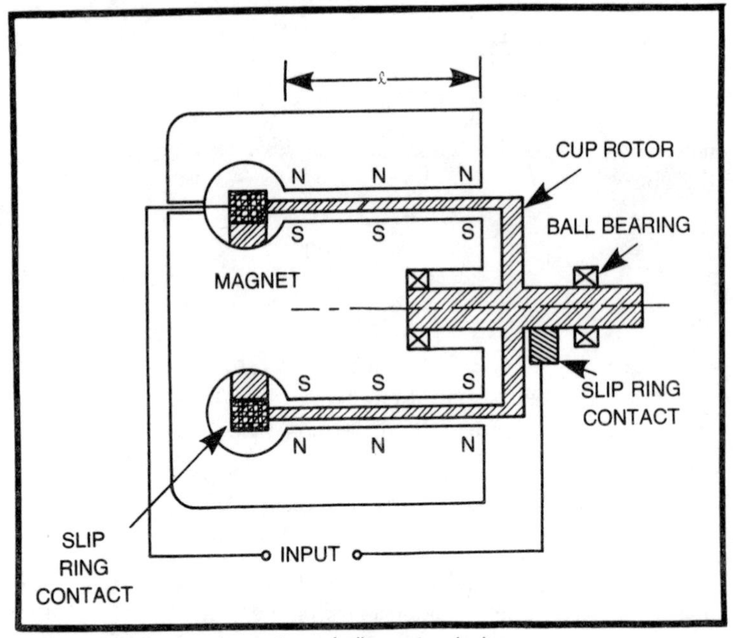

Fig. 9-4. The homopolar motor, a hollow rotor design.

to speed. Certain of the smaller servo motors are capable of accelerating from a dead stop to 3900 rpm in less than one millisecond and in less than 12° of shaft rotation. This implies an acceleration of 2.34×10^7 deg/sec^2 which in turn implies either a fantastic amount of torque or a very low inertia.

One of the most obvious ways to obtain a very low inertia is to arrange the motor so that the iron does not turn, as in a D'Arsonval galvanometer, which simply has the coils immersed in the magnetic field between the outer poles and a fixed central slug. Unfortunately, the D'Arsonval instrument has the central slug supported by posts which prevents the coil from making a full turn.

The matter of motor geometry is, therefore, a matter of selecting a design which will permit the rotor to operate within the field of two stator parts. One of the oldest of such designs is the *homopolar motor*, illustrated in Fig. 9-4. This motor has the unique distinction of being the only dc design that directly produces dc and requires no commutation or rectification. The original Faraday motor was a homopolar design.

As shown in the figure, the rotor is a hollow cup (shown here in cross section) which is situated in an annular magnetized gap. The field in the annular gap is essentially uniform and is shown with the south pole on the inside and the north pole on the outside. This motor tends to be an inherently low voltage, high current type since, in effect, it has only one turn on the armature. The voltage of the rotor is given by:

$$V_g = Blr\omega \qquad \text{volts} \tag{9.20}$$

Where

B = flux density in webers
r = radius in meters
l = length of the field in meters
r = rotational velocity in radians/sec

If we consider an example where $r = 0.1\,m$, $l = 0.1\,m$, $B = 0.4$ weber/meter^2, and the input voltage is 1.63 volts we can solve for the speed where the V_g just cancels the input voltage, neglecting brush drop and windage losses. This yields a result of 408 rad/sec or 3900 revolutions per minute.
The torque is given by:

$$\text{Torque} = Blri \qquad \text{newton meters} \tag{9.21}$$

where

i = current in amperes

For the same motor the torque would be 4×10^{-3} newton meters per ampere, or 2.58×10^{-3} lb-ft per ampere. This is precious little for such a large motor. In order to develop one hp at 3900 rpm, the motor would have to draw 450 amperes! And this is before allowance for windage loss.

This form of motor or generator is not very practical in small sizes. The brushes for the slip rings do not have to be slightly resistive to reduce the circulating current as in a commutating motor and may therefore be made of silver-graphite or other low loss contact material; however, the brush drop will still remain in the 100 to 200 mV/brush region which would mean a significant power loss for a 1.63V motor. On the other hand, if the size were scaled up by a factor of 10 in radius and length the machine would develop a V_g = 163 volts at 3900 rpm and the torque at 450 A (without windage and friction losses) would go to 180 newton-meters (132 ft-lb) which at that speed would amount to 98 hp. This is an impressive amount of power, but the machine would be over 8 feet in outside diameter. The performance of the machine in terms of acceleration would be truly impressive since the rotor could be made of aluminum and would have relatively little inertia. A low inertia aluminum rotor for this machine could be constructed with a weight of about 35 pounds

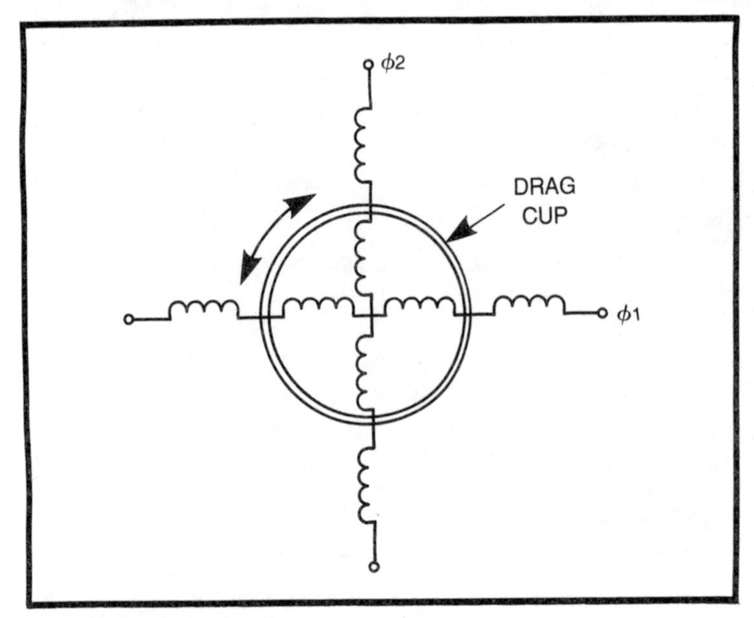

Fig. 9-5. The drag cup motor.

and a moment of inertia of about 11 slug-ft^2. This would accelerate at the rate of 23.04 rad/sec at constant current of 450 amperes and would reach 3900 rpm after 17.7 seconds, having turned through 574 revolutions. If the current were allowed to climb to 1000 amperes the machine could reach 3900 rpm in about 8 seconds, having turned through only 117 revolutions. The drum on a motor of this size would easily handle several thousand amperes in a starting surge. The principal limitation would be the shaft strength and the current handling capability of the slip rings.

In actual practice, the homopolar machine has been used principally for dc generation where very large currents and relatively low voltage is required. These applications include such items as electroplating of ship propellers and ship hulls, and degaussing of battleships.

A variation of this theme, which is used more practically in small sizes, is the *drag cup motor*. This device is generically related to the dragcup speedometer shown in Fig. 4-2 except that the rotating bar magnet has been replaced by a rotating magnetic field. Physically, the motor is arranged in fashion very similar to the homopolar machine shown in Fig. 9-4 except that the fixed permanent, or electromagnets, have been replaced by a pole structure very much like the pole structure of an induction motor. Fig. 9-5 shows a two-phase arrangement, but a three-phase arrangement could just as easily be used. In operation, the machine is very similar to a squirrel cage induction motor. In practice, these motors are usually built in small ratings and designed for two-phase servo operation. Because of the transformer action, the stationary windings have little difficulty including the required very large circulating currents in the aluminum drag cup. The motor also has the advantage over its dc counterpart in that no slip rings are required since the eddy currents are induced in the drag cup. These motors usually have a dual field structure arranged so that a portion is inside of the drag cup and a portion is outside.

Another variation of the theme is the *shell motor*, or related *basket motor*. One winding for this form of motor is shown in Fig. 9-6. The general shape of the winding is shown. This winding could be thought of as occupying the same surface as the drum of the homopolar motor. This motor format has a great advantage over the homopolar in small sizes in the fact that both V_g and torque are

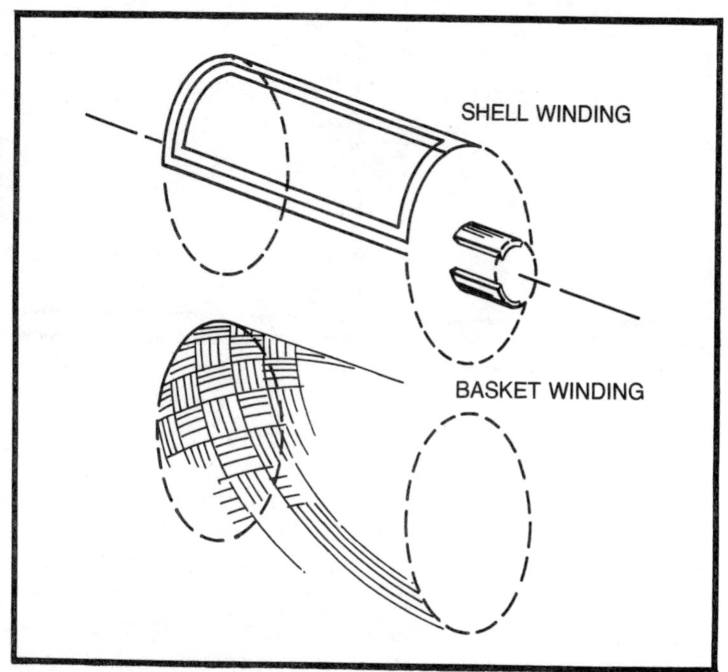

Fig. 9-6. The dc shell motor winding and a variation, the basket rotor. Only two commutator bars and one winding of a shell of a shell motor are shown for simplicity. The actual motor would have a number of such windings to form the shell, or basket. In the basket type rotor, the wave-wound shell is woven like a basket for mechanical support.

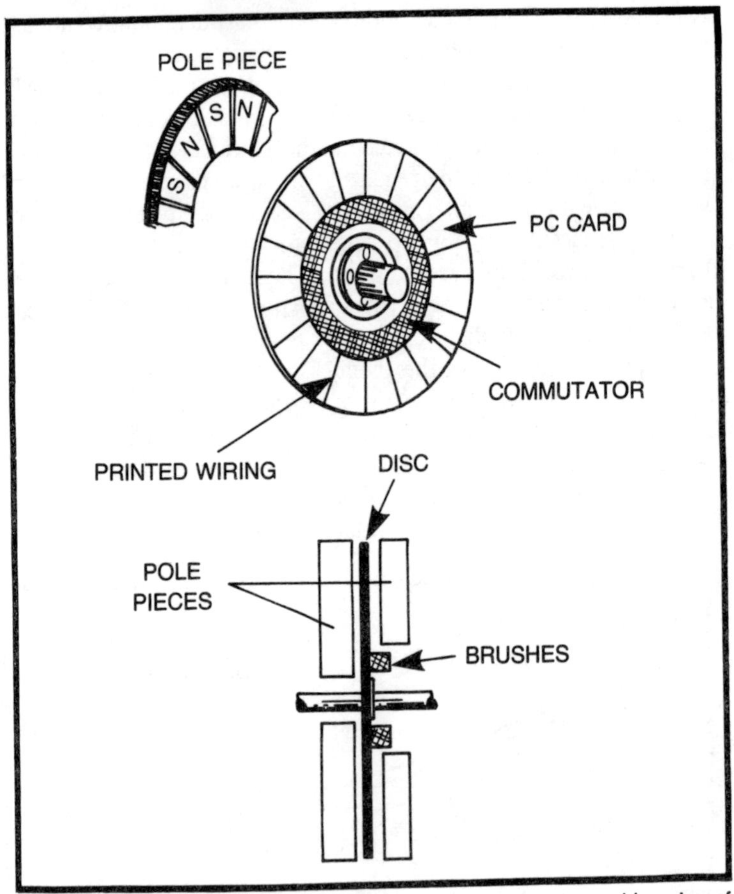

Fig. 9-7. The pancake motor. The unit shown is broken apart to provide a view of the rotor and pole pieces.

multiplied by twice the number of turns. In small sizes, therefore, these motors can be made to operate on more practical values of voltage and current. The motor is sometimes described as a basket motor since in a skewed pole design, the windings are woven like a basket. This actually has a certain mechanical advantage because of the strength that the weaving imparts. The windings on small sizes are frequently simply impregnated with epoxy and left to be self-supporting. Unfortunately, this is a very expensive way to build an electric motor since the windings must be very true and must be secured against the centrifugal force which tends to expand the open end of the basket. However, these motors will give the outstanding acceleration performance quoted earlier. This can sometimes be worthwhile as in the case of digital computer tape transports and in memory disc head positioning devices where the utmost in mechanical agility is required in order to avoid excessive delay for the digital circuitry.

A somewhat different configuration is the *pancake motor* illustrated in Fig. 9-7. In this case the geometry of the rotor has been transformed into a single flat disc with a shaft through the center. Some of these motors are fabricated upon a printed circuit card, with the wiring etched on the card and the commutator etched on as well. The card is double-sided with the return wiring on the far side of the card from the commutator. Like the basket rotor motor this unit enjoys an advantage of twice the number of turns in V_g, and torque compared to the homopolar. Compared to the basket motor, the pancake motor enjoys another advantage in the fact that the disc does not have to resist the efforts of centrifugal force to deform it with respect to the air gap. It is usually not possible to print as many turns upon the disc as it is to weave into the basket. These motors are, therefore, usually lower voltage/higher current affairs. However, a significant advantage in cost of rotor preparation exists. The pancake motor has been employed for tape drives and similar uses, like the basket motor, and is usually built in larger sizes than basket types. One version of the pancake motor employs a minimum amount of rotating iron to provide a support for more conventionally wound wiring.

PRO AND CON

When built completely without iron in the rotor, these motors have the advantage compared to conventional permanent magnet types in that there is no detenting or cogging torque when the motor is not energized. This can be a distinct advantage in some applications since the cogging effect can quantize or granularize the response of a control system. There is also the extremely rapid start and stop property which is the principal advantage.

On the negative side, there is the fact that these motors are usually a great deal more expensive than a conventional type and they are also usually less powerful and less efficient than a conventional motor of the same size. This latter property stems from the fact that they must generally have from three to five times the amount of air gap in order to permit passage of the drum or winding between the pole pieces. Even if the winding proper were of zero thickness, they would have twice the air gap. Since the air is typically 5000 times less magnetic than the pole pieces, the larger air gap usually means that the flux density is considerably smaller or else that the energizing magnets or winding must be a great deal larger.

In addition, the very rapid response of motors of this type in the smaller sizes effectively precludes the use of pulse width and pulse frequency modulated switching type speed regulators since the motors are often fast enough to follow the pulsations and to modulate the output shaft speed. If there is any significant "wind-up" in the shaft and gearing it can lead to severe torsional oscillation in the system. It is generally necessary to either cause the system response to fall off to less than unity gain at the mechanical oscillatory frequency or to provide mechanical damping to accomplish the same end.

Motors of this high speed response are usually best driven with constant-current supplies in order to obtain the best and smoothest response. Unfortunately, this requires the use of the control transistors in the linear rather than the switching mode which means that far larger regulator transistors are required. In passing it should be noted that a constant current type drive for an ordinary iron rotor PM motor will do wonders for the response rate.

The Choice

These factors combine to restrict the use of the hollow rotor motors to those applications where the response requirement outweighs the other considerations.

10
Feedback Mechanisms

As noted in the discussions of Chapter 1, many types of controls consist of or contain a full closed-loop servo. The purpose of the loop closure is, naturally, to inform the system to what measure the assigned command has been completed or perhaps, at what rate it is being completed. In our "oversimplified power steering" example of Chapter 1 the feedback sensor consisted of the switch rod and contacts A and B. As a matter of fact, a good many controls have nothing much more complicated than a switch to sense the direction of the command and a deadband adjusted to permit the motion to damp out.

It is noteworthy, however, that the deadband error must be great enough to allow the damping of the nonlinear system to provide stability when the system is running at full speed and in maximum "coast" condition. This may not be accurate enough for some applications. In addition, when the command input is small, the system will not get up to speed and will thus tend to err on the negative side; that is, it will be overdamped for small input commands. A better and more sophisticated type of system is required to sense the speed and perhaps the load and to adjust the deadband accordingly.
We shall discuss some of these features in this chapter.

THE BRIDGE FOLLOWER

One of the simplest and easiest feedback mechanisms to use is the potentiometer. Chart-and-pen recorders frequently employ a potentiometer to tell the system where the pen is on the paper. The

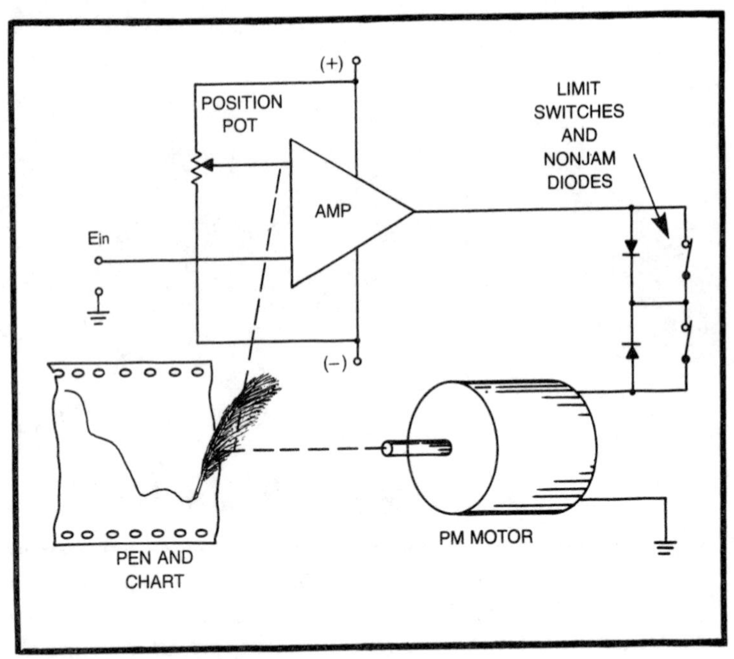

Fig. 10-1. The bridge follower.

simplified schematic of Fig. 10-1 illustrates this arrangement. An op amp with a balanced supply is arranged to drive the motor in the direction to null the difference between E_{IN} and the existing position potentiometer setting. As shown, the op amp has no electrical feedback, so it will tend to run to saturation with very small differences of potential between the inverting and noninverting input.

A very important feature is shown in this figure in the limit switches and the non-jam diodes. In systems with limited mechanical motion, it is important that some form of limit switch be provided to prevent the motor from crashing into the frame or whatever limits the travel. Once there, with the limit switch open, it must be possible for the motor to be driven away from the switch, otherwise the system would sooner or later stall in one of the limit postions. The diodes across the switches will permit the PM motor to back away from the limit switch but prevent further travel toward the switch. If the error signal remains toward the switch the diode holds off the current and prevents motor burnout.

The circuit of Fig. 10-1 lacks several features required of a chart recorder. First of all, with zero input (that is with E_{IN} shorted) the pen would be in the center of the chart. It is often desirable to be able to "zero" the pen anywhere on the paper. The second item is the fact that most op amps will not follow or retain their common mode rejection ratio closer than 1 or 2 volts to the supply rails. The circuit of Fig. 10-2 illustrates simple techniques for accomplishing these ends. Note, however, that E_{IN} is no longer referenced to ground, but rather floats between the offset potentiometer tap and the op amp input.

The arrangement is still not complete since there is no dead-band adjustment. Without "someplace to put the screwdriver" the chances of obtaining a stable and responsive system are almost negligibly small. The circuit of Fig. 10-3 illustrates a mechanism for obtaining the requisite control. When the output of U1 goes more positive than the forward drop of CR-1, U2 output saturates near the positive rail, thereby saturating Q1. Conversely, when the output of U1 goes more negative than the forward drop of CR-2, U3 saturates

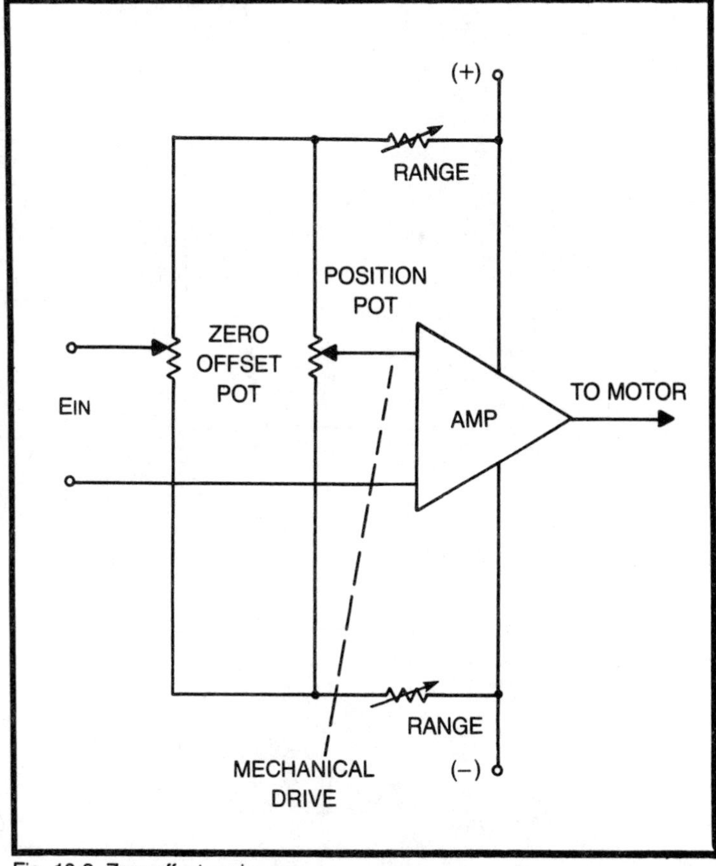

Fig. 10-2. Zero offset and range.

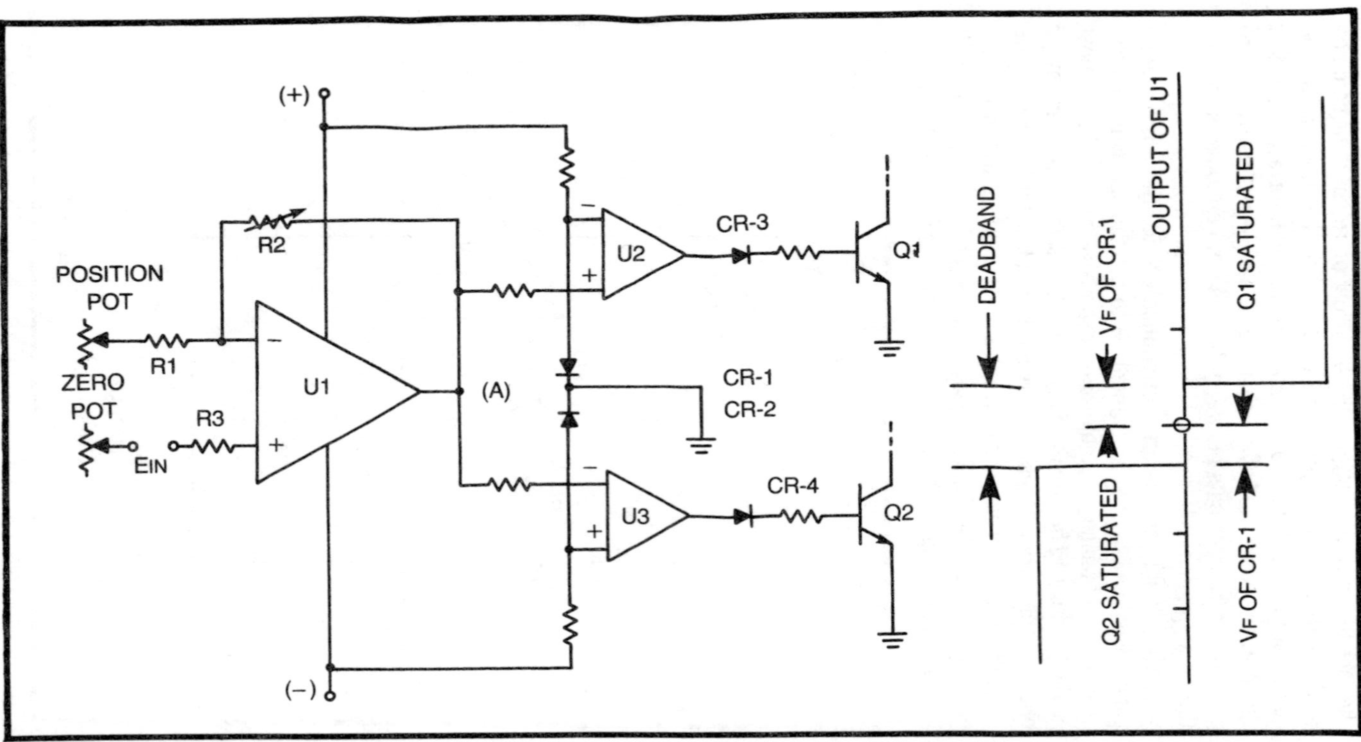

Fig. 10-3. The deadband control.

in the positive direction thereby saturating Q2. When the output of U1 is between the forward drops of CR-1 and CR-2, both Q1 and Q2 are cut off. Here we have mployed the forward drops of CR-1 and CR-2 to establish a relatively stable deadband, as measured at point (A). However, related to the input, the gain of op amp U1 enters the picture. The deadband measured at the input terminals is essentially:

$$\Delta E_{\text{DEADBAND}} \cong \frac{(E_{CR\text{-}1} + E_{CR\text{-}2})\, R_2}{R_1} \tag{10.1}$$

Since R_2 is variable, we see that the condition for an electrically adjustable deadband is readily obtained. Mechanisms for making use of the saturated condition of Q1 or Q2 to drive the motor will be shown in a subsequent chapter.

You will note that Figs. 10-1 and 10-2 took no particular note of the inverting and noninverting inputs of the op amps. This can generally be done as a matter of electrical convenience since the sense of the final mechanical feedback can be reversed simply by reversing motor leads or the leads to the position pot.

The offset biases of CR-1 and CR-2 obviously increase with temperature. This effect can be eliminated in a variety of ways. However, it is noteworthy that the looseness and "coast" of a gearbox is usually proportional to temperature. Also, the looser the box, the broader the required deadband. A certain degree of deadband compensation for temperature can be obtained by placing CR-1 and CR-2 in physical contact with the gearbox.

One of the principal advantages of the bridge follower system is that the input/output relationship is determined almost entirely by the accuracy of the potentiometer. Potentiometers with a very precise and stable relationship between shaft rotation (or linear travel) and tap voltage are readily obtainable. The law can also be tailored to be logarithmic, etc. Since the electronic amplifier enters in only in driving the potentiometer to a null (almost) the system accuracy can be stably tied to the potentiometer accuracy. The potentiometer can be purchased with accuracy traceable to the Bureau of Standards.

A system of the sort shown in Fig. 10-3 is usually started with R2 set for minimum gain. It is nearly always possible to increase the gain setting until the motor chatters, or "hunts," in a relatively fixed position. In the case of high frequency, or rapid chatter, the usual cause is play or lost motion in the drive, which is larger than the deadband. Resistor R2 can be reduced slightly to just stop the chatter by broadening the deadband. A step input can then be applied by rapidly slewing the zero offset or introducing a step voltage. The slewing will result in a response which may be either slightly over or under-damped and the deadband adjusted for best system operation.

COMMON MODE REJECTION

The suggestion that the output position was *almost* entirely due to the potentiometer is due to the fact that many common op amps have very limited common mode rejection ranges. For example, when operated with *both* inputs at more than either 1/2 of the ± supply voltages (and not closer than 3 volts to either) the output of something like a 741 will not be zero by a factor at least as large as the forward drop of a diode. This shifting-zero effect will show up as a false null for the potentiometer, and the shift will be impressed as an error on the system. To avoid these effects, one of the high-common-mode-rejection-ratio amplifiers is generally employed. In such amplifiers, one or more additional op amp is usually employed to cancel the common-mode signal. The circuit of Fig. 10-4 is relatively effective and enjoys the benefit that the unit can be made to cancel common-mode voltages much larger than the supply voltage.

When a voltage with a sizable common-mode component is applied between terminals A and B, say for example −7 V at A and −6.9 V at B, amplifier U4 will alter its output to drive the net input current at the inverting input to zero. The voltage at D will be +0.69 V, if the typical values for R and R are employed. Note that the amplifier is operating at a negative gain. Now also note that an essentially identical current is driven into the inverting input (summing input) of U5. The total current at the summing input of U is:

$$i_A + i_{FB} + i_{FA} = 0$$

where

$$i_A = i_B + \Delta i_B$$

$$i_A = -i_{fB}$$

therefore:

$$\Delta i_B + i_{FA} = 0$$

The output of U5 responds only to the difference between A and B. In this case the output voltage of U5 measured at (E) will be +0.01 V. In real op amps, the current sensitivity will vary slightly between units; therefore, the CMRR trim is supplied. All R_A and R_B resistors should be closely matched. A setting can usually be found for minimum deviation from zero output at (E) with inputs A and B tied together and cycled from positive to negative limits. With the values

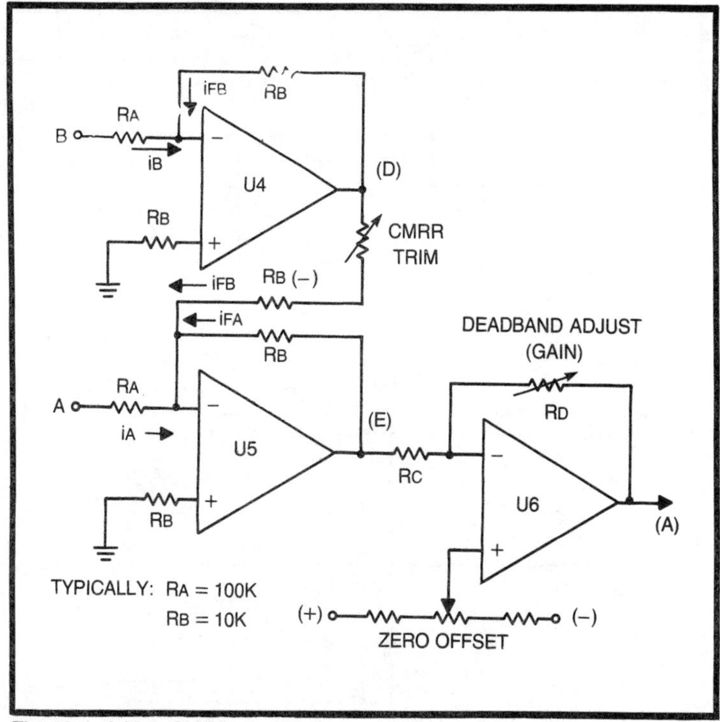

Fig. 10-4. A high CMRR amplifier.

of R_A and R_B shown, the unit should function up to nearly ten times the supply voltage.

This unit presents an input impedance of $2R_A$. In a similar arrangement with the inputs A and B applied to noninverting amplifiers a much higher input impedance is obtained, but the actual impedance is dependent upon the op amps and thus is not defined by resistors. If multiple scales are required, a resistor-defined input impedance is an advantage in terms of calibration stability.

Op amp U6 replaces op amp U1 in the previous circuits with a single significant difference, the inclusion of the zero offset circuit. If the circuit is employed as a linear voltmeter, it will be found that this control will permit the zero setting to be placed within about −1.5 volts of the plus rail and +2.5 volts of the minus rail for common op amps such as the 741. In servo service, the zero should naturally be set for essentially zero output at zero difference between A and B. The zero offset can be employed to offset small differences in the forward drops of CR-1 and CR-2. This zero offset mechanism usually shows more stability than internal offset control of the op amp.

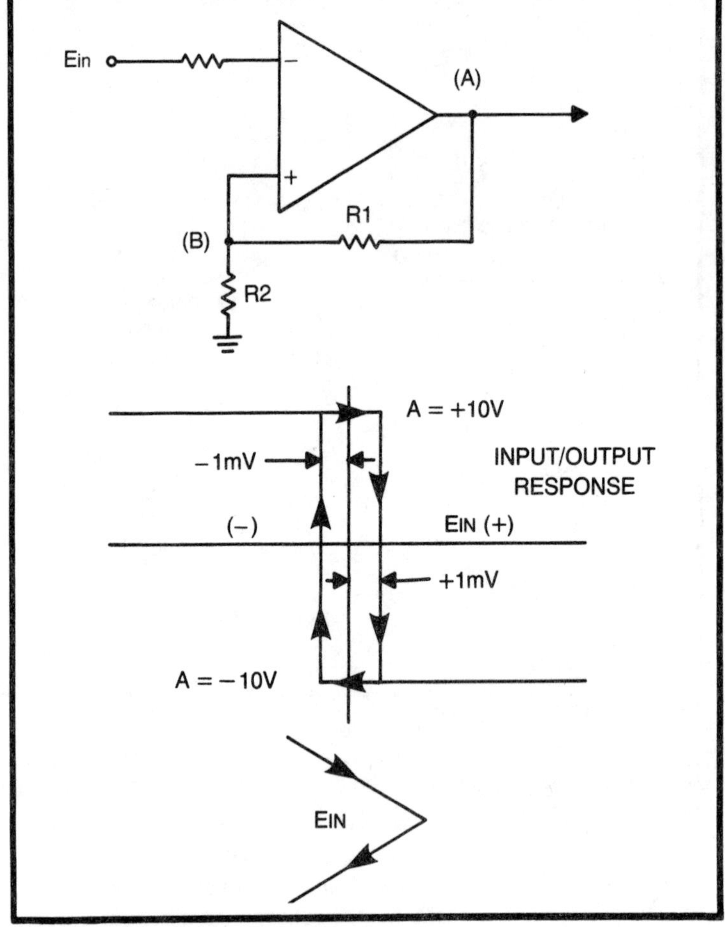

Fig. 10-5. The latch comparator.

THE UNI-DIRECTIONAL BRIDGE

In certain cases, particularly with temperature controls, the control function is uni-directional; that is, the heater can only supply heat and not cold. The deadband is nonetheless important since otherwise the output control device could wind up by operating in the linear range and would probably be destroyed. A small amount of "toggle" is generally employed to prevent this. This is obtained through the use of a certain amount of electronic positive feedback. An example is shown in Fig. 10-5. The op amp will saturate in the negative (−) direction whenever E_{IN} exceeds the voltage at (B) and in the positive (+) direction whenever E_{IN} is more negative than (B).

The voltage at (B) is, of course, not a fixed quantity but rather is dependent on the voltage at (A) and, therefore, on the previous history. If we assume:

$$(A) + (\text{sat}) = +10\ V$$
$$(A) - (\text{sat}) = -10\ V$$
$$R1 = 1\ M\Omega$$
$$R2 = 100\ \Omega$$

Then

$$(B)\ /\ (A) = \frac{10^2}{10^6} = 10^{-4}$$

and (B) can assume one of two values, $+1\,mV$ and $-1\,mV$. The $2\,mV$ gap is the hysteresis deadband. Suppose E_{IN} proceeds on a sawtooth from -2 volts to $+2$ volts, and back. From -2 V to zero (A) will be $+10$ V. At $+1\,mV$ (A) will switch abruptly to -10 V whereupon (A) will abruptly switch to $+10$ V again. The action is shown on the accompanying input/output curve.

The amount of this hysteresis is, of course, determined by the ratio of R1 and R2. This can and should be adjusted so that any noise on E_{IN} is smaller than the deadband. Any significant amount of noise on E_{IN} will give the unit the appearance of a "soft" turn-on characteristic. This is to be avoided if any mechanical switching is involved. If a small region of quasi-linear control is desired, a small triangle wave added to the op amp inverting input will yield a duty-factor modulated output pulse signal which is linear on the time average. The op amp may be either a common op amp or a comparator. The comparator is usually stabilized for output voltage and also optimized for slew rate and will, therefore, have a somewhat better performance. However, an otherwise unused part of a multiple op amp IC functions very well. The circuits of Fig. 10-6 illustrate several temperature control type configurations along with several output configurations. As noted the bridge sensors may be absolute, as in the thermistor bridge, or differential, as in the silicon diode bridge. The thermistor bridge takes advantage of the negative temperature coefficient of the device. The diode bridge takes advantage of the 2 $mV/°c$ coefficient on the forward drop of the diodes. The strain gauge bridge was thrown in to illustrate that this general circuit can be employed for mechanical and pressure measurements as well, although the latter are generally bidirectional servo's.

SYSTEM HYSTERESIS

In the final analysis most systems have some feedback hysteresis or "slop" or spring. In the Potentiometer Follower this may

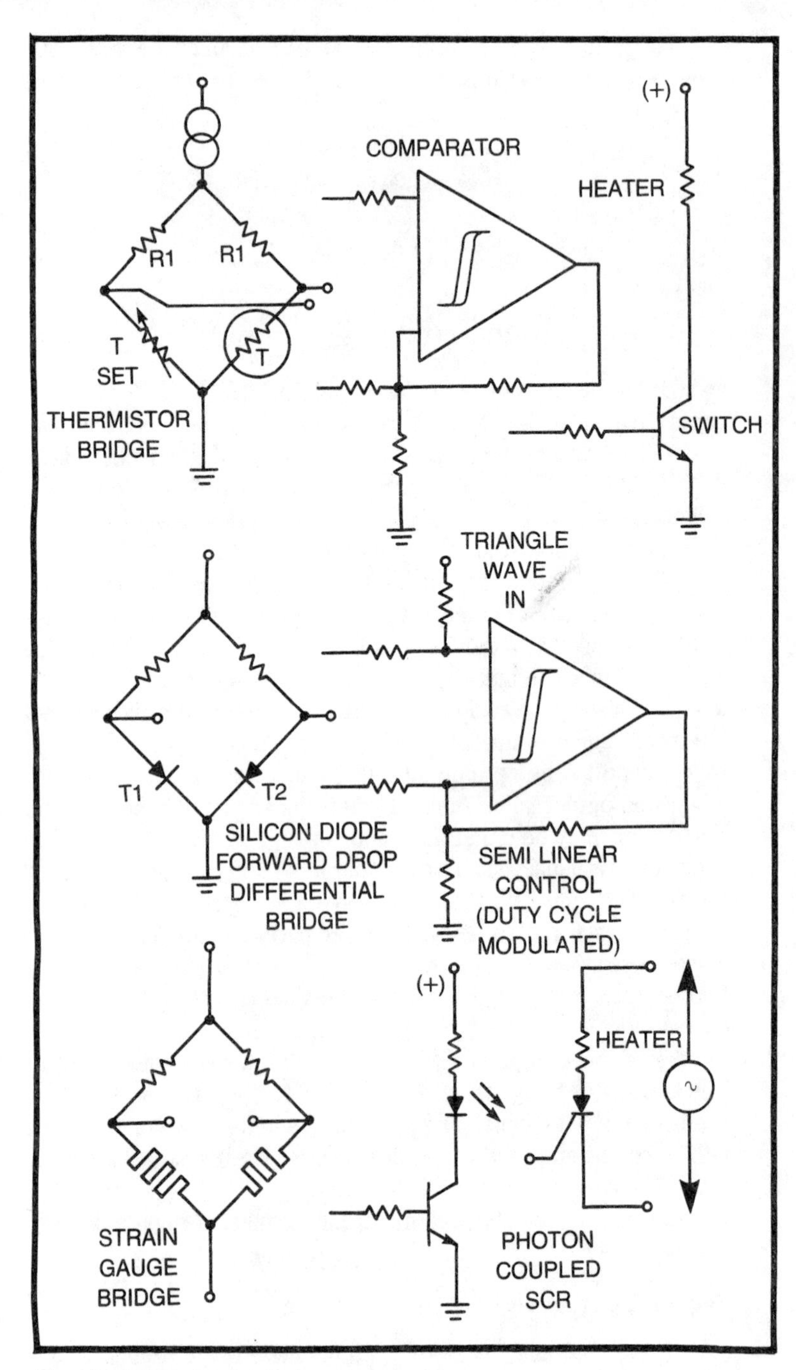

Fig. 10-6. Bridge feedback elements.

take the form of play or lost motion in the gear train or of elasticity in the belt or drive cord. In the thermal Uni-Directional Bridge, it can take the form of thermal inertia and thermal drop between the feedback element and the heat source. By the time the feedback element senses that the work load (oven, aquarium, etc.) is up to temperature the heater or cooler (which had to be hotter or cooler than the limit in order to warm or cool the load) is too hot (or cold) and there must always be some overshoot. In an analogous manner the elasticity in a gear train or timing belt drive sets a lower limit on the overshoot. Play in a gear train is analogous to time delay between application of heat and sensing of temperature. The latter might occur when the sensor measures the objects emerging from a conveyor oven and the system controls oven temperature. With only modest care in the design of the electronic control, the system play and elasticity become the chief source of control error in slow-moving systems. In systems with fast changing input commands inertia is generally the culprit.

VELOCITY OR FREQUENCY SENSORS

Probably the least expensive and perhaps the most common velocity sensor is the PM motor, which has a no-load output directly proportional to shaft speed. This velocity output may be used in a variety of ways.

The circuit of Fig. 10-7 illustrates one way in which a PM motor can be used as a generator to control deadband width. In this circuit comparator pair $U2$ and $U3$ function just as in Fig. 10-3, with the exception that the deadband is established by the voltage drop across $R1$ and $R2$. The circuit made up with $U7$ and $Q2$ is a negative feedback servo intended to keep point (F) at ground potential. If (F) goes below ground, the output of $U7$ goes high and $Q2$ draws more current, thus restoring (F) to ground. Conversely if (F) rises above ground the output of $U7$ goes low and $Q2$ draws less current.

Let us first assume that the system is stationary and the generator develops no voltage. Then $-E_{REF}$ pulls $Q1$ into a fixed level of conduction determined by $R3$ and the forward drop of the bridge. As the generator begins to spin in *either* direction a net negative voltage appears in series with the (–) supply, and $Q1$ conducts more heavily thus raising the drop across $R1$ and $R2$; the deadband is increased. The $U7$ and $Q2$ circuit does its thing and the deadband voltage remains centered at ground.

This circuit does draw a few milliamperes from the motor, but in a brush contact machine that is not necessarily a bad idea, since a little current will tend to keep contact resistance down. A com-

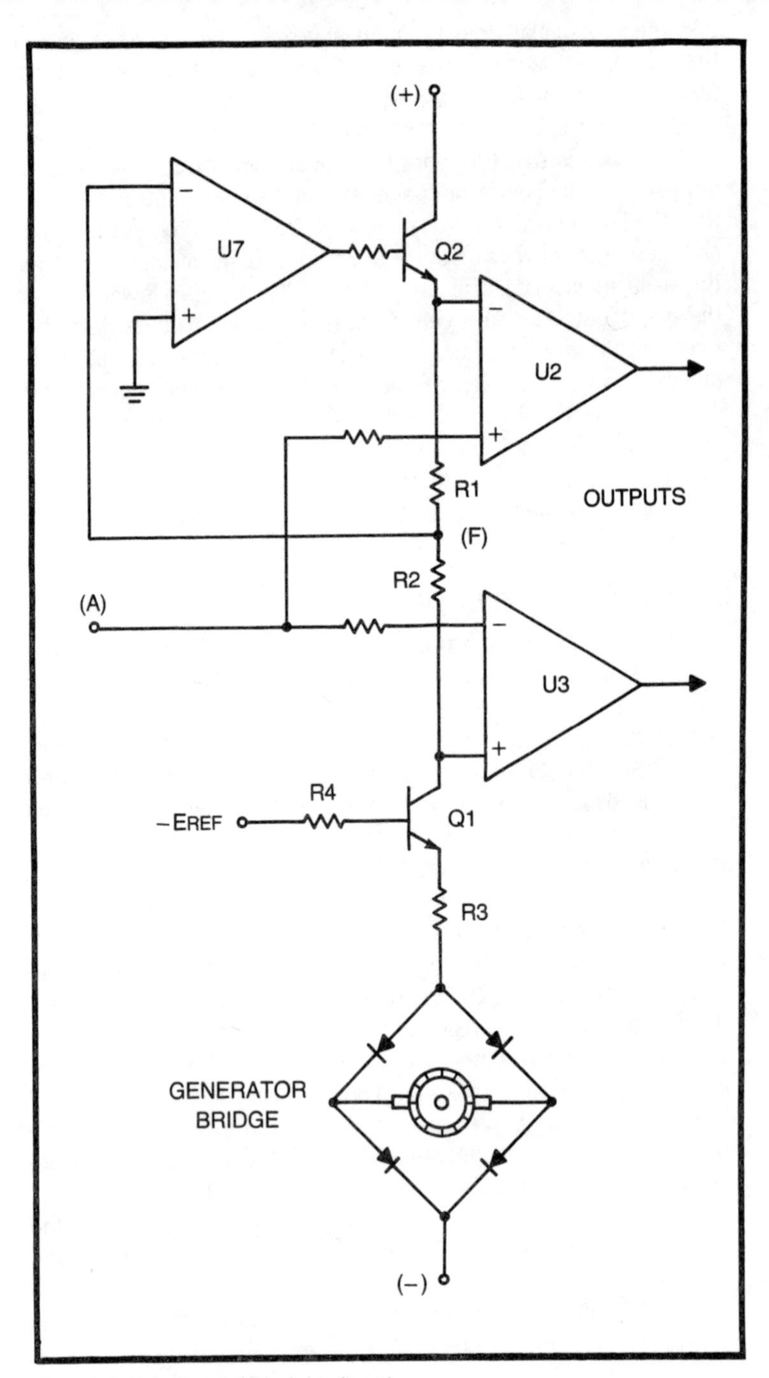

Fig. 10-7. Velocity control of deadband.

promise between $-E_{REF}$, $R3$, and the motor voltage is necessary in order to make the system have appropriate high speed and low speed deadbands. A somewhat more complex system uses inverting and noninverting op amps to develop the deadband reference voltages. In the circuit of Fig. 10-8, $U8$ simply inverts the generator voltage. The negative deadband reference is developed at the output of $U9$ and the positive at the output of $U10$. Under static conditions the deadband reference is controlled by the ratio of $R1/R2$. Under dynamic conditions, the ratio is set by $R1/R3$. This circuit allows nearly independent control of static and dynamic deadband at the expense of some additional complexity. For systems exhibiting viscous damping, this is only an approximation, but it will considerably improve static accuracy.

CEMF Sensing

When a PM motor or a shunt-wound motor with fixed field excitation is employed in a system, the counter electromotive force (CEMF) may be sensed in order to determine the motor speed. Figure 10-9 illustrates one technique by which such speed sensing is performed. The bridge is arranged such that the two voltages are equal with the motor in the stalled rotor condition. Under these circumstances, as shown in the figure, the difference between V_A and V_B is exactly the CEMF. This CEMF value can be level translated using the High CMRR amplifier described earlier and employed in a circuit similar to that shown in Fig. 10-7 to obtain a motor velocity feedback for deadband broadening. Resistors $R1$ and $R3$ can be much smaller than resistors $R2$ and $R4$ for the sake of efficiency and response. A somewhat similar scheme was shown in Fig. 8-2 for a series wound universal motor. For the series wound motor this technique does not really work as well since the field is a function of the motor load, and the CEMF is, therefore, not a linear function of motor speed.

This technique is fairly often employed in speed control systems since it does not require any additional mechanical connections and is relatively cheap to implement. It can be used to good advantage to stabilize the capstan speed on a dc operated tape recorder and can be used for deadband control in positioning systems.

THE TACHOMETER MOTOR

High quality servo motors are frequently built with a tachometer incorporated into the motor frame. This can be either a brush/commutator type PM motor or a permanent magnet alternator. The latter type is more prevalent in high-performance types because of

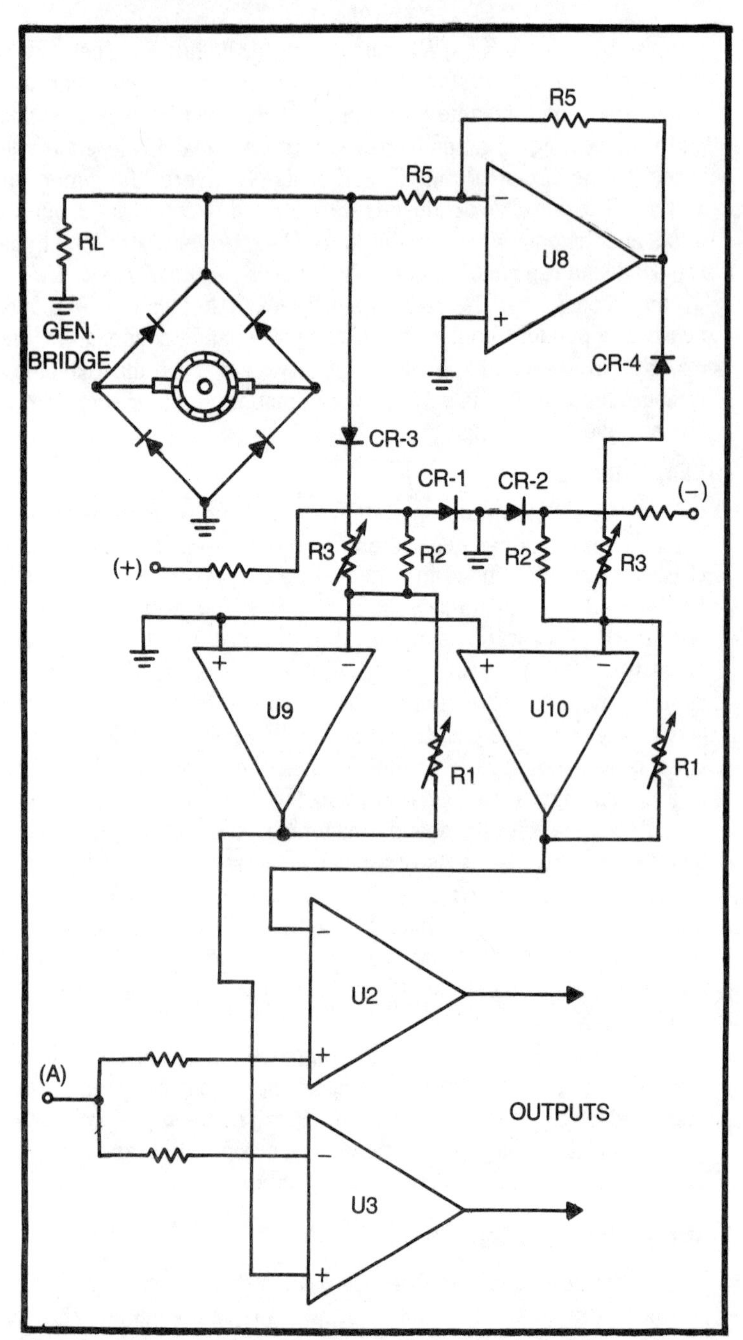

Fig. 10-8. Velocity control of deadband with independent static and dynamic control.

the brushless feature which tends to give the tachometer nearly unlimited life. In the case of the permanent magnet alternator, several things can be done to utilize the speed signal. The simplest, cheapest, and also least accurate is to simply rectify and filter the voltage output. As shown in equation 3.1, the peak voltage of the rotating machine is proportional to its angular velocity. If we rectify this voltage with a bridge or other rectifier we therefore can obtain an output voltage which is proportional to the rotational velocity of

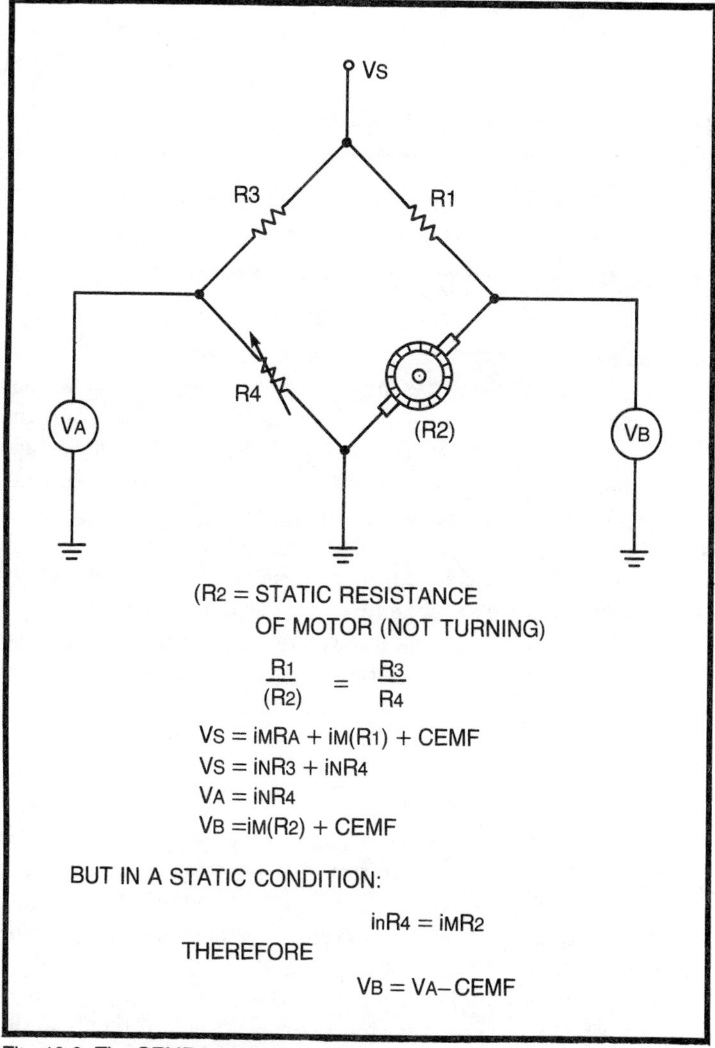

Fig. 10-9. The CEMF bridge.

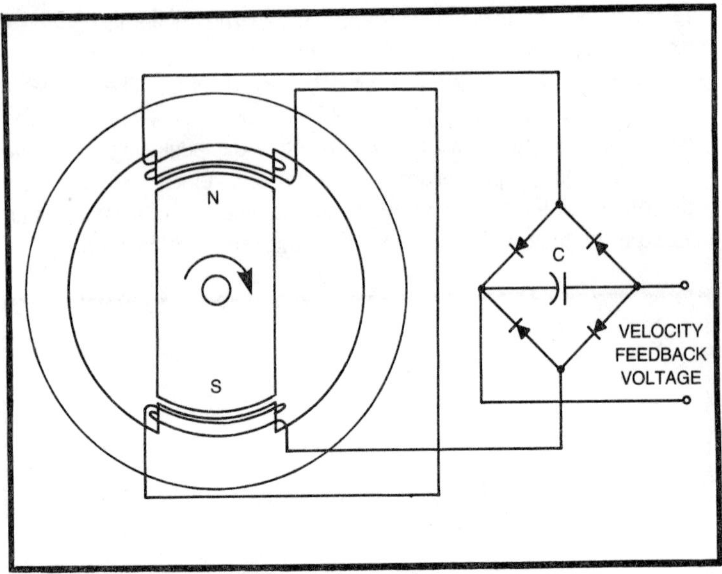

Fig. 10-10. The brushless PM velocity sensor.

the machine. This arrangement is illustrated in Fig. 10-10. The small brushless velocity sensor has the advantages of being less noisy than the brush type system and does not have to accommodate the vagaries of the brush resistance. Furthermore, it has the advantage of eliminating the brush wear and friction load. It should be noted however that a PM alternator of this type has a distinct "cogging" or detent type torque since the magnet would much rather be between the poles than crosswise. This uneveness of torque is very noticeable when rotating a small motor of this type by hand.

The output voltage of this system is, of course, somewhat subject to variation with temperature since the forward drop of the diodes is subject to temperature variation. In the final analysis, it is this factor which tends to limit the accuracy of a system of this type.

THE ONE-SHOT TACHOMETER

There is a large class of devices which I shall lump together in the class of **one-shot tachometers**. In essence all of these devices operate on the principle of providing a fixed quantum of energy into an integrating circuit for each increment of angular rotation. The integrated output is then a function of rotational velocity (or other parameter). Figure 10-11 illustrates this device in its simplest form. The battery charges the capacitor up to a fixed level every time the switch closes in that direction and then discharges it through the meter in the opposite position. **The average current through the meter is proportional to the number of closures per second.**

The circuits of Fig. 10-12 illustrate a variation of this scheme in which a slotted disc is used as a photon interruptor. It breaks the light path between the LED and $Q1$. When $Q1$ is illuminated, it saturates and delivers a short spike to the base of $Q2$—through $C3$ and $R3$, which have been chosen have a time constant much shorter than the shortest illumination period. Amplifier $U1$ is arranged as an integrator, or low-pass filter, to average out the charge transfer in much the same way that the inertia of the meter movement did in Fig. 10-11. This is not a true integrator in the mathematical sense but rather a low-pass averaging filter whose time constant is determined by the values of $R1$ and $C1$. If a second such active filter with a longer time constant is used as a reference, the angular acceleration of the unit is obtainable from the difference between the output terminals. In the configuration shown, the output of $U1$ would run between $-(V+)$ at zero velocity and nearly $-(V+/2)$ at maximum speed when $C3R3$ was nearly equal to the on time of $Q1$. Up to nearly the maximum limit, the unit is quite linear in output.

Also shown in the figure is a magnetic pickup made up of a bar magnet with a winding. The zener diode will transform the output pulse into a nearly square wave above some minimal speed by clamping the negative swing to nearly ground and clipping the positive swing. This action limits the output voltage to an essentially constant voltage wave, but the duty cycle is also constant. It is

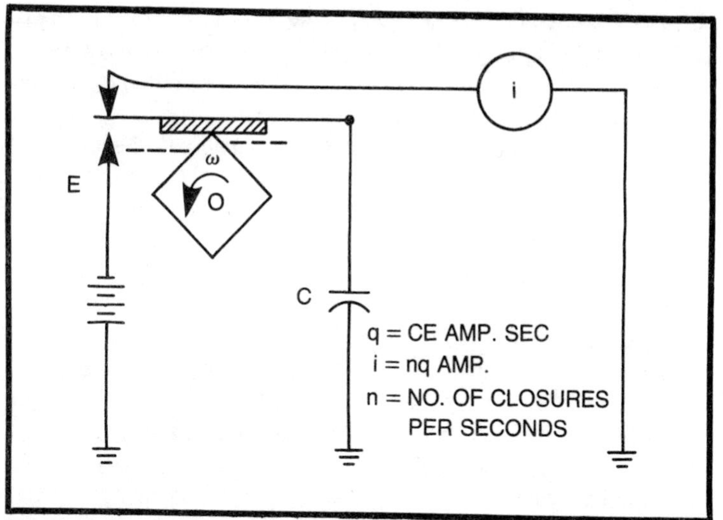

Fig. 10-11. The one-shot tachometer.

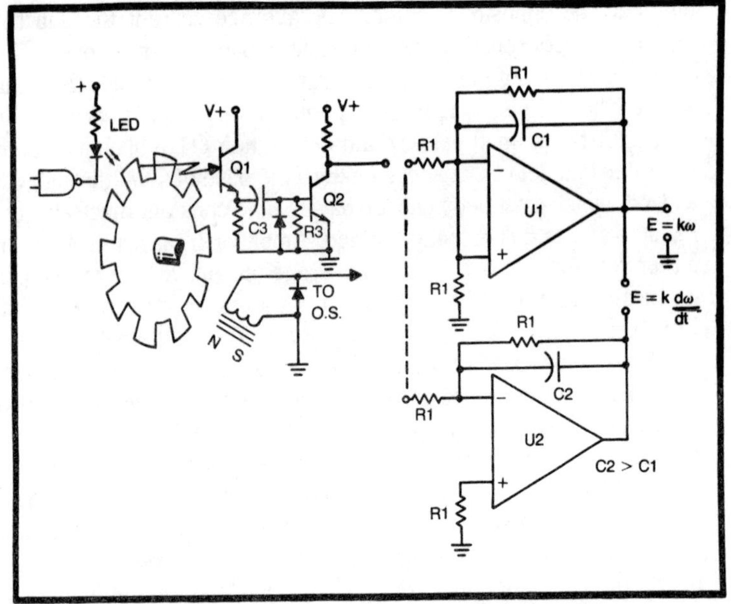

Fig. 10-12. The optical tachometer and the magnetic tachometer.

therefore necessary to drive a one-shot of some form to obtain a constant output pulse width. A circuit similar to the $Q1$ and $Q2$ arrangement will suffice.

Since the accuracy and linearity of the one-shot tachometer is directly related to the stability and precision of the one-shot, the use of one of IC one-shot devices such as the N555 or the 74121 or 74122 is to be recommended. These devices have been carefully designed to provide a stable and repeatable output with variation in temperature, supply voltage, etc.

The Photo Interruptor can be either a made up assembly such as the General Electric H-11 series or may be fabricated from separate LED and photo transistor elements. It is noteworthy that the LED can be driven by TTL logic as shown, and the one-shot mechanism can be implemented by turning the LED off after the initial pulse.

Both types of units have the advantage of offering no friction load although the magnetic pickup may have a detenting torque. The photo transistor arrangement is workable down to the lowest speeds, including zero but must be shielded from stray light. The magnetic pickup is inoperative below some minimum speed but can often be easily implemented by placing the pickup near an existing iron or steel (but not stainless) gear. An earphone magnet makes a very good pickup however suitable magnetic pickups are available commercially. Hall effect magnetic pickups, which are workable down to zero velocity, are also available. These units often feature direct TTL output.

These tachometer techniques are essentially analog in nature. They can be made to be stable and repeatable to better than one percent with a little care and can be improved to 0.1 percent with a great deal of care. For applications requiring even better control, other circuits will be discussed in Chapter 12, where phase-locked loops will be discussed.

11

Tracking Systems

One of the fundamental applications of motor controls is to be found in tracking systems. In their more sophisticated forms, such systems steer ships, submarines, airplanes, guided missiles, and a host of other craft. The controls are characterized by servo systems which will contain a multiplicity of closed loops, each of which must be independently stable. However, the largest loop usually must contain some form of vehicle or device which is the ultimate object of the control. This device is usually nonelectronic and is sometimes subject to natural and nonelectronic perturbations, which must be overcome by the system. The actions of wind, waves, and current, for example, will tend to divert a ship from the course set into the autopilot. To function properly, the system must be arranged to compensate for these effects.

A detailed consideration of these effects is beyond the scope of this text, but this chapter will attempt to set forth some of the problems and considerations involved to provide a general understanding of the topic. For those wishing to delve deeper I would recommend:

John D. Trimmer RESPONSE OF PHYSICAL SYSTEMS John Wiley and Sons, Inc., NY

BASIC CONSIDERATIONS

Figure 11-1 illustrates the way in which some of these natural forces can affect the system operation. Nearly everyone is familiar with the sight of a light plane traveling across the sky crabwise on a windy day, the airplane obviously not traveling in the direction in which the nose is pointed. The airplane engine is only capable of causing the airplane to move with respect to the air surrounding it. If the body of air is moving with respect to the earth, then the motion of the airplane with respect to the earth is the vector sum of two

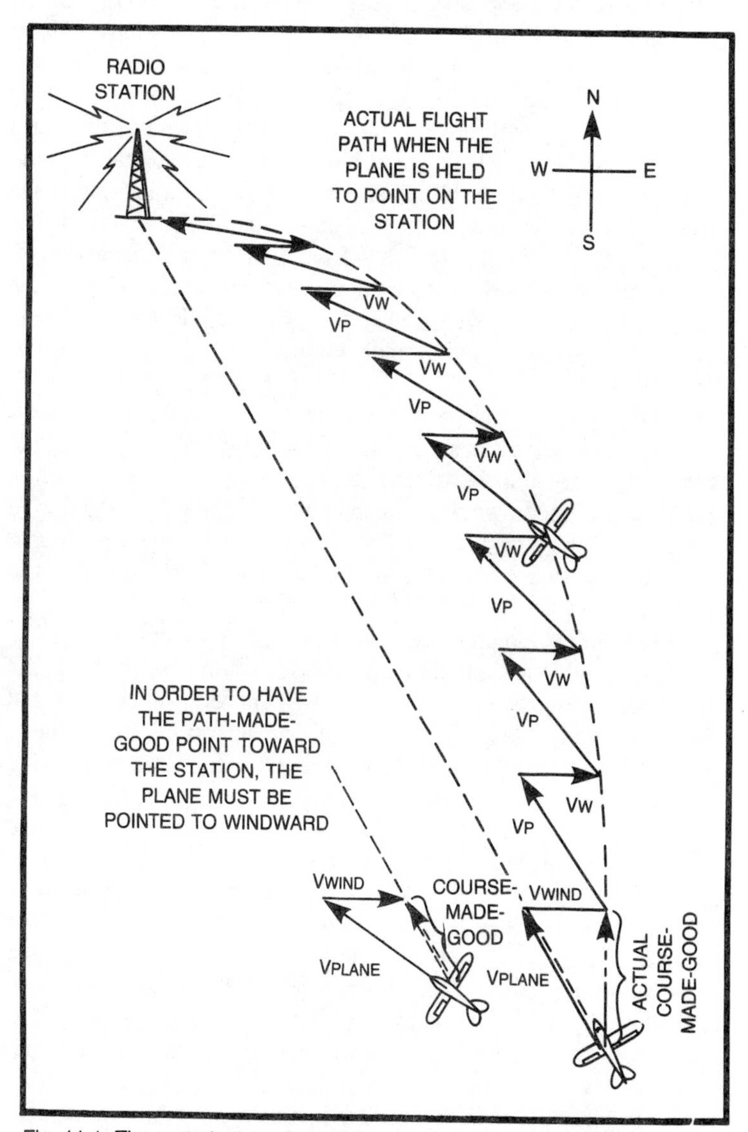

Fig. 11-1. The perturbation of an airplane course due to wind drift.

motions: the motion of the airplane with respect to the air and the motion of the air with respect to the earth.

In the illustration, an airplane is presumed to be navigating in the direction of a radio station, with the aid of a radio compass which tells the pilot the direction of the radio station with respect to the axis of the airplane. If the pilot arranges to point the airplane always in the direction of the radio station, the airplane will be carried in the direction of the wind off of course and will continually have to have course changes with respect to north in order to stay pointed toward the station. On the last part of the trip the airplane wll be approaching the radio station nearly straight upwind and from the east rather than from the southeast. The flight path would have taken it far from the straight line path from the initial position to the station and would have taken a significantly longer time (about 17 percent in the example) than a direct path would have. The inset in the lower left illustrates that in order to make the true path to the station good, the pilot would have to hold the nose enough to windward of the true course so that the wind would blow him exactly back onto the course.

The object of the example is not to teach air navigation but rather to indicate that many tracking systems are required to include offset capability in order to function effectively. The same diagram could have been drawn with a ship in place of the airplane and wind drift and current drift vectors added to the ship velocity. Ships, sailboats, airplanes, and guided missiles will seldom travel in the direction in which they are pointed and will generally require an offset in the pointing direction of a tracking apparatus.

A second requirement for offset arises when a moving vehicle is supposed to intercept a second moving vehicle. The illustration of Fig. 11-2 shows this effect. An airplane flying in a straight line is attacked by a missile which is traveling 1.5 times as fast. If the missile is kept pointed toward the airplane, in the example, it will take nine seconds to catch the plane. If the missile assumes an angle θ so that the velocity vector in the direction of the airplane travel is equal to the airplane velocity, the pursuit time is reduced to 6.26 seconds. Note that in both this and the radio compass case, the course of the missile (or airplane) with respect to north is not changed throughout the flight on the optimum course. In both cases, if the heading is adjusted so that neither the bearing with respect to north nor the bearing to the target is changing, the course is optimum.

The missile case is obviously oversimplified. If the airplane pilot were to see the missile on an unvarying bearing he would alter his

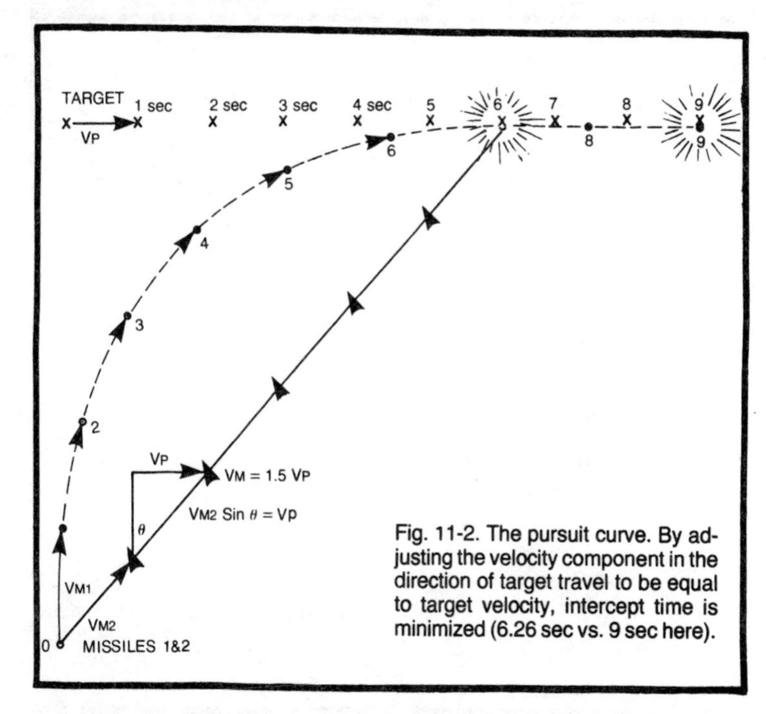

Fig. 11-2. The pursuit curve. By adjusting the velocity component in the direction of target travel to be equal to target velocity, intercept time is minimized (6.26 sec vs. 9 sec here).

course and not fly on a straight path. Human beings respond instinctively to an approaching object whose bearing does not change. It is naturally recognized as being on a collision course. It is the technique of avoidance that usually fails when a collision results.

Tracking systems are also characterized by the number of axes in which they operate. In some cases it is only necessary to correct the direction left and right while in others, an up and down correction is also required. For our first example, we shall consider a tracking system which operates with only a single-axis correction and employs incoherent radiation.

THE SOLAR BOILER

Among the various schemes which are frequently implemented to extract energy from sunlight, the solar boiler is one of the best working techniques. With this device a reflector is used to focus the rays of the sun upon a boiler tube which becomes very hot. The tube is cooled by water, which may be superheated into steam or simply used as a source of hot water for household or other purposes. The scheme has the advantage that the concentrated sunlight can be used to make steam hot enough to use with ordinary heat engines such as turbines or reciprocating steam engines. Because of the

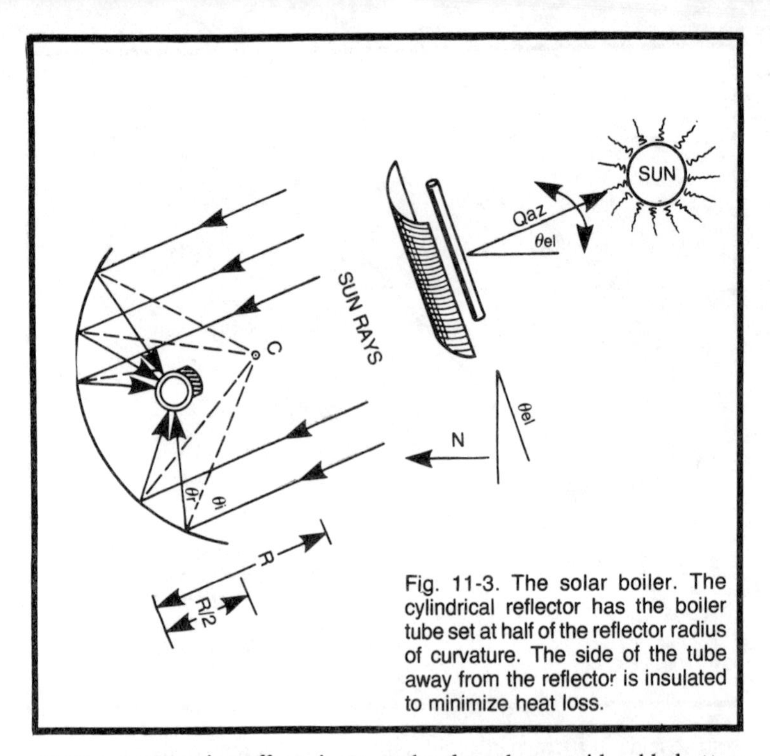

Fig. 11-3. The solar boiler. The cylindrical reflector has the boiler tube set at half of the reflector radius of curvature. The side of the tube away from the reflector is insulated to minimize heat loss.

great concentrating effect, it can make the tube considerably hotter than the boiling point even on somewhat cloudy days and in cold winter weather.

Figure 11-3 shows the general arrangement of the boiler. A shiny, reflecting cylinder is arranged with a boiler tube running down parallel to the cylinder axis and at half of the radius of curvature. As seen from the ray diagram $\theta_i = \theta_r$ and the rays concentrate on the boiler tube. The tube absorbs the concentrated heat and light, which is transferred to the water. The tube collects heat from the reflector faster than it would unaided by very nearly the ratio of the reflector area to the tube cross section area. If the water is moved into the tube relatively slowly the output will be very hot. Faster flows will yield a cooler output.

Without going into the thermodynamic aspects of the boiler, it should be obvious that one of the principal difficulties with this system is the requirement that it must be pointed pretty much right at the sun in order to work. The apparent position of the sun when viewed fom the earth lies in a plane which tilts with the season, and its angle to the horizontal is a function of latitude. If however the system axis is tilted to the north by an angle such that the sun is centered at noon, the tilt adjustment for the system will only have to be adjusted about once a week in order to keep the system reasonably true. In the plane normal to this tilt axis, the sun moves at approximately 15° per hour. Since the sun subtends an angle of about a half degree, the system would work effectively for only about a thirtieth of an hour if not moved to track the sun. It should be noted that the system is much more sensitive to errors in the focusing plane than in the plane containing the axis of the cylinder, which tends to aggravate the problem.

The axial inclination and the azimuth direction for the sun can be calculated with great precision for any time and any position on the face of the earth, so it is possible to use a pre-programmed position system and control the boiler pointing angle with cams and clockworks or with computer device. A simpler system, though, employs a tracker as shown in Fig. 11-4. In this unit the shadow of the boiler tube or a fin falls on a pair of solar cells. Whenever the apparent motion of the sun carries it sufficiently ahead of the boiler

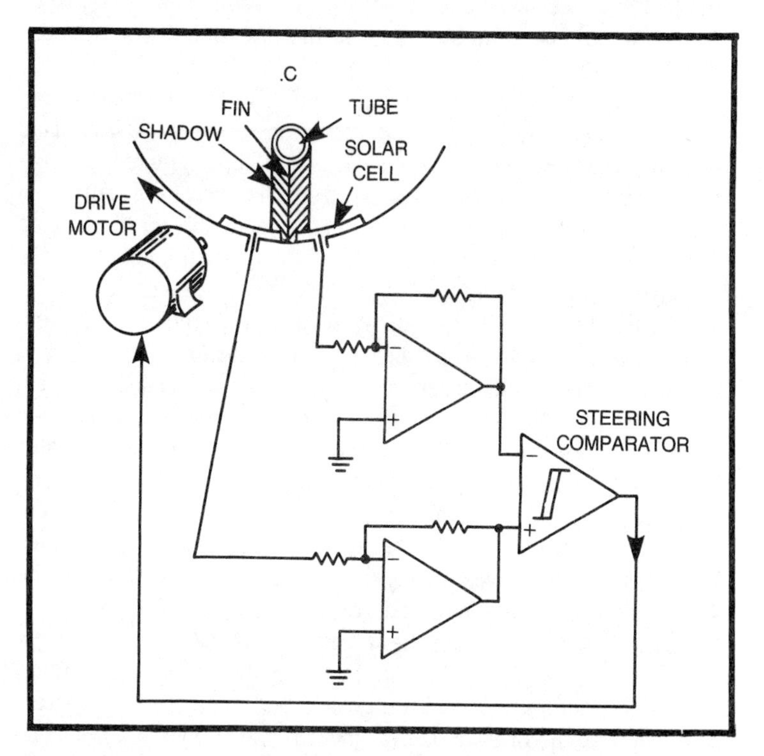

Fig. 11-4. The sun-tracking servo. The output of the comparator operates the motor switch to reduce the differences in the solar cell output.

direction, the comparator toggles and the motor rotates the reflector in a clockwise direction until the comparator untoggles, thereby deenergizing the motor. The system then waits for the sun to get ahead again whereby the process repeats.

This is a relatively crude system, but it is simple to implement and requires only an on-off control of the motor in a single direction. It is relatively cheap and easily stabilized. With the motor arranged to drive the reflector at 1° per minute, a pointing accuracy of better than a half degree can easily be maintained. This is more than adequate for efficient operation of the boiler.

Naturally, one of the more elegant servos could be employed to bidirectionally steer the system, and the axis tilt correction could also be automated. But this very simple system will provide adequate performance at minimum cost.

THE LOBE-SWITCHING OR CONICAL SCAN TRACKER

The very earliest radio automatic direction finders (ADF) employed a lobe-switching technique to derive a drive signal to swing the antenna around to point in the direction of the received signal.

The antenna patterns illustrated in Fig. 11-5 show the patterns of the two antennas superimposed upon one another. The cardioid antenna pattern gives a response which is approximately proportional to the cosine of half the angle from the system. This type of pattern is relatively easy to achieve in an electrically small antenna; therefore, it is employed for the ADF. Two such antennas were employed with their patterns superimposed as shown.

It may be seen that signals coming in from the right excite a stronger signal in the right hand antenna than in the left hand antenna. If a pair of identical receiver detectors were used as shown on the bottom of the page their detected dc difference could be used to switch a motor to run the antenna clockwise when R is greater than L and counterclockwise when L is greater than R. If you think about this for a while you will see that it will wind up with the system 0 axis always pointing toward the signal source. There is an unstable point about the 180° axis; however, any perturbation would send the system slewing around to the zero degree point since R exceeds L in the entire right hand hemisphere, and vice versa.

The system shown in Fig. 11-5 employs a pair of identical receiver-detectors to derive the error signal. As a practical matter, at the time of the original development in the late 1920s, it was no simple task to produce such a pair of matched receivers. The ADF had to cover the marine and aircraft beacon bands from 180 to 400

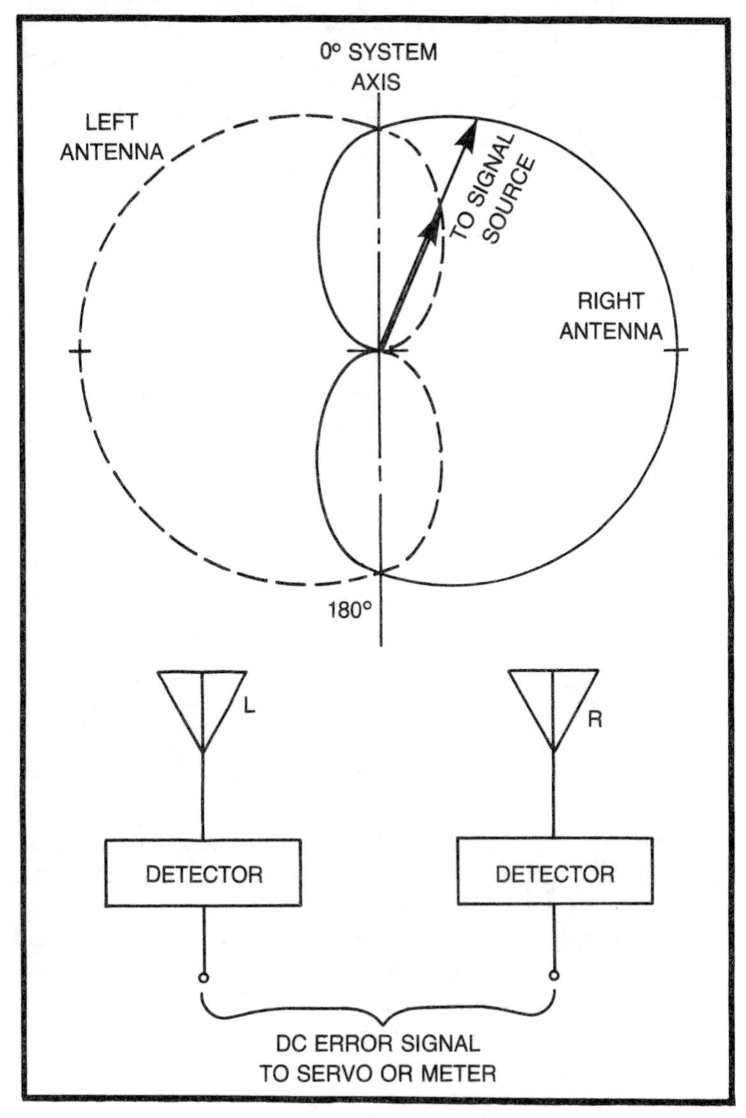

Fig. 11-5. The ADF system.

kHz and the broadcast band from 550 to 1550 kHz. Furthermore, the signals could range from very weak to very strong as the station was approached. The presence of any gain error between the receivers would manifest itself as a bearing error.

Accordingly, the systems were actually built using a lobe-switching scheme in which a single receiver was switched first to one

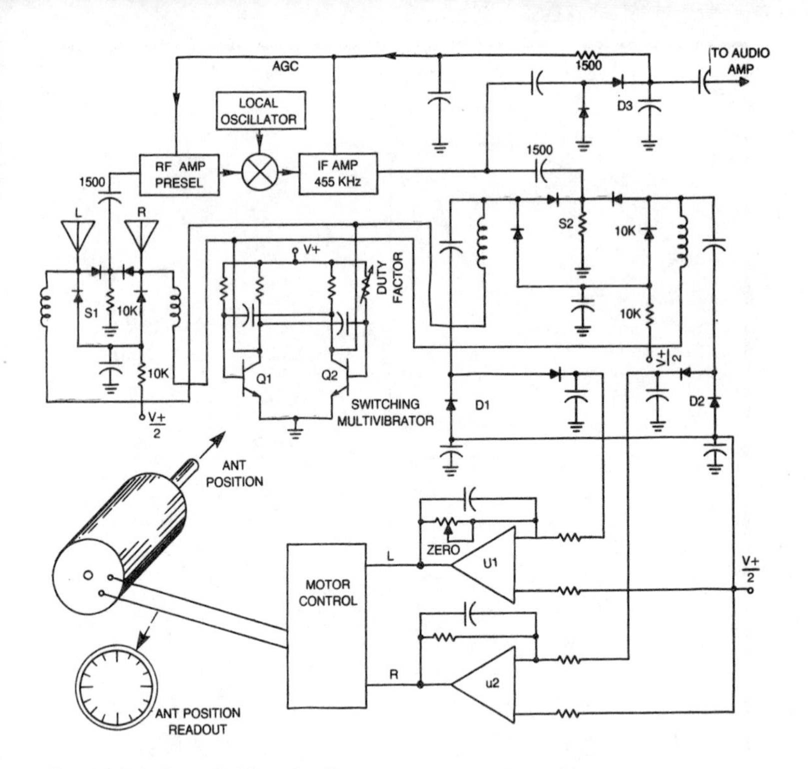

Fig. 11-6. Lobe switching circuit.

antenna and then to the other and the output decommutated into two separate detectors which provided the error signal. In the original systems the switching was electromechanical, but modern systems generally employ electronic switching.

A system of this type is illustrated in Fig. 11-6. Switches $S1$ and $S2$ each consists of a radio frequency diode bridge given by the multivibrator made up of $Q1$ and $Q2$. One of the transistors is cut off and the other is saturated at all times except during transition. The bridges are driven by the transistor collectors through rf chokes.

When the driven node of the bridge is high the diode connected to the center tap conducts through the 10 kilohm resistor, which rises to nearly $V+$. The diode connected to $V+/2$ is back biased, thus open circuited, and the antenna connected to the on-biased diode anode is switched into the circuit through the relatively low on-biased forward impedance of the diode.

Simultaneously, the opposite, or low-driven, node back biases the diode to the centertap thus disconnecting the antenna while the diode connected to $V+/2$ is forward biased thereby grounding this antenna. This process of connecting one sensor and grounding the deselected sensor is required to obtain the required isolation in the presence of a strong signal. Switches of this type should have a forward current of several milliamperes in the *on* diodes and a cutoff voltage of several volts in the *off* diodes. This is required in order to minimize the effects of intermodulation in the presence of strong signals. For very high frequencies, PIN diodes would be the best choice, but for the low frequencies used in an ADF, the storage time of PIN diodes is not adequate and the use of ordinary high-speed switching diodes, such as the 1N914, is satisfactory.

The circuit is provided with an audio output to permit the operator to identify the station being tracked by listening to the call sign. If the switching is done at a relatively slow rate and the switching transition is fast, the switching modulation can be held to a band of frequencies below the audio when the antenna is nearly nulled. Then the spikes caused by phase discontinuities will fall above the audio band. For example, a switching rate of 5 Hz with a transition time of a few microseconds will produce relatively little objectionable modulation of the audio if the audio amplifier is equipped with a 50 to 5000 Hz bandpass filter.

Signal detectors $D1$ and $D2$ must be passed through integrators in order to smooth out the switching modulation. These units are shown here as voltage doubler types since they are easily arranged with the rest of the circuitry to form a basic single-supply scheme. This is not necessarily required as the scheme could have probably been built with a smaller parts count for use with a bipolar supply.

A duty factor control is shown on the multivibrator since any departure from a 50 percent duty factor in the switching waveform will show up as a bearing error. A superior arrangement would employ a divide-by-two and an oscillator—to be discussed in Chapter 12 on phase-locked loops. The zero control shown on the integrator is used to correct *minor* errors in the zero. This is also not too good since it tends to unbalance the time constants of the two halves if any sizable offset is achieved. This balancing function would be better performed with the zero offset adjustment of the deadband control. The relative sluggishness of the arrangement due to the integrators is not undesirable since the system works on AM signals which vary in power, and the error signal requires considerable smoothing to eliminate the flutter due to the signal modulation and local reflections.

THE CONICAL SCAN TRACKER

The conical scan system is simply a variation of the lobe switching system. This technique was used on some of the earliest tracking

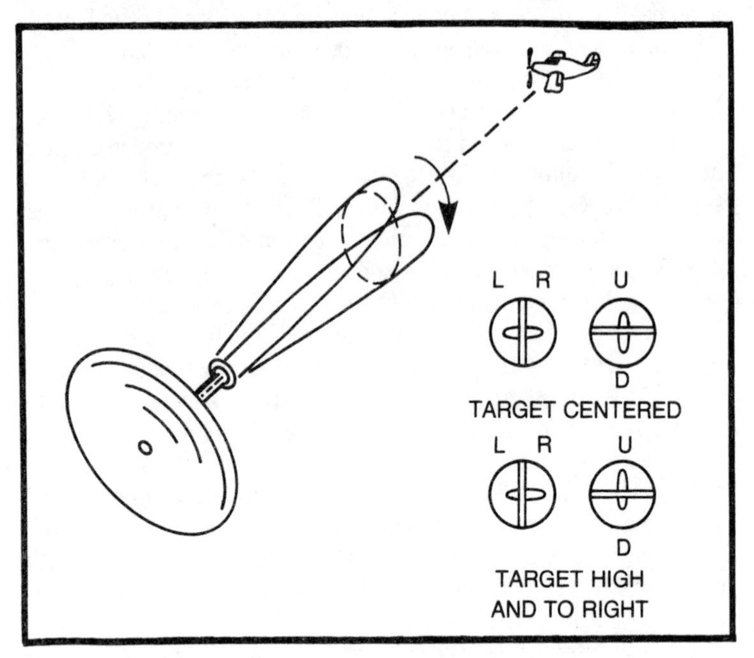

Fig. 11-7. Conical scan tracking.

radars, such as the Sperry TPL which was employed to point searchlights for night antiaircraft fire and in the GE/Westinghouse SCR-584 antiaircraft gunfire control radars. Figure 11-7 illustrates basic technique.

The antenna system was equipped with a rotating feed which was slightly offset so that the flashlight-like beam of the radar was caused to wobble in a small circle. In the TPL there was a set of "point" and "train" oscilloscopes which presented the target blips, as shown in the inset, into left hand, right hand, up and down halves of the respective oscilloscope screens. When the target was centered and left and right halves were of equal height and the up and down halves were also of equal height. When the target strayed from the axis of the system the beam would point more directly at it in one part of the wobble, and the signal would be stronger. Conversely, on the other side of the swing the beam would point less directly at the target, and the signal would fall. The lower portion of the inset shows this effect. On this particular type radar, there were two operators charged with operating the handwheels to bring the target back onto the axis of the screen.

The operation of the signal processing involved can be understood by a study of the optical wobbling mirror tracker of Fig. 11-8.

In this system, an optical mirror is attached to a rotating shaft slightly off normal so that it wobbles. This causes the reflection from the mirror to wobble in a circle. As shown in the lower half of the figure a four-segment detector is employed, and the image of a point source of light will describe a circle on the detector. The series of four op amps is arranged so that the summing junction of each is connected to two of the segments. When the target is aligned with the axis of

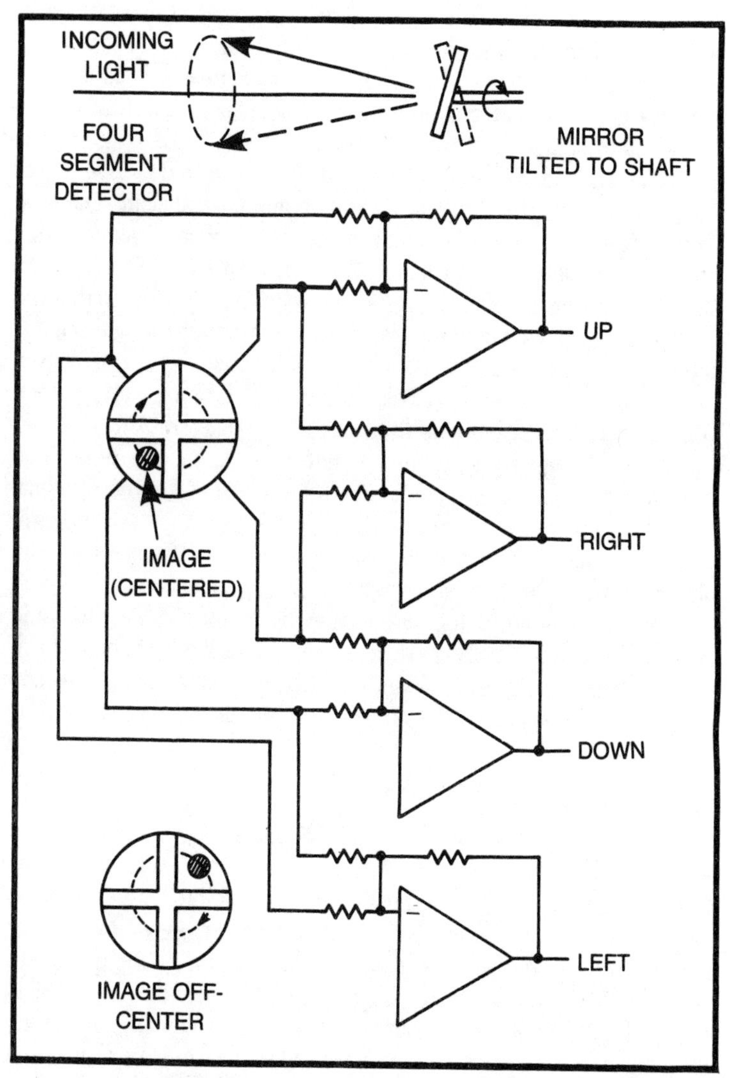

Fig. 11-8. The wobbling mirror.

the mirror shaft the target spends an equal period of time in each of the sectors and the up and down outputs are equal on the average (after integrating). The left and right outputs are similarly equal.

In the case where there is an error angle between the target image and the shaft, the image spends more time in the segments in the direction opposite to the error. The integrated outputs of the sensors are thus no longer equal, and the error signal to drive the positioning systems may be derived from the difference.

This is not exactly the same as the conical scan radar system in which the incoming video was de-commutated into up-down and left-right detector pairs using a pair of quadrature signals derived from the antenna beam position sensor. However, the overall operation is quite similar and the principles relatively alike.

The wobbling plate system can also be used to describe one of the principal faults of the lobe-switching and conical scan schemes. The curves of Fig. 11-9 illustrate the effect which is obtained with a noisy or scintillating target where the signal varies over an appreciable range during the course of a single scan. It may be seen that the up-down signals would not average out to zero over a single scan. If the system is very snappy, the error would cause the servo system to jitter rather considerably. This used to be a common feature of the conical-scan radars, such as the SCR-584. While tracking, the antenna would tend to jitter very slightly nearly all the time. The propeller on a propeller-driven plane was a great source of signal jitter. In an optical system other factors can cause amplitude jitters in a similar fashion. Since the conical scan was usually a mechanically driven scheme, it tended to operate at speeds consistent with mechanical motions of the target that were roughly of the same period. Some of the modern track-while-scan radars have come back to the use of a conical scan system, but the scan rate is electronically controlled and far faster than mechanically induced target scintillations.

THE MONOPULSE TRACKER

In order to cure some of the unsteadiness of the conical scan system a great effort was put forth in the early 1950s to build radar trackers that could take the position error reading from each pulse individually so that the instantaneous bearing error would be completely free of bearing errors due to modulation of the signal. In actuality, the detection scheme shown on the bottom of Fig. 11-5 is an amplitude monopulse scheme since it provides instantaneous bearing error readings with the two antennas and two matched receivers.

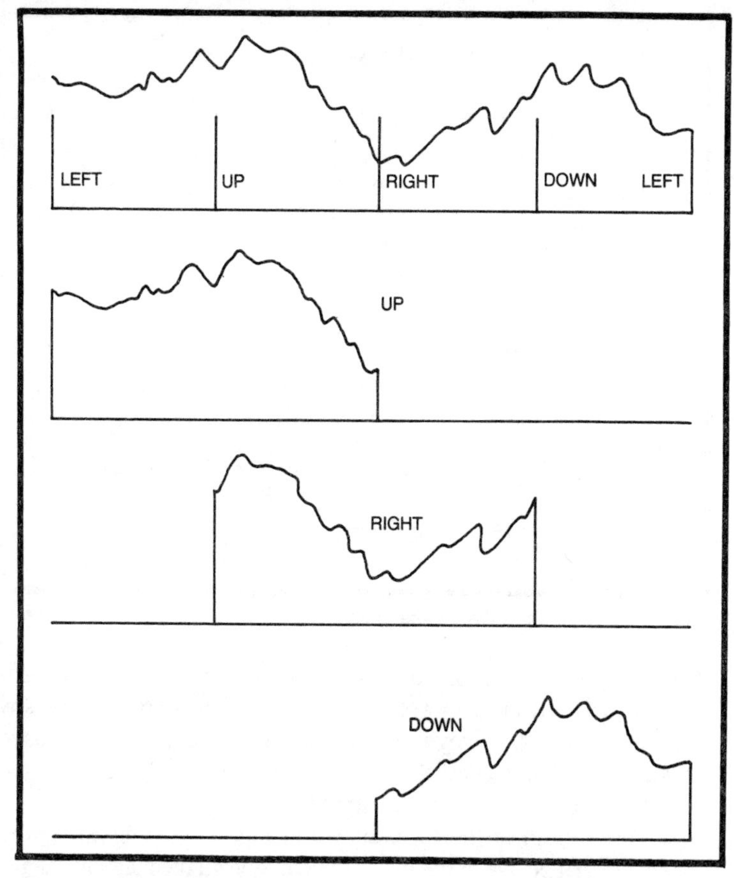

Fig. 11-9. Noise jitter.

If the signals are modulated the result goes up and down in both channels simultaneously. If the target is exactly on the axis, the resulting error signal is always zero regardless of modulation. However, if the target is off the zero axis the signal strengths in the two channels will vary with the ratio of the antenna pattern strengths. If the servo drive signal is taken from the absolute difference of the two signals rather than the ratio of the signal strengths, the error signal will be modulated. That is, a strong signal will produce a large difference whereas a weak signal will produce a small difference, at the same bearing error.

Since the introduction of analog dividers, it is relatively simple to make a system of this sort respond only to bearing error and not to signal strength. It should be noted that in a highly nonlinear bang-

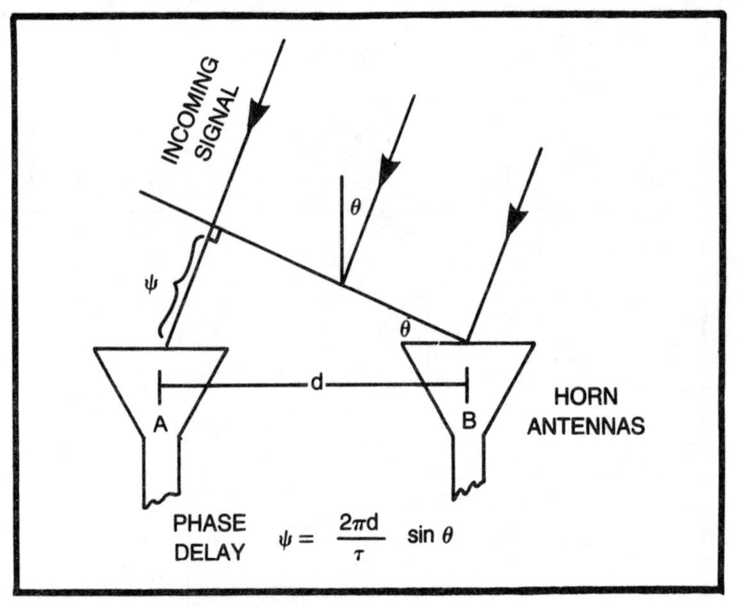

Fig. 11-10. Angle of arrival phase delay.

bang servo this factor is somewhat less important than in a linear system since the drive merely saturates for signals above some low threshold. It does, however, mean that the overall system deadband is wider for weak signals than for strong, and the positioning error thus becomes larger for weak signals.

The system of the form shown in Fig. 11-5 is an *incoherent* system in that no use of the phase information on the incoming signal is made. This is a technique which can be used with any form of incoherent radiation. There are other systems, however, which employ the phase information obtained from the incoming radiation. These are usually referred to as *coherent monopulse* techniques.

Figure 11-10 shows a pair of identical waveguide horn antennas with an incoming signal arriving at an angle to the plane of the antenna mouths. It can be seen that it takes a little longer for the wave to get to A than it does to B. This means that a target off the axis of the system will induce signals with different phase angles in the two antennas, with the phase difference being given by the formula. If these two signals are applied to a device called a hybrid junction (Fig. 11-11), the output will be signals containing the sum and the difference of the two.

The sum signals and the difference signals may be applied to a pair of matched superheterodyne receivers and converted to intermediate frequency without loss of the phase angle information. The output of the *sum* channel is then applied to a phase detector, and the difference induced in quadrature in the detector, which produces a differential output proportional to the phase difference between the signals. This is used as our error voltage to drive the servo. The operation of this and other types of phase detectors is explained in some detail in Chapter 12, where phase-locked loops are described.

The particular monopulse shown is a *phase monopulse*. However, the system may be made to operate in essentially the same manner utilizing the differences in signal strength obtained from not having the antenna patterns exactly overlap. One or the other of these schemes is employed in nearly every modern radar or satellite ground station system.

There are also a number of applications in which the actual carrier radiation information is destroyed, but in which the modulation information is preserved. This type of system is typical of certain of the laser trackers in which extremely short pulses on the order of picoseconds may be used. In effect, these systems merely measure the time-of-arrival of the reflection at two separated detectors. The servo system then drives this difference to zero. Obviously, a system of this type must have modulation encoded upon the carrier with a wavelength which is small with respect to the distance between the antennas or photodetectors if the system is to have any accuracy. In a laser tracker this implies that the light packet is shorter than the distance between the antennas, or at least has a rise time which is short with respect to the time it would take light to pass

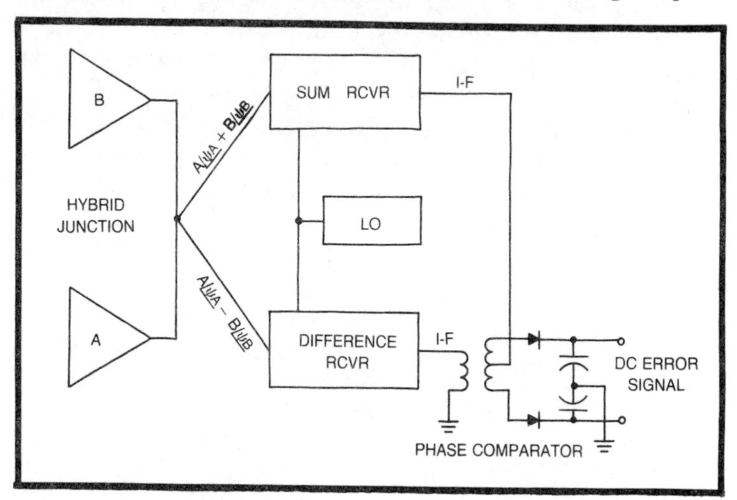

Fig. 11-11. The coherent monopulse.

between the antennas. For a detector separation of one meter this implies a time of 3.33×10^{-9} seconds. The electronic circuitry following would have to be able to respond to something approaching 10^{-10} seconds rise time or about 10 GHz.

AUTOPILOTS

Referring back to the system shown in Figs. 11-5 and 11-6, you should notice that it is relatively easy to manually adjust the antenna to some fixed angle with respect to the ship, aircraft, or missile frame; then the motor control could be applied to operate the rudder control so that the entire vehicle was turned to null the antenna. The operator would merely have to supply the offset angle, and the system would maintain the vehicle heading in the desired direction. This is the form of control generally used with air-to-air missiles. Actually, the control is a good deal more sophisticated than that in most cases since other things must be considered. For example, most airplanes will not turn properly with a control input applied only to the rudder. The ailerons must be used to bank the airplane as well as in order to make a proper turn.

In the case of the guided missile with a heat-seeking or laser or radar tracker, there must also be some predictive system to speed the intercept. In general these vehicles are equipped with an inertial or other form of autopilot, and the tracking information is superimposed upon the existing steering error signal system to provide an override function. Frequently in systems of this type the antenna system is gimballed and allowed to track the target. The autopilot is then excited to drive the error angle between the antenna (or sensor) gimbal frame and the vehicle frame to the desired offset angle.

The actual operation of these very sophisticated systems is well beyond the scope of this text; however, the basic principles of the error sensing mechanisms usually fall within one of the various categories described.

12
Phase Locked Loop Controls

In a previous chapter we dealt with certain types of feedback mechanisms, including one-shot tachometers for speed control. However, when a really precise speed control is required, the choice usually goes to a phase locked loop. There are a variety of reasons for this choice but the principal one is the fact that there are few things in this world which can be measured with the accuracy that can be obtained in frequency and time measurements. A rather modest quartz watch will keep time to within about a minute per month, which represents a speed error of about one part in 43,200 or 2.3×10^{-5}. With a little bit of care, a nonovened 5 MHz quartz oscillator can be synchronized with National Bureau of Standards WWV radio transmission to one part in 10^6 or better. With the addition of an oven, or other temperature control, a stability of one in 10^8 is obtainable. If you had a scales this accurate, you could take two successive weighings of the aircraft carrier Enterprise and tell whether the Captain had carried his sextant aboard! If you want to get really precise you can purchase a cesium or rubidium standard with a guaranteed accuracy over the long term of ± 3 in 10^{-12} which is settable to one in 10^{13}. For the long term rate, a cesium clock set at the time of the birth of Christ could by now have accumulated an error of 0.0062 seconds (if it hadn't given up long ago). And one of the best parts of this is the fact that the accuracy of frequency standards can be verified against the National Bureau of Standards by radio, anywhere in the world, using equipment available to the most modest laboratories.

MAINTAINING ACCURACY

The techniques for maintaining accuracy are beyond the scope of this text; however, the reader is directed to:

Frequency-Time Services
National Bureau of Standards
Boulder, Colorado 80302
Tel: 303-447-1000

The whole point of this discussion is to note that it is possible to develop a standard of comparison and to cause at least small mechanical devices to rotate with tremendous accuracy if the speed of the device can somehow be keyed to the frequency of a quartz crystal. This is the function of the phase locked loop control.

LONG AND SHORT TERM STABILITY

At about this point in the discussion someone usually says, "Well heck, we have lots of electric clocks around and they always keep time to better than a minute a month. Why not just use a synchronous motor and forget about all the fancy stuff?"

The answer to this question lies in the question of long and short term stability. A little discussion of power systems may help provide us with the answer.

We noted earlier that the ac synchronous motor (or generator) will develop a torque which is either positive or negative depending upon the phase angle (see Figs. 5-1 and 5-2, and the discussion). This property is employed by the power companies on a continuing basis.

My laboratory is south of Rochester, N.Y., and I pay the electric bill to the Rochester Gas and Electric Co. However, RG&E is a part of the New York State Power Pool, which is in turn connected to the New England Power Pool and the Pennsylvania Power Pool. Each of the companies in the NY Power Pool is connected to the central dispatch location through use of an elaborate computer setup and telephone and telegraph lines. The participants are continually buying or selling power from one another, on a load-sharing basis. This power is simply dumped into the pool, and I cannot distinguish whether the electricity I use came from the Adam Beck Hydro plant at Niagara Falls or from the RG&E Gannet Nuclear plant.

When RG&E has excess spinning capacity up and they decide to sell, they open the throttle slightly on one of the turbines and it pushes the rotor a little harder; the phase angle of that unit advances with respect to the system, and it is then delivering power into the pool. The long-term average frequency is locked up to the Bureau of Standards at precisely 60 Hz to avoid disastrous unplanned power flow reversals (remember the blackout). For this reason, the *long term average frequency* is probably somewhat more accurate than one in 10^8, as attested by the fact that my clock does not need to be reset except on the rare occasion when the power is interrupted.

On the other hand, the phase jockeying amounts to a short-term frequency jitter. A check using a microsecond timing circuit controlled by a photointerruptor mounted on the shaft of a two watt synchronous motor running the output shaft at 30 rpm will typically reveal a time varying between 2.005375 and 1.994351 seconds on adjacent timings perhaps ten seconds apart. Averaged over a whole day the timing comes out to 2.000000 seconds.

Now, it could be argued that the timing jitter might stem from an eccentricity in one of the reduction gears. A measurement of the zero crossing times will convince one that this is not the entire problem. These typically will measure between 8307 and 8373 microseconds on timings taken about 5 seconds apart, as compared to 8333 for a perfect 60 cycles. The instantaneous line frequency has a *short-term stability* not much better than four parts per thousand. A set of 100 such zero crossing readings showed a stability of 2.0 parts per thousand which is noticeably better than the 2.7 parts per thousand error found for the clock motor, which lends some credence to the thought that a portion of the error may be in the gear train.

The outcome of this story is simply that the line frequency is simply not terribly accurate on a cycle-to-cycle basis. A motor with a large inertia or an added flywheel will not be so sharply affected by the phase jitter on the power line. Still, speed variations of a few parts per thousand over a period on the order of several seconds are still to be expected.

A PHASE LOCKED LOOP MOTOR CONTROL

The simplified block diagram of Fig. 12-1 illustrates the basic components of a *phase locked loop motor control*. The motor shaft carries a photointerruptor similar to the one discussed in Chapter 10. Actually, this could be any form of rotation sensor which outputs a modulated signal with a frequency proportional to the shaft speed, such as the small PM alternator, the cam and breaker, the magnetic pickup, etc. A referenced signal suitably scaled from the quartz crystal oscillator or another suitably stable frequency source is compared for phase with the chopped signal from the photosensor.

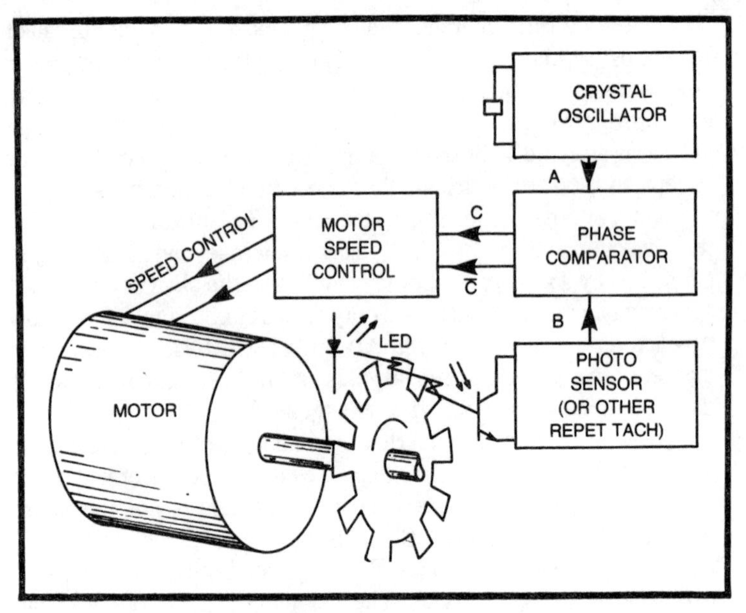

Fig. 12-1. A phase locked loop speed control.

If the motor starts to get ahead, a signal is sent to the speed control, which calls for a slower speed, probably by reducing the voltage or average current. If the motor starts to lag, the converse correction is applied.

The comparison between this form of system and the one-shot tachometer control is about the same as the comparison between a Wheatstone bridge and an ohmmeter. In the one-shot system, any variation in the capacitance or the period of the one-shot will give a corresponding shift in the motor speed. Similarly, any drift in the zero setting will be reflected as a change in motor speed. These are probably not large errors, but they can be eliminated in the phase locked loop system. Here, the interruptor phase is continuously being compared with the phase of the crystal derived frequency. The system has its own kinds of errors which will be discussed shortly, but they can be made to be orders of magnitude smaller than the one-shot system errors.

How does one compare the phase of the two signals, and for that matter, what is the meaning of the phase of an on and off digital signal? Let us suppose that the signals at A and B are both perfect square waves (implying a 50 percent duty cycle). Figure 12-2 shows the use of an EXCLUSIVE OR gate as a phase comparator. As noted from the truth table, this unit gives a true, or high, output only when the signals are different and not when they are alike. From this you can see that when A and B are perfectly in phase the output of the gate will be always zero, and when A and B are exactly opposite, or 180° out of phase, the output will always be one. For points in between, the unit tends to produce a pulse output whose average value is a function of the phase angle. The average value of output C is shown in the accompanying graph.

Now, the output of a typical logic gate is typically not voltage controlled with any great precision, so the use of a voltage average of the output waveform is not a particularly accurate reference. However, EXCLUSIVE OR gates usually come more than one to a package and they *generally* will track one another pretty well with temperature. In this case if A is inverted, it will be seen that the output of the second EXCLUSIVE OR will tend to have the opposite slope. In this case a high CMRR amplifier will be able to derive the sense signal from the pair with minimal error.

Note: This is a case of employing an uncontrolled parameter, namely the output one and zero voltages of a device. This is never a good engineering practice for any circuit to be produced in quantity.

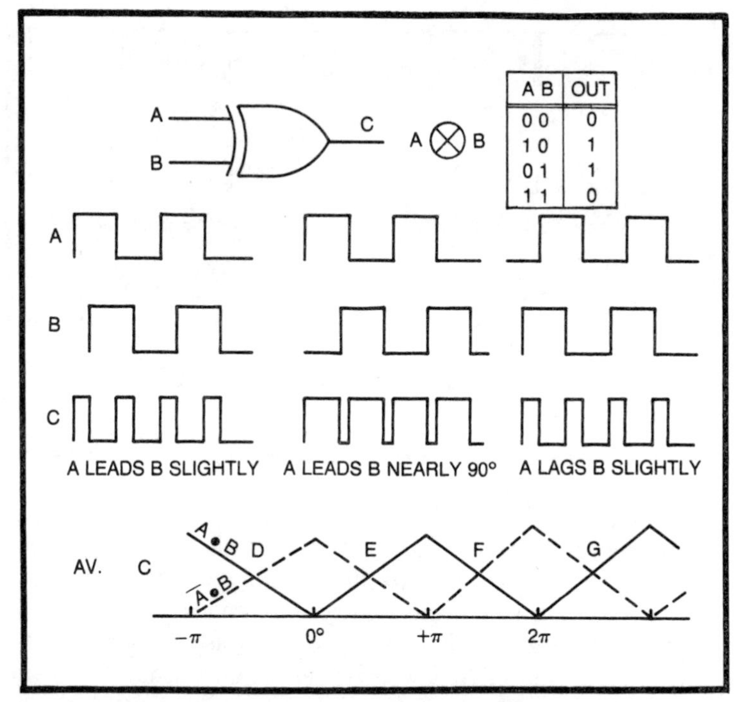

Fig. 12-2. The EXCLUSIVE OR digital phase detector.

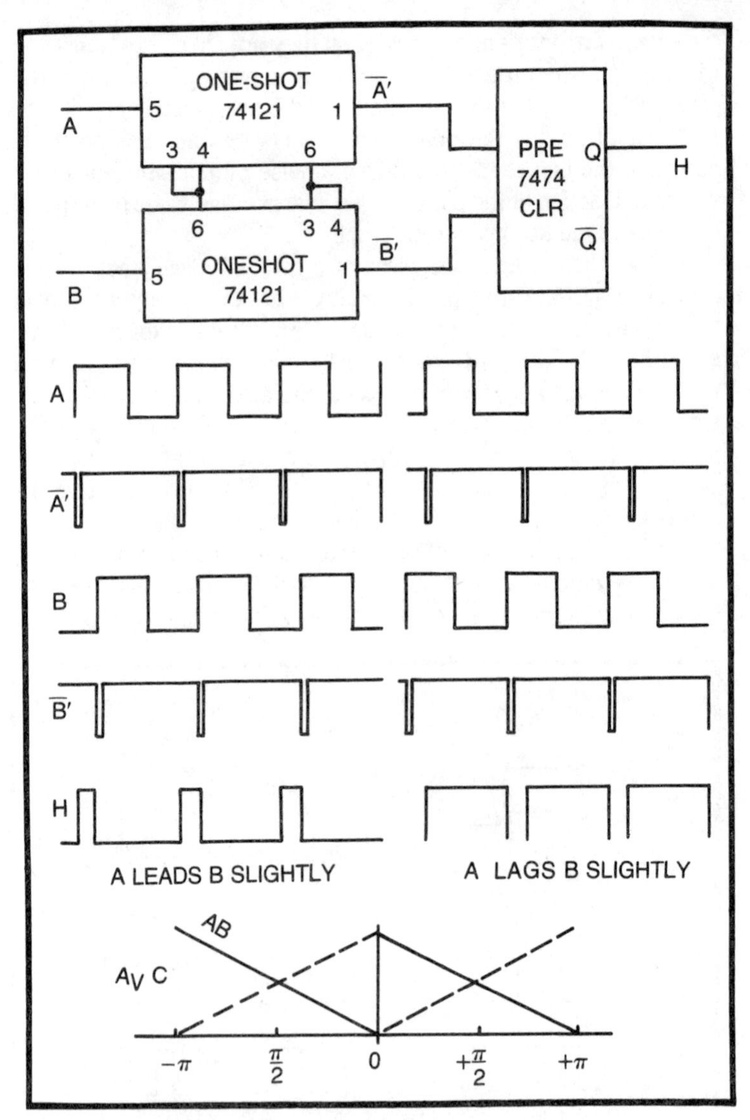

Fig. 12-3. The dual one-shot and flip-flop digital phase detector.

A far better practice is to supply an amplifier with precisely clamped zero and one levels to stabilize the slope of the error signal.

Let us suppose that the polarity of the control is such that the system *looks up* at point E; in other words a phase shift less than 90° causes the motor to slow down. If the motor were suddenly to lose a part of its load and accelerate past point F it would be on an unstable part of the curve, with the signals reversed. Thus it could not lock up at F or slightly beyond but would proceed on to G which is again stable. However, the chances of it locking at G if it left F with any significant rate are slight and the system would be unlocked—and not under control. The signals are in the direction to lock the system all the way between D and F. However, outside of the range between 0° and 180° the corrective action rapidly diminishes. For this reason a system forced beyond this range will generally come unlocked. The control has to be sufficiently effective between the 0° and 180° points to account for all transient disturbances.

A circuit with certain advantages is shown in Fig. 12-3. In this circuit a pair of one-shots is employed to set and reset a flip-flop. The one-shots have the advantage that a Schmitt trigger is frequently incorporated in the one-shot package. This tends to make the system less noise sensitive with slow rising waves. This system can be employed with sine waves on either or both inputs. This circuit works well except in the region of 0° and 180°, where we find the preset and clear inputs being keyed simultaneously. The cross connection between the one-shots will restrict this to a few nanoseconds period, since the first one-shot will inhibit the other. Only in the event that the crossings are within the one-shot delay will there be a conflict.

Figure 12-4 illustrates another advantage of this form of phase detector if one of the waves has other than a perfect 50 percent duty

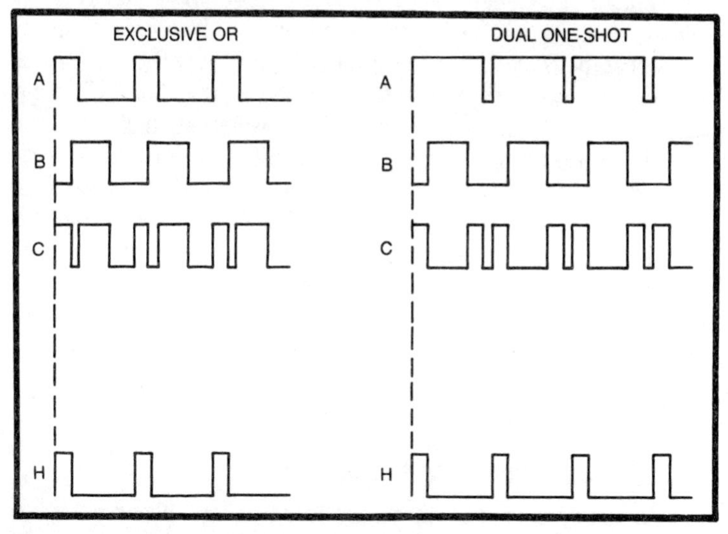

Fig. 12-4. Harmonic distortion in the EXCLUSIVE OR phase detector versus the dual one-shot flip-flop phase detector.

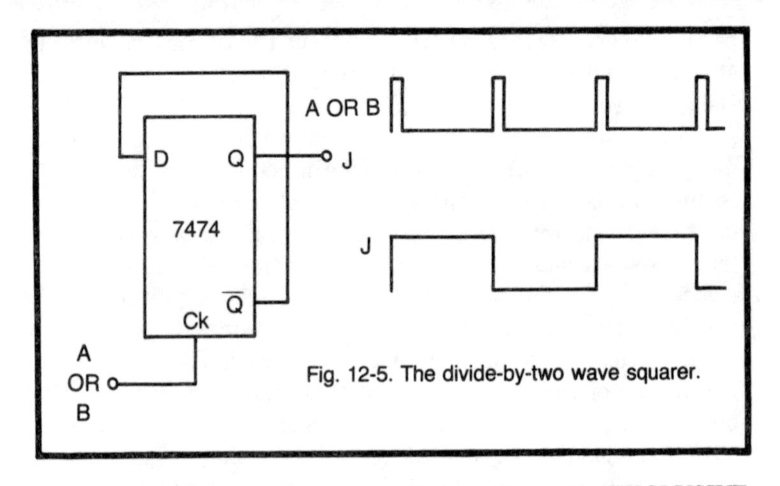

Fig. 12-5. The divide-by-two wave squarer.

cycle. As may be seen in the top half of the figure, the EXCLUSIVE OR type phase detector interprets the difference between waves A and B quite differently even though the rise-time relationship between the left and right hand examples is the same. This difference is sometimes referred to as harmonic error for reasons that will be discussed in the next chapter, which pertains to Fourier analysis. The lower waveform shows that the one-shot flip-flop phase detector is not subject to this kind of error, since it responds only to the rising edge of the two waves and ignores the length of the on time.

Figure 12-5 illustrates the divide-by-two wave squarer. The 7474 is classified as an edge-triggered flip-flop. It transfers the data present at the D input to the Q output on the rising edge of the clock. If at the start we assume that Q is low and $\overline{Q}$ (read "not Q") is high then on the first rising edge of the clock Q will be set high, and $\overline{Q}$ set low. The next rising clock edge finds the low at D and sets Q low again. This circuit will produce perfect 50 percent duty cycle square waves at one-half of the frequency of the clock input—provided only that the clock input is regular. This is, of course, not always the case. An irregularity of this sort represents a *phase jitter*, in any event. The EXCLUSIVE OR type phase detector is nearly always preceded by a pair of dividers to square up the wave to prevent an action of the type shown in Fig. 12-4. The circuit of Fig. 12-6 shows this arrangement.

THE SINE WAVE PHASE DETECTOR

In some cases it is necessary to detect the phase difference between two sine waves. The most commonly used detector in this case is the same unit used for years in FM radios. Figure 12-7 shows

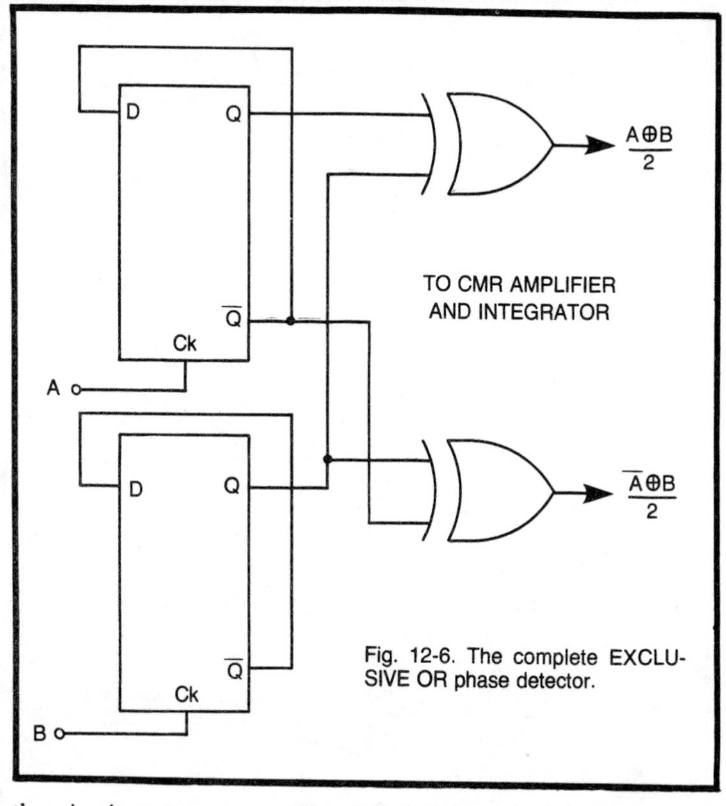

Fig. 12-6. The complete EXCLUSIVE OR phase detector.

the circuit arrangement. We shall not examine this circuit too exhaustively since it is treated in great detail in nearly all radio texts. The diagram of Fig. 12-8, however, shows a special case for this device.

It should be noted in Fig. 12-7 that the action of transformer *T*1 is such that it provides a phase-inverted output to points L and M

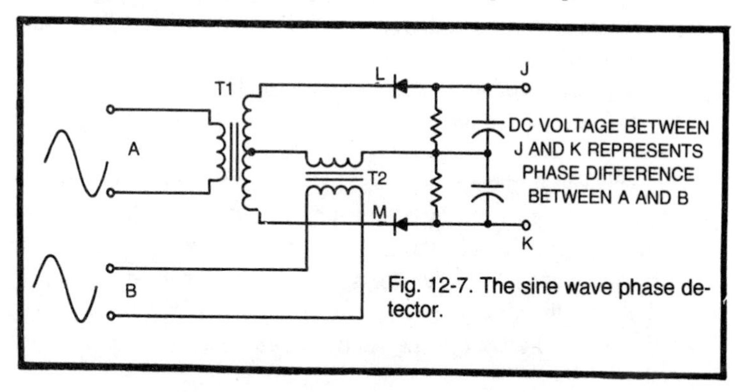

Fig. 12-7. The sine wave phase detector.

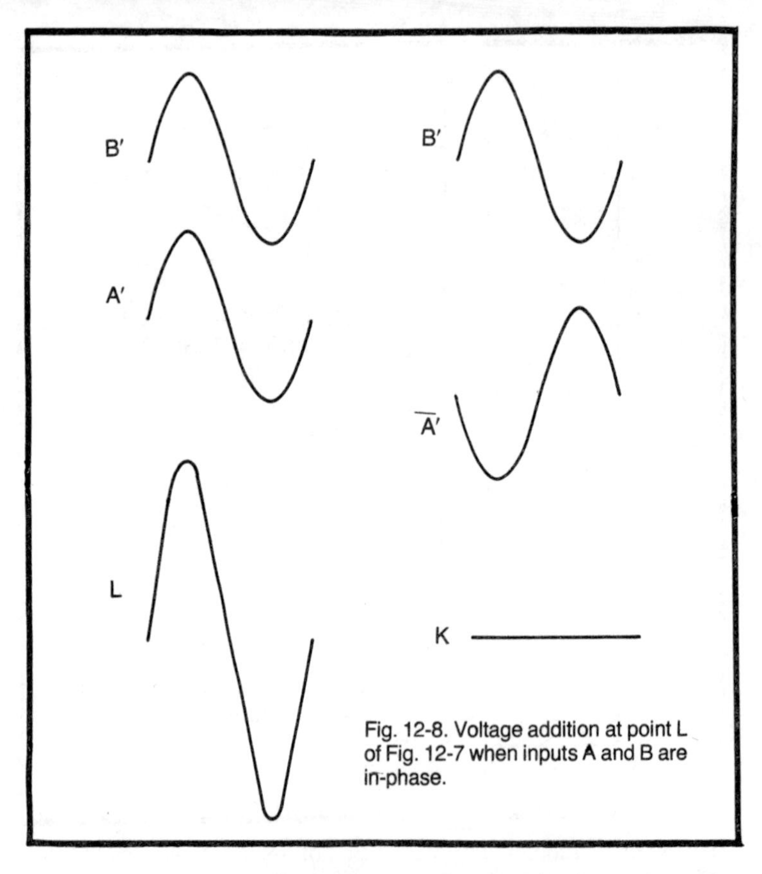

Fig. 12-8. Voltage addition at point L of Fig. 12-7 when inputs A and B are in-phase.

whereas $T2$ feeds L and M in-phase. In Fig. 12-8 the special case for inputs A and B exactly in-phase and equal in amplitude is shown. It may be seen that the induced components, which are designated by the primed symbols, add up to double the voltage at L, and to cancel it at K. Obviously, the rectified voltage at J will be strongly negative, and the rectified voltage at K will be zero. If the voltage at B were reversed in phase, the opposite case would be obtained—with K strongly negative and L zero. If both diodes were reversed the output voltage would go strongly positive at the more favored terminal.

It will be stated without proof that the sum of two sine waves of the same frequency but with any phase and amplitude is always another sine wave of the same frequency. Therefore, this circuit works equally well with the voltage at either terminal smaller. It should be relatively easy to convince yourself that the circuit winds up with the rectified output at J and K equal when the inputs are in quadrature or have a 90° phase difference. The circuit will also work well with any waveform which is symmetrical and has equal positive and negative halves.

THE IC PHASE LOCKED LOOP

A fairly large number of IC phase locked loop packages are offered by various manufacturers. These devices usually offer a resistance-capacitance (RC) tuned oscillator which may be voltage or current controlled in frequency, a phase detector, and an integrating amplifier, all in a single IC package. In frequency modulation (fm) detector service, the RC oscillator is adjusted to about equal the center frequency with no signal applied to the detector. When a signal is applied, the oscillator is forced into phase lock with the incoming signal, and the control voltage or current then represents a precise analog of the frequency or phase deviation of the incoming signal. Most of these devices are extremely sensitive and will track a signal of a millivolt or less.

Usually these devices have been carefully designed to minimize the frequency shift with temperature and supply voltage. Temperature stabilities on the order of a few to a few tens of parts per million per °C can be obtained. This is not as stable as a crystal oscillator but it is not bad either. Some, but not all, of these devices can be used in phase locked loop motor controls.

In the phase locked loop motor control, the object of the game is to hold the oscillator signal as steady as possible and to force the motor to track it. This is the opposite of the fm detector where one wants the oscillator to track the incoming signal. On some of the IC packages, the control signal to drive the oscillator is connected outside of the package, and therefore, it is possible to break the feedback loop. This is the type which is generally preferred for this type application. A generalized block diagram of an IC PLL is shown in Fig. 12-9. Each of these blocks is usually supplied. The elements for the low-pass active filter or integrator are generally external to the package, but in order to save pins, the link between A and B is sometimes made internal, and the link between the voltage controlled oscillator (VCO) and the phase detector is sometimes internal. As shown, the oscillator runs at a constant frequency, and the output of the integrating amplifier is used for motor speed control. The reference signal from the motor may be either a sine or square or even other wave shape, but it is usually best if this is kept strictly symmetrical. The unit offers the flexibility that the motor speed may be adjusted by varying R, C, or the voltage or current control. For high accuracy applications the reference link may be broken and a crystal oscillator substituted, leaving the interval VCO unused.

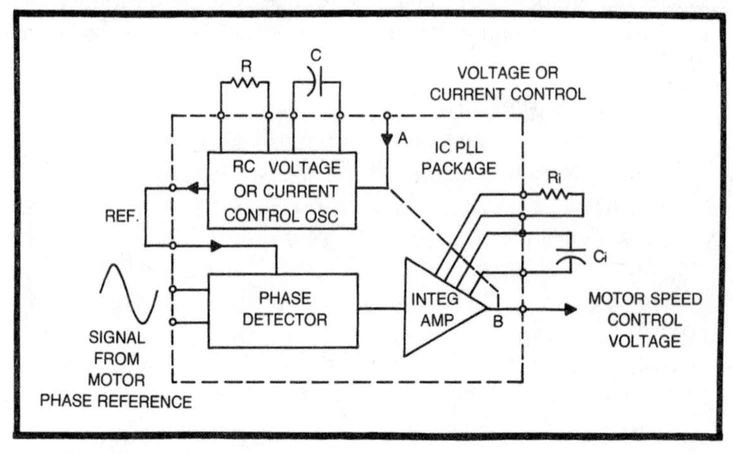

Fig. 12-9. The IC phase locked loop.

If link A-B is completed, the oscillator will track the input frequency over a very wide range, perhaps more than 2 : 1 with some units. The output voltage at B may then be applied to a *potentiometer balancing servo* in place of the reference potentiometer voltage and made to track a speed control voltage at the signal input. This system has a very wide lock range, but the accuracy is lower than is achievable with the fixed-frequency oscillator.

FREQUENCY DIVISION

The stability of the crystal oscillator is very advantageous in many applications. On the other hand, there are other uses where it would be nice to have the advantage of a variable frequency as well. Probably the most successful means to obtain such flexibility is with a *digital synthesizer*. The discussion of the varieties of these units is well beyond the scope of this text; however, a simple programmable divider is shown in Fig. 12-10. In this unit the crystal reference input frequency is applied to a chain of one or more conventional digital counters. The current count is compared to a reference word in a digital magnitude comparator. The reference word is set in an array of switches with an open switch considered to be a 1. The comparator is arranged so that the appropriate A versus B output condition is 1 and the others are 0. When the count equals the number set in the switches, one-shot 74121 fires long enough to ensure that all of the counters clear. Without the one-shot a race condition could develop and only the fastest counter would clear. The one-shot should be arranged to fire for something less than a half of an input cycle in order to avoid missed counts. Because of the

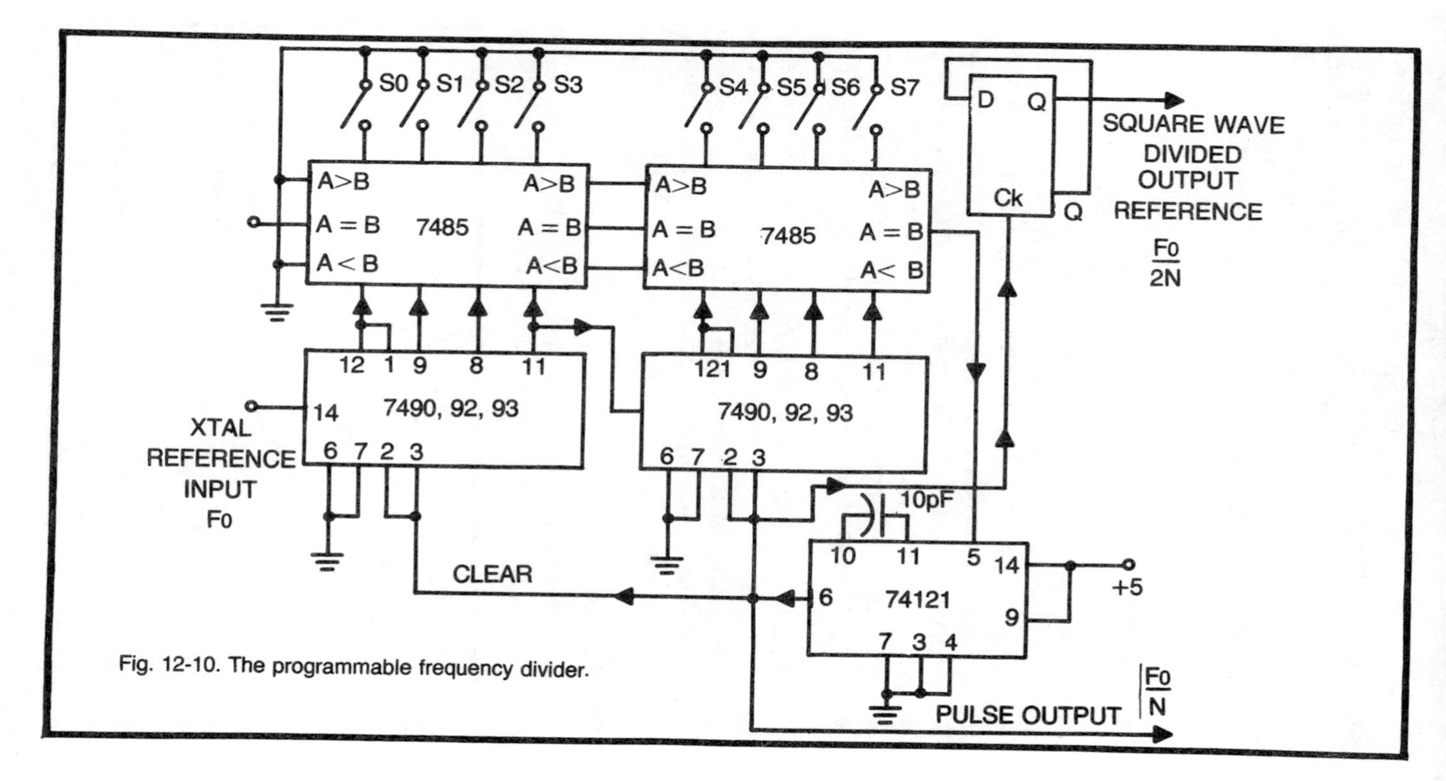

Fig. 12-10. The programmable frequency divider.

shortness of the one-shot output the wave will be extremely asymmetrical. This can be used directly in a dual-one-shot-flip-flop phase detector or it may be squared up in a divide-by-two circuit. If binary counters such as the 7493 are used, the count may be set for any value from $N = 1$ to $N = 266$ with the switches. With a 10 pF timing capacitor on the 74121 this circuit can operate with input frequencies in excess of 5 MHz.

OSCILLATORS

There are a great many types of stable oscillators that may be constructed, and most engineers have some favorite ones. This again is a topic which we cannot more than touch upon here. The circuit shown in Fig. 12-11 is very simple to implement in TTL logic. It can be made to operate with a stability on the order of one in 10^6 per month with a good 1 to 7 MHz crystal. The trimmer capacitor will generally permit the unit to be "rubbered" or tweaked over a range of 20 to 50 kHz for synchronization, and in a fairly stable temperature environment the short term stability is probably better than one in 10^7.

Crystals cut for the 3.585 MHz frequency used for the color burst in TV have the distinct advantages of being cheap and being evenly divisible by 59750 to yield 60 Hz. Since the divisor is an even number, the final divide-by-two circuit may be used to ensure a square wave output.

PHASE JITTER

In a well constructed PLL system it will usually turn out that the principal source of the phase jitter is due to mechanical inaccuracy in

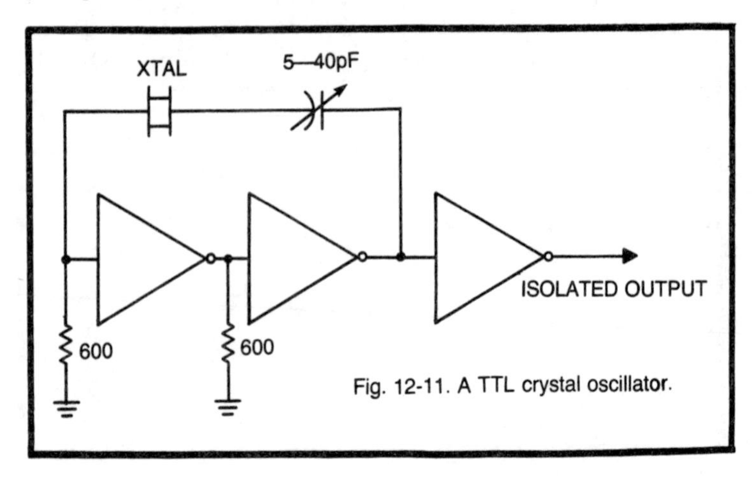

Fig. 12-11. A TTL crystal oscillator.

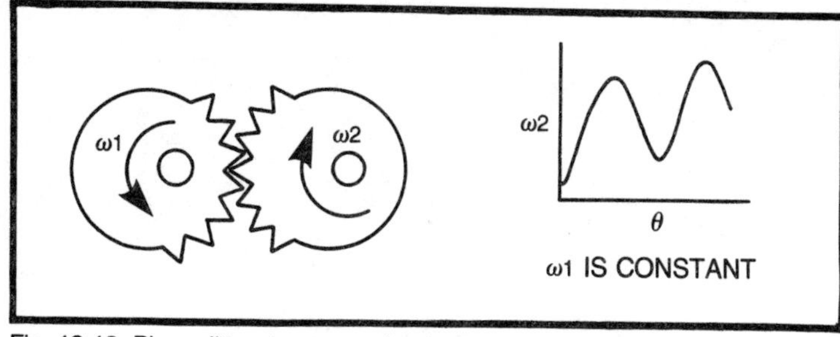

Fig. 12-12. Phase jitter due to mechanical eccentricity.

the system. Figure 12-12 shows two eccentric gears meshed together. As shown, the effective radius to the pitch circle of the first gear is smaller than the effective radius of the second gear therefore the second will be turning slower than the first. One half turn later, the situation will be reversed and the second gear will be turning faster than the first. In between, the velocity will be equal. Although this illustration is somewhat exaggerated, the presence of any mechanical inaccuracy of eccentricity will, in general, force a modulation upon the output velocity. When straight tooth spur gears are used in a gear train, the point of contact of the involute teeth changes radius throughout the range where the opposing teeth are in contact thus imparting a tooth-rate modulation to the output gear. This effect is worse in straight tooth bevel gears. The effect is minimized in spiral, helical, herringbone, and hypoid gears, which accounts for the popularity of the later types in propulsion machinery where the smoothness and quiet operation is at a premium.

The photo interruptor, magnetic sensor, and other rotation pickups are similarly modulated by eccentricity. While these items can be very easily fabricated to modest precision, the fabrication of units with minimal phase jitter is a very precise business. In high precision PLL applications, these units are usually purchased items. It is noteworthy that the phase jitter accuracy of the finest unit can be completely destroyed by a slight bit of misalignment in the mounting shaft and the subsequent modulation by a sliding oldham's coupling or universal joint.

13
The Fourier Transform and Wave Analysis

In the earlier chapters, we touched briefly upon the fact that some of the waves we dealt with would not be simple sine waves. The next two chapters describe a variety of controls in which the current is often deliberately mutilated by switching devices in order to obtain some form of control of a motor or other equipment. In this chapter we shall deal with some of the mathematical techniques that facilitate calculations with some of these odd or mutilated waveforms. Rather surprisingly, it turns out that the mathematics for handling this type of thing are a great deal older than the general case mathematics required for ac circuit calculations.

FOURIER'S CONTRIBUTION

Jean Baptiste Fourier (1768–1830) rose from a humble birth to become a mathematician, statesman, and confidante of Napoleon. However, his greatest single contribution is a mathematical procedure and technique which bears his name. In 1807, Fourier announced his discovery that almost any function of a real variable can be represented by a series of sines and cosines of integral multiples of the variable. This sounds fancier than it actually is. It simply means that almost any curve which can be drawn on the face of an oscilloscope or a sheet of graph paper can be synthesized by adding up a series of sine and cosine waves. Furthermore, if the curve is one that repeats itself, we need only consider waves with periods that are integral number fractions of the period of the wave being synthesized, like a half, a third, a fourth, a fifth, etc. This is the same

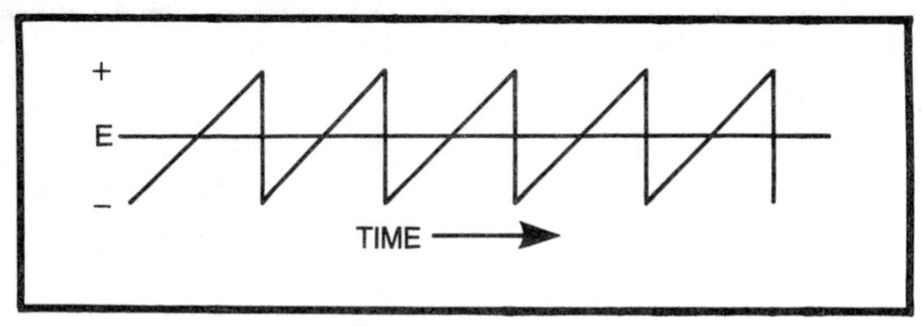

Fig. 13-1. The common sawtooth wave.

as saying that they have frequencies twice as great, three times, etc.

For example, consider the sawtooth wave shown in Fig. 13-1. This is a rather commonly encountered wave in electrical apparatus and is often used to supply the sweep for an oscilloscope etc. This wave may be simulated by the judicious choice of a series of sine waves as shown in Fig. 13-2. In the illustration we have selected only four components and yet as you see, we are able to construct a reasonable approximation of the sawtooth wave using only the fundamental and the second, third, and fourth harmonics. You will note that the sharp corners are absent. We will see later that this is because of the lack of higher harmonic terms. It is fairly easy to visualize that a slight decrease in the amplitude of either E_3 or E_4 would bring the dip above the arrow closer to the desired curve but this would drive the other points farther away.

The physical significance of the analysis is rather great. Suppose that we wanted to build an oscilloscope that used the sawtooth wave for the horizontal sweep, and suppose furthermore that the amplifier we were using to drive the horizontal sweep gave up at the fourth harmonic of the sweep frequency. The wave on the bottom would be as close as we would get to having the sawtooth sweep and the display would be rather badly distorted.

Before getting too mathematical, we might further investigate some of the properties that can be deduced about the addition of sine waves.

The effects of phase are not so readily guessed. The curves of Fig. 13-3 illustrate what happens when a fundamental and the third harmonic are added for the cases where the third harmonic is in-phase and where out-of-phase. *The waves are said to be in-phase when they cross the zero axis going in the same direction at the same instant*. The in-phase addition shown at *C* yields roughly a square wave. This is the sort of thing that can happen when an amplifier is saturating while trying to pass sine waves.

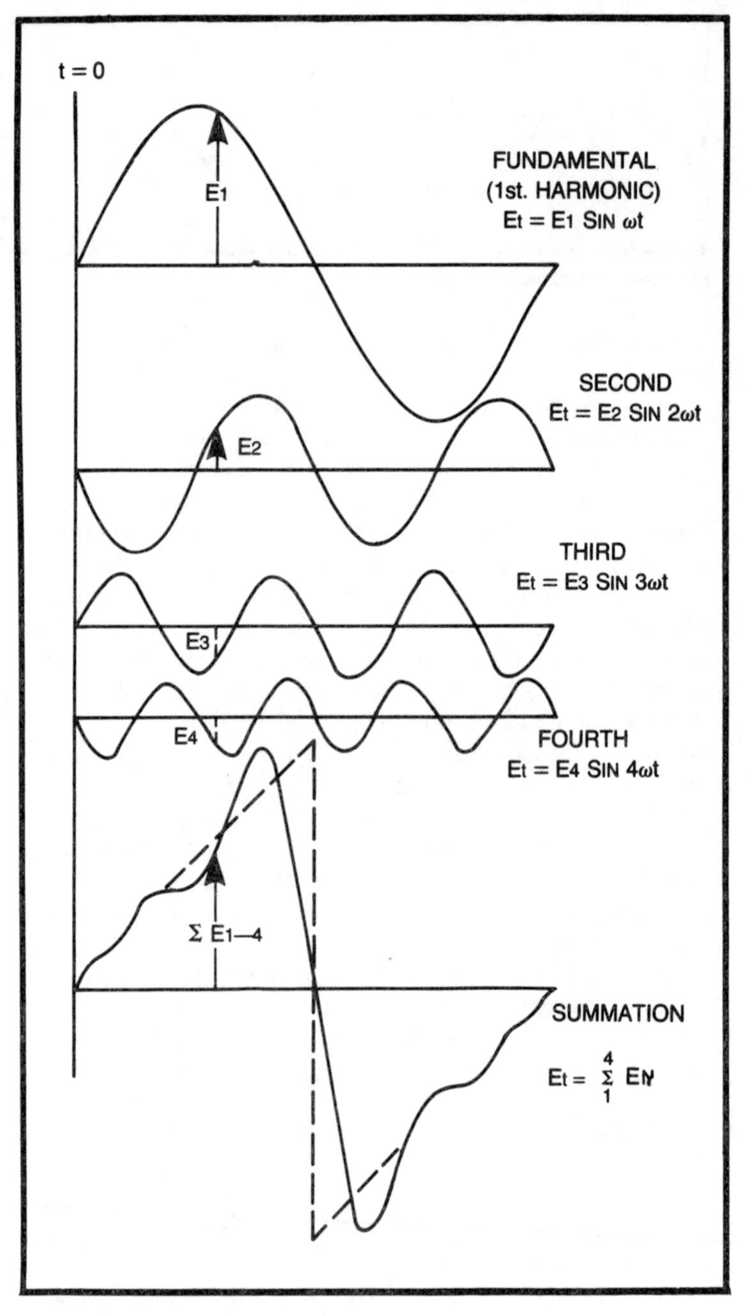

Fig. 13-2. The fourth harmonic resultant provides a good approximation of a sawtooth.

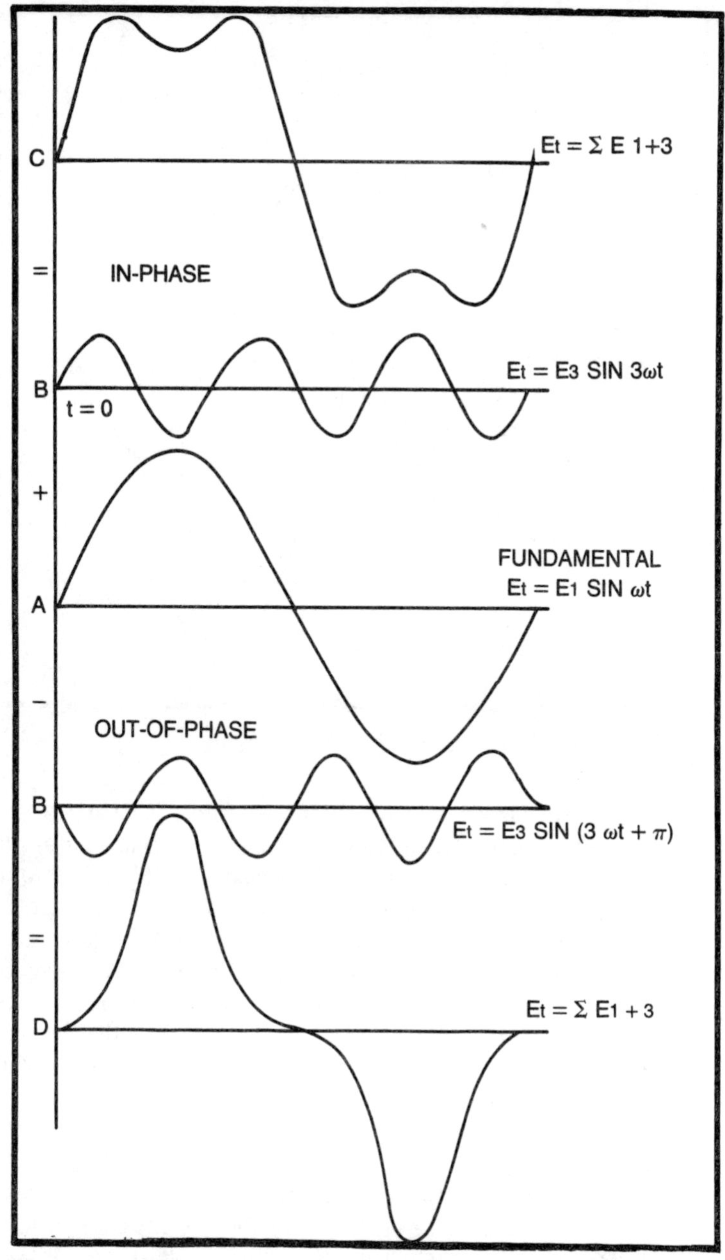

Fig. 13-3. The effect of phase relationship when a fundamental and the third harmonic are combined. When in-phase, a near square wave results, as shown at the top. When out of phase, the resultant is sharply peaked, as at the bottom, resembling a $\sin^2$ function.

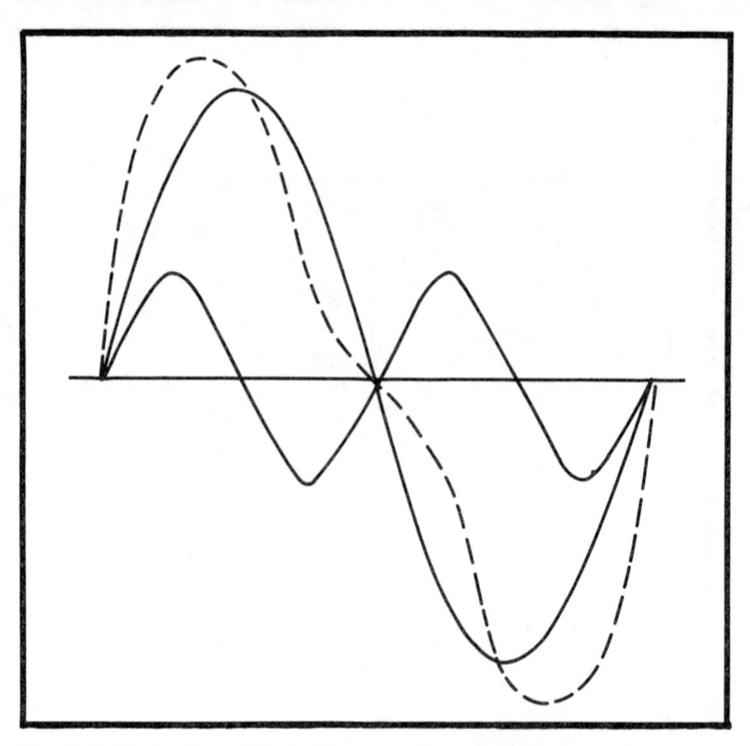

Fig. 13-4. The fundamental plus the second harmonic.

The wave shown at D is sharply peaked and nearly a $\sin^2$ function. This is the sort of thing that can happen when a class B amplifier has a crossover distortion, meaning that both sides of a push-pull, or complementary, amplifier are biased off too far. It always comes as a bit of a shock that two such different looking waves could have the same harmonic content. Most wave analyzers and spectrum analyzers do not show phase relationship, so the display would be identical for both C and D. Also note the fact that both C and D are symmetrical, not only about the zero crossings but also about the midpoint between the zero crossings. This is generally characteristic of the absence of second and other even harmonics.

Figure 13-4 shows that the symmetry about the midpoint between zero crossings can be destroyed by the presence of a significant second harmonic component. Figure 13-5 illustrates that an arbitrarily phased second harmonic component can even destroy the symmetry about the zero voltage axis. This is the sort of wave that one can obtain from an overdriven single-ended amplifier which is "topping out." Figure 13-6 shows the harmonics which are not even can also destroy the symmetry about the midpoint between zero crossings.

It can be shown that a waveform which is symmetrical about the zero crossings and the midpoints between zero crossings has no even-order harmonics, and all odd-order harmonics have a phase of either zero or π.

THE ANALYSIS OF WAVEFORMS

Next let us consider the analysis or the means by which we may arrive at the judicious choice of the harmonic phases and amplitudes. To begin with let us consider the recurring voltage waveform of Fig. 13-7. This wave repeats itself with a period T; therefore, we would expect that there would be no components with a period greater than T since this would cause a variation from cycle to cycle. We may write an expression for the wave in the form:

$$E_{(t)} = \frac{E_o}{2} + \Sigma_1^n A_n \cos \frac{2n\pi t}{T} + B_n \sin \frac{2n\pi t}{T} \quad (13.1)$$

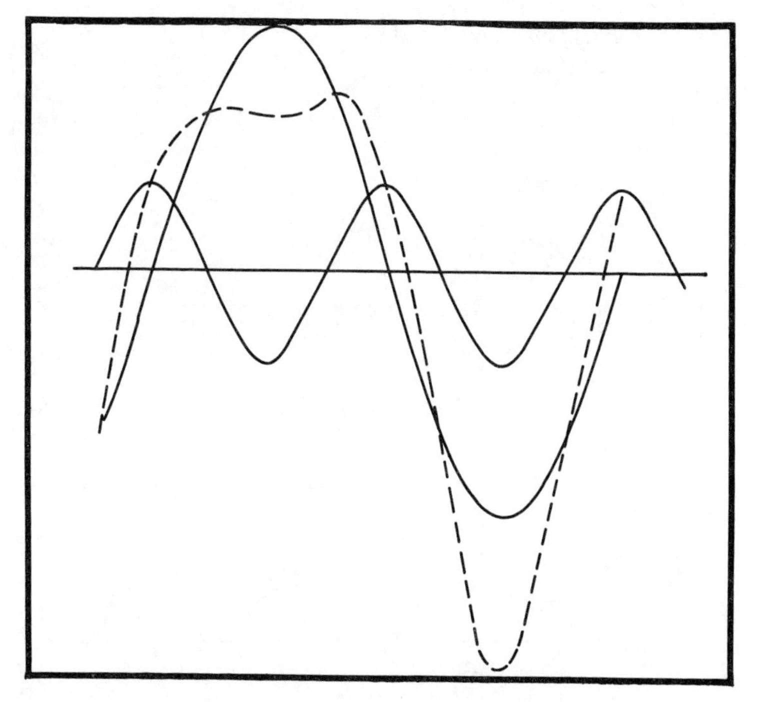

Fig. 13-5. An arbitrarily phased second harmonic component can destroy the symmetry about the zero voltage axis.

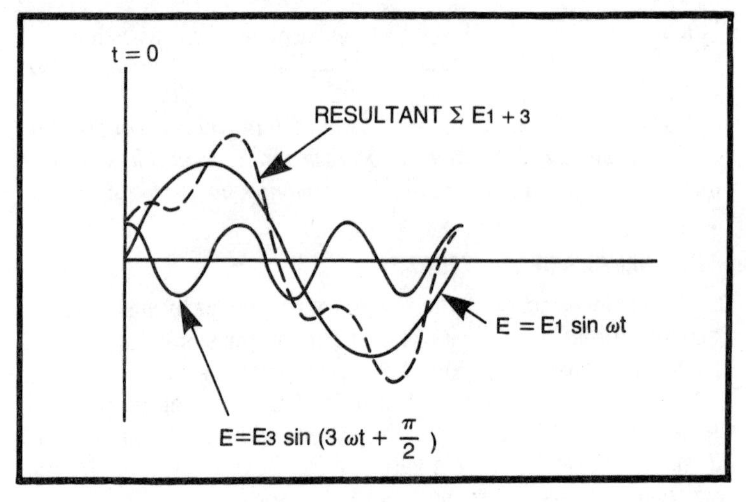

Fig. 13-6. Harmonics that are not even can also destroy the symmetry about the midpoint between zero crossings.

where

A_n is the amplitude coefficient of the n'th harmonic given by:

$$A_n = \frac{2}{T} \int_0^T E_{(t)} \cos \frac{2n\pi t}{T} \, dt \tag{13.2}$$

Similarly:

$$B_n = \frac{2}{T} \int_0^T E_{(t)} \sin \frac{2n\pi t}{T} \, dt \tag{13.3}$$

and E_0 is the average value, or dc value, given by:

$$E_o = \frac{1}{T} \int_0^T E_{(t)} \, dt \tag{13.4}$$

For the examples of Figs. 13-1 through 13-4 the positive and negative halves of the wave cancel and E_o is zero. This case occurs when

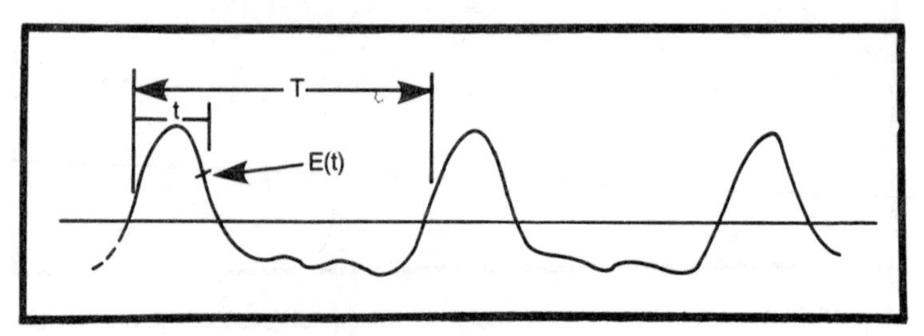

Fig. 13-7. A recurring voltage waveform.

there is no dc bias, as for example when the wave has been passed by a capacitor or transformer.

The double symmetry case (about zero crossings and mid-points) may be resolved into two conditions. When

$$E_{(t)} = -E_{(-t)} \tag{13.5}$$

the function is said to be odd and all of the A_n (or cosine) terms cancel. This case is illustrated in Fig. 13-3. If the time axis is located in the center of a loop, when

$$E_t = E_{(-t)} \tag{13.6}$$

or if the figure posseses mirror symmetry about the time axis as shown in Fig. 13-8 then the function is said to be even and all of the B_n or cosine terms vanish.

A large number of the frequently encountered cases have been solved analytically and are presented in the literature. Most of these are idealized geometric waveforms. The general spectrum of a particular wave can sometimes be synthesized by summing up the solutions for known waveforms.

One of the best known of these solutions is for the rectangular pulse-train waveform. This is very frequently encountered in digital and control work since it simply represents a regular on and off function, in a repetitive pattern. A tabulated solution is shown in ITT-Reference Data for Radio Engineers, Edition IV, Chapter 33, Fig. 6A, represented here as Fig. 13-9. This is shown as an even function and all of the sine terms (B_n's) vanish. It can readily be seen that the average value is:

$$A_{av} = A\frac{t_o}{T} \tag{13.7}$$

and the coefficient of the M'th term is

$$A_n = 2A_{av}\left[\frac{\sin\frac{n\pi t_o}{T}}{\frac{n\pi t_o}{T}}\right] \tag{13.8}$$

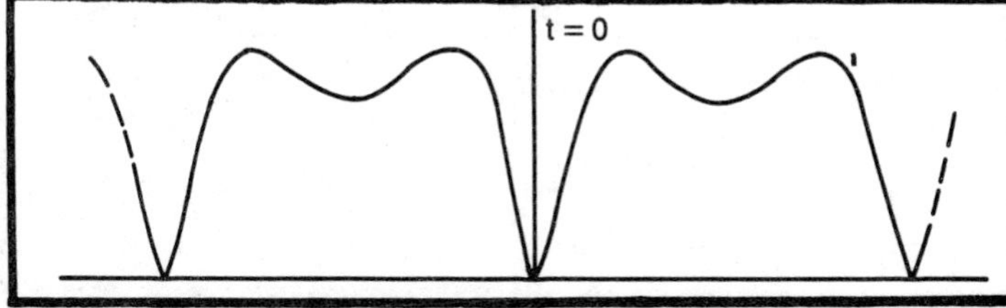

Fig. 13-8. Mirror symmetry about the time axis wave shape.

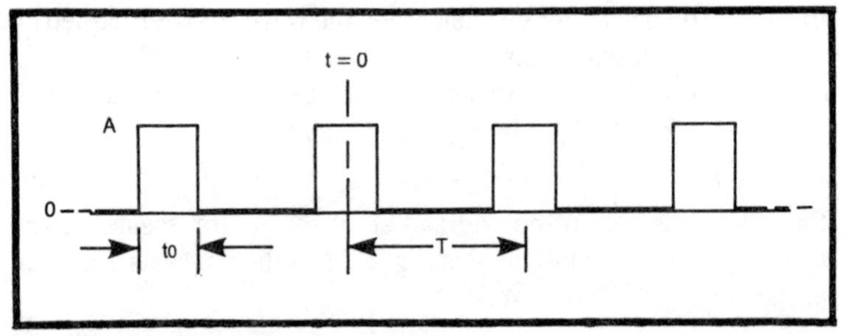

Fig. 13-9. Rectangular impulse waveform.

The function sin x/x may be found tabulated in a number of handbooks. The general form is illustrated in the curve of Fig. 13-10. This curve describes the envelope or locus of values of the harmonic and the actual expression for the wave may be written following equation 13.1.

$$E_{(t)} = 2A_{AV} \left(\frac{t_o}{T} + A_n \text{ COS } \frac{n\pi t)}{T} \right. \tag{13.9}$$

Actually one is usually interested in the envelope of the harmonics, i.e., the spectrum of A_n's. We find the sin x/x function recurring in many of the geometric wave shapes.

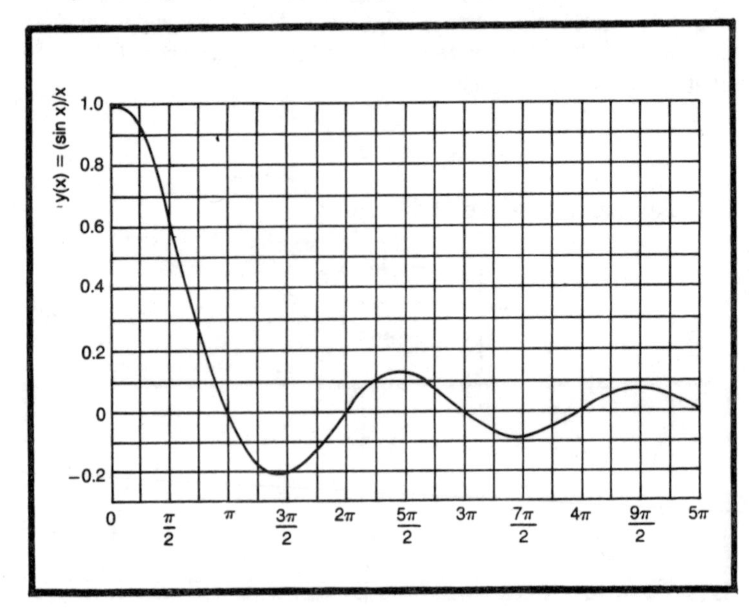

Fig. 13-10. Sin x/x function waveform.

If one were to take a tuned voltmeter or a calibrated spectrum analyzer and tune it across the spectrum he would find a multiplicity of signals each of which was spaced from the fundamental or $1/T$ signal by $1/3T$, $1/5T$, $1/7T$, etc., for this wave. The response of a complex circuit could then be calculated by considering the response to each of these components individually and then summing the results.

For the cases in which the analytical solutions are available, the technique is quick and painless. However, much of the real power of the Fourier transform is the ability to apply it to the analysis of waveforms for which there is no known expression. Long chains of recorded data may be reduced to relatively simple expressions, the performance of which may be predicted with great accuracy.

NUMERICAL FOURIER ANALYSIS

Until recent years the engineer was always reluctant to launch into a Fourier analysis of a waveform because of the tedious nature of the work and the amount of calculation required.

The methods given in one source (Kerchner and Corcoran. *Alternating Current Circuits*. John Wiley & Sons, 1938. P 155–159) requires the reading and recording of 72 sample voltages, looking up the sine and cosine of 144 different angles, 432 multiplications, 292 additions, and a full sheet of 8 1/2 × 11 inch paper to write down the answers—for each separate component! A 16-component analysis would require 2304 sines and cosines, 6912 multiplications, 4672 additions, a scratch pad an inch thick, and an inordinate amount of persistence.

The advent of the high speed digital computer has considerably altered the outlook toward this technique because all of the arithmetical manipulation can be performed without error or boredom by a tireless computer in the blinking of an eye.

Accordingly, this tool, which was once employed only "in extremis," is finding progressively increasing and widespread use in the analysis of data streams. A number of firms are offering hard-wired Fast Fourier Transform (FFT) machines which can listen to the music from a symphony orchestra and plot the individual notes and harmonics in "real" time (i.e., with only a few millisecond delay), or sort the sounds of a submarine from the whistles of shrimp, the chirping of whales and porpoises, and the rumble of waves.

That blink of the eye required for the summation of one of the component terms is not to be taken too literally if the work is being performed on a desktop calculator. It takes these relatively slow machines a fair amount of time to run a very lengthy summation;

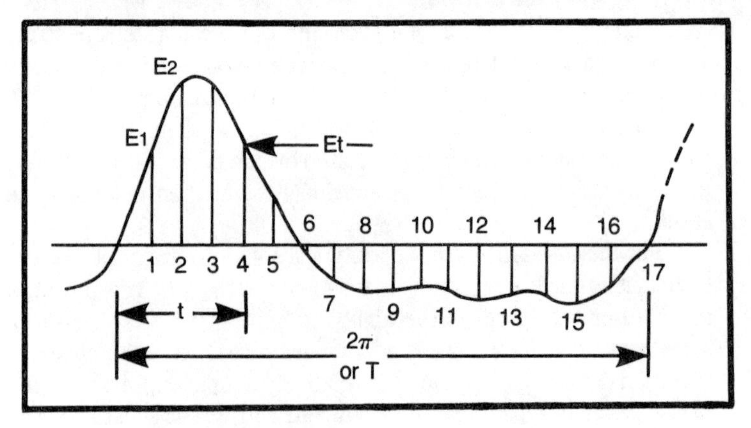

Fig. 13-11. Wave sample diagram.

computation times on the order of 30 seconds to a minute per component are not unusual.

In generalized functions, such as Fig. 13-7, the vibrations of a single cylinder engine, electrocardiograms, etc., none of the terms or harmonics may be neglected. For these functions an analytic expression is seldom known and would be difficult to fit; therefore, a sampling or graphical method is usually employed when digital analysis is desired. This technique also lends to computer solution. In this technique the period of the wave is set equal to 2π radians and the wave is divided into a series of (generally equal) samples (M).

The values of $E_{(t)}$ are sampled at each of the ordinates (17 in Fig. 13-11). The coefficients are then determined by

$$E_o = \frac{\Sigma_1^M E_t}{M} = \frac{E_1 + E_2 + E_3 \ldots E_M}{M} \tag{13.10}$$

and

$$A_n = \frac{1}{M} \Sigma_1^M E_t \cos \frac{2n\pi t}{T} \tag{13.11}$$

For example, for 17 equal steps and to determine the third harmonic, A_3:

$$A_3 = \frac{1}{17} \Sigma E_1 \cos \frac{6\pi}{17} + E_2 \cos \frac{12\pi}{17} + E_3 \cos \frac{18\pi}{17}$$

$$+ \ldots + E_{17} \cos \frac{102\pi}{17} \tag{13.12}$$

Similarly:

$$B_n = \frac{1}{M} \Sigma E_t \sin \frac{2n\pi t}{T} \tag{13.13}$$

The labor of Fourier analysis of a detailed function may be appreciated from the example. The magnitude of the n'th harmonic is:

$$E_n = \sqrt{A_n^2 + B_n^2} \tag{13.14}$$

and the phase is:

$$\angle \phi = \arctan \frac{A_n}{B_n} \tag{13.15}$$

From this we observe the power of this technique which can serve as a digital analogue to a spectrometer. If a sampling voltmeter of sufficient speed could be employed to digitally sample the electric field in (polarized) sunlight then the Fourier-analyzed output would provide a complete (voltage) spectrum with the added benefit of providing phase information. In actual optical spectrometers and most electronic spectrum analyzers, the power spectrum only is provided. (This corresponds to equation 13.14.)

$$P_n = E_n^2 = A_n^2 + B_n^2 \tag{13.16}$$

Some of the additional flexibility of digital spectrum analysis will become evident in the subsequent examples.

The numeric technique has the limitation that it cannot predict harmonics any higher than the number of ordinates used. For example with the 17 voltage samples employed, harmonics no higher than the 17th may be computed, and really, the system loses accuracy at about half of the number of samples.

PHASE CONTROL WAVEFORMS

In recent years, with the advent of the SCR, one of the techniques used for controlling the average current in a motor or other device is to use an SCR to switch on the current sometime after the start of the cycle. The waveform is shown in Fig. 13-12.

Also shown in this figure is a printout of the various values of B_n. Note that the changing sign of the sine function tends to change the sign of the tabular values. This happens more rapidly for the higher harmonic terms, so they generally tend to decay in summation. The summation term at the bottom lists the voltage magnitude of the component, and the negative number beneath gives the relationship in decibels (20 log V_1/V_n) of the component to the fundamental.

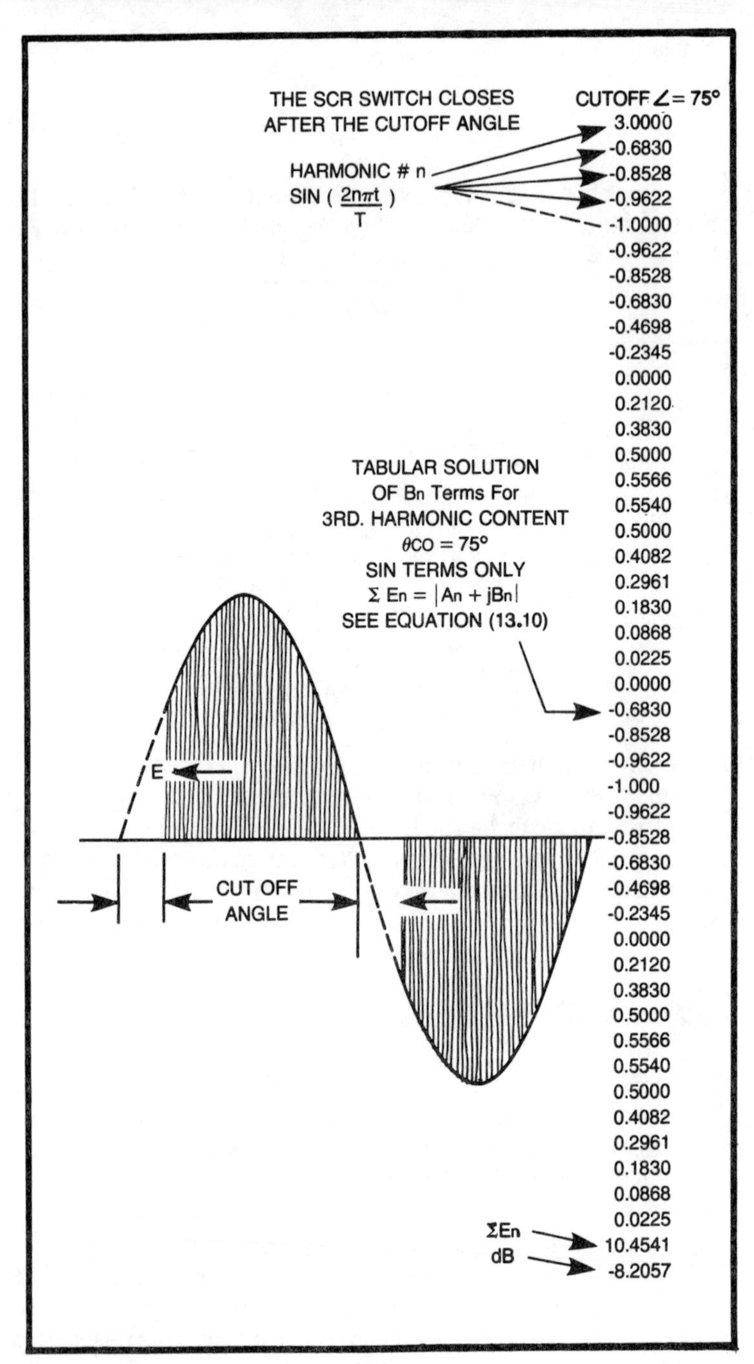

Fig. 13-12. The phase control waveform.

The data tabulated on Table 13-1 show the harmonic content of various cutoff angles. The labels on the first page apply to the entire table. The set was run for steps of 5° in cutoff angle with the value of the voltage taken in the center of the interval. The value labeled Fundamental Voltage is the absolute magnitude of the fundamental component. The value for any angle compared to the value for zero angle gives the relative magnitude of the fundamental component. The value labeled ΣE represents a measure of the average value of the waveform for that cutoff angle. This value compared to the value for zero cutoff gives a measure of the average value of voltage compared to the unmutilated wave.

The table entries are printed in triads showing harmonic number, the absolute magnitude of the harmonic and the number of decibels the harmonic is down from the fundamental.

It should be noted that since the voltage samples were taken *between* the 5° divisions, the values carry a slight error. For example, it will be found that the average value for 90° compared to the average value for 0° works out to 0.5218 and the value for 95° works out to 0.4782, indicating that the half average value point is at 92.5° instead of 90° where it should be. This is an artifact of the program and can be allowed for.

If we examine the tables we find that by the time the wave is about half cutoff (between 90° and 95°) the third harmonic is only about 5.5 dB down from the fundamental. This means that for a 60 Hz system the third harmonic, or 180 Hz component, is more than half as large as the fundamental. By the time that the cutoff angle reaches 155°, the third harmonic is nearly as large as the fundamental and the 11th harmonic (660 Hz) is nearly half as great.

The tables are not normalized so that the summation harmonic voltage may be compared to the fundamental for any degree of mutilation. For instance, in the case of the 155° cutoff, the 11th harmonic voltage is 1.1831 compared to the unmutilated fundamental of 35.9999, or −29.6657 dB. Since motors are often much more lossy on the harmonic (and higher) frequencies the distinct possibility exists that a motor can be burned up by increased core losses at a reduced power setting. Most of the transformer irons used have a characteristic of increasing loss with frequency; this is due in part to eddy current losses due to the thickness of the laminations and to the hysteresis loop of the iron itself. A typical 60 Hz induction or shaded pole motor will use 0.032 inch iron laminations, whereas a good 400 Hz motor will usually have laminations less than 0.005 inch in thickness. It is not unusual to find that the inductance of motors and transformers is a function of frequency above some limit. For this

Table 13-1. Harmonic Content of "Phase Control" Sine Waves With Various Cutoff Angles.

0.0000	5.0000	10.0000	15.0000
35.9999 45.8075	35.9999 45.8075	35.9852 45.6332	35.9282 45.2859
	3.0000 7.631107-12 -253.4742	3.0000 0.1743 -46.2959	3.0000 0.5176 -36.8281
	5.0000 3.654216-12 -259.8701	5.0000 0.1743 -46.2959	5.0000 0.5106 -36.9467
	7.0000 6.562202-12 -254.7850	7.0000 0.1743 -46.2959	7.0000 0.5001 -37.1262
	9.0000 1.865261-12 -265.7112	9.0000 0.1743 -46.2959	9.0000 0.4864 -37.3683
	11.0000 2.787436-12 -262.2219	11.0000 0.1743 -46.2959	11.0000 0.4695 -37.6757
	13.0000 1.781375-12 -266.1109	13.0000 0.1743 -46.2959	13.0000 0.4496 -38.0515
	15.0000 3.546829-13 -280.1292	15.0000 0.1743 -46.2959	15.0000 0.4270 -38.4999
	17.0000 4.180000-12 -258.7025	17.0000 0.1743 -46.2959	17.0000 0.4019 -39.0255
	19.0000 1.698264-12 -266.5259	19.0000 0.1743 -46.2959	19.0000 0.3747 -39.6335
	21.0000 2.047876-12 -264.8999	21.0000 0.1743 -46.2959	21.0000 0.3459 -40.3291
	23.0000 1.751370-12 -266.2584	23.0000 0.1743 -46.2959	23.0000 0.3159 -41.1160

20.0000	25.0000	30.0000	35.0000
35.8049	35.5952	35.2827	34.8550
44.7682	44.0842	43.2390	42.2390
3.0000	3.0000	3.0000	3.0000
1.0195	1.6647	2.4337	3.3030
-30.9107	-26.6006	-23.2256	-20.4668
5.0000	5.0000	5.0000	5.0000
0.9851	1.5644	2.2075	2.8691
-31.2087	-27.1407	-24.0732	-21.6902
7.0000	7.0000	7.0000	7.0000
0.9350	1.4223	1.8990	2.3047
-31.6618	-27.9675	-25.3804	-23.5929
9.0000	9.0000	9.0000	9.0000
0.8710	1.2485	1.5430	1.7039
-32.2775	-29.0998	-27.1834	-26.2163
11.0000	11.0000	11.0000	11.0000
0.7955	1.0555	1.1831	1.1831
-33.0654	-30.5583	-29.4908	-29.3848
13.0000	13.0000	13.0000	13.0000
0.7113	0.8587	0.8717	0.8717
-34.0367	-32.3501	-32.1436	-32.0377
15.0000	15.0000	15.0000	15.0000
0.6221	0.6768	0.6680	0.8068
-35.2010	-34.4176	-34.4554	-32.7096
17.0000	17.0000	17.0000	17.0000
0.5320	0.5320	0.6012	0.8328
-36.5591	-36.5081	-35.3700	-32.4339
19.0000	19.0000	19.0000	19.0000
0.4464	0.4464	0.6165	0.8116
-38.0829	-38.0318	-35.1525	-32.6580
21.0000	21.0000	21.0000	21.0000
0.3718	0.4243	0.6323	0.7211
-39.6715	-38.4740	-34.9324	-33.6842
23.0000	23.0000	23.0000	23.0000
0.3159	0.4397	0.6103	0.6103
-41.0862	-38.1633	-35.2391	-35.1331

con'd on following page

con'd from page 243

40.0000	45.0000	50.0000	55.0000
34.3032 41.0918	33.6219 39.8062	32.8090 38.3920	31.8654 36.8599
3.0000 4.2463 -18.1463	3.0000 5.2349 -16.1542	3.0000 6.2387 -14.4179	3.0000 7.2273 -12.8868
5.0000 3.5029 -19.8181	5.0000 4.0648 -18.3516	5.0000 4.5167 -17.2233	5.0000 4.8298 -16.3877
7.0000 2.5911 -22.4367	7.0000 2.7320 -21.8026	7.0000 2.7320 -21.5901	7.0000 2.6359 -21.6476
9.0000 1.7265 -25.9633	9.0000 1.6731 -26.0617	9.0000 1.6731 -25.8491	9.0000 1.8420 -24.7602
11.0000 1.1655 -29.3761	11.0000 1.2932 -28.2985	11.0000 1.5803 -26.3447	11.0000 1.8777 -24.5935
13.0000 1.0372 -30.3891	13.0000 1.3199 -28.1212	13.0000 1.5329 -26.6090	13.0000 1.5780 -26.1041
15.0000 1.0653 -30.1565	15.0000 1.2328 -28.7145	15.0000 1.2328 -28.5020	15.0000 1.1995 -28.4859
17.0000 1.0000 -30.7066	17.0000 0.9999 -30.5324	17.0000 1.0000 -30.3198	17.0000 1.1737 -28.6749
19.0000 0.8390 -32.2304	19.0000 0.8390 -32.0561	19.0000 1.0128 -30.2088	19.0000 1.1587 -28.7862
21.0000 0.7047 -33.7454	21.0000 0.8501 -31.9422	21.0000 1.0002 -30.3173	21.0000 0.9762 -30.2751
23.0000 0.6897 -33.9330	23.0000 0.8632 -31.8100	23.0000 0.8632 -31.5974	23.0000 0.8954 -31.0259

60.0000	65.0000	70.0000	75.0000
30.7945	29.6024	28.2972	26.8892
35.2216	33.4896	31.6770	29.7976
3.0000	3.0000	3.0000	3.0000
8.1706	9.0399	9.8089	10.4541
-11.5243	-10.3032	-9.2024	-8.2057
5.0000	5.0000	5.0000	5.0000
4.9872	4.9872	4.8454	4.5973
-15.8122	-15.4693	-15.3281	-15.3413
7.0000	7.0000	7.0000	7.0000
2.5320	2.5320	2.7057	3.0238
-21.6998	-21.3569	-20.3890	-18.9804
9.0000	9.0000	9.0000	9.0000
2.1583	2.4949	2.7312	2.8012
-23.0871	-21.4851	-20.3076	-19.6443
11.0000	11.0000	11.0000	11.0000
2.0491	2.0491	1.9417	1.8807
-23.5378	-23.1948	-23.2710	-23.1049
13.0000	13.0000	13.0000	13.0000
1.5098	1.5098	1.6818	1.8937
-26.1906	-25.8477	-24.5193	-23.0451
15.0000	15.0000	15.0000	15.0000
1.3271	1.5412	1.6332	1.5603
-27.3110	-25.6690	-24.7738	-24.7271
17.0000	17.0000	17.0000	17.0000
1.3472	1.3472	1.2833	1.3760
-27.1802	-26.8372	-26.8677	-25.8186
19.0000	19.0000	19.0000	19.0000
1.1305	1.1305	1.2905	1.3683
-28.7039	-28.3610	-26.8190	-25.8673
21.0000	21.0000	21.0000	21.0000
1.0122	1.1770	1.1944	1.1357
-29.6635	-28.0103	-27.4912	-27.4856
23.0000	23.0000	23.0000	23.0000
1.0572	1.0572	1.0412	1.1791
-29.2861	-28.9432	-28.6841	-27.1599

con'd on following page

con'd from page 245

80.0000	85.0000	90.0000	95.0000
25.3900	23.8132	22.1730	20.4852
27.8657	25.8961	23.9037	21.9037
3.0000	3.0000	3.0000	3.0000
10.9560	11.2994	11.4737	11.4737
-7.3001	-6.4752	-5.7224	-5.0347
5.0000	5.0000	5.0000	5.0000
4.2996	4.0279	3.8637	3.8637
-15.4246	-15.4346	-15.1764	-14.4887
7.0000	7.0000	7.0000	7.0000
3.3871	3.6921	3.8637	3.8637
-17.4964	-16.1907	-15.1764	-14.4887
9.0000	9.0000	9.0000	9.0000
2.7064	2.5186	2.3662	2.3662
-19.4451	-19.5129	-19.4354	-18.7477
11.0000	11.0000	11.0000	11.0000
1.9873	2.2032	2.3662	2.3662
-22.1276	-20.6752	-19.4354	-18.7477
13.0000	13.0000	13.0000	13.0000
1.9723	1.8803	1.7434	1.7434
-22.1935	-22.0517	-22.0883	-21.4006
15.0000	15.0000	15.0000	15.0000
1.4933	1.5948	1.7434	1.7434
-24.6103	-23.4818	-22.0883	-21.4006
17.0000	17.0000	17.0000	17.0000
1.5320	1.5320	1.4142	1.4142
-24.3876	-23.8306	-23.9062	-23.2185
19.0000	19.0000	19.0000	19.0000
1.2855	1.2855	1.4142	1.4142
-25.9113	-25.3544	-23.9062	-23.2185
21.0000	21.0000	21.0000	21.0000
1.2446	1.3166	1.2207	1.2207
-26.1922	-25.1471	-25.1838	-24.4961
23.0000	23.0000	23.0000	23.0000
1.1791	1.1167	1.2207	1.2207
-26.6616	-26.5773	-25.1838	-24.4961

100.0000	105.0000	110.0000	115.0000
18.7659 19.9113	17.0321 17.9417	15.3007 16.0099	13.5890 14.1305
3.0000 11.2994 -4.4063	3.0000 10.9560 -3.8322	3.0000 10.4541 -3.3084	3.0000 9.8089 -2.8313
5.0000 4.0279 -13.3657	5.0000 4.2996 -11.9566	5.0000 4.5973 -10.4439	5.0000 4.8454 -8.9570
7.0000 3.6921 -14.1218	7.0000 3.3871 -14.0285	7.0000 3.0238 -14.0830	7.0000 2.7057 -14.0179
9.0000 2.5186 -17.4440	9.0000 2.7064 -15.9772	9.0000 2.8012 -14.7470	9.0000 2.7312 -13.9365
11.0000 2.2032 -18.6063	11.0000 1.9873 -18.6597	11.0000 1.8807 -18.2076	11.0000 1.9417 -16.8999
13.0000 1.8803 -19.9828	13.0000 1.9723 -18.7256	13.0000 1.8937 -18.1477	13.0000 1.6818 -18.1482
15.0000 1.5948 -21.4129	15.0000 1.4933 -21.1423	15.0000 1.5603 -19.8298	15.0000 1.6332 -18.4027
17.0000 1.5320 -21.7617	17.0000 1.5320 -20.9196	17.0000 1.3760 -20.9213	17.0000 1.2833 -20.4966
19.0000 1.2855 -23.2854	19.0000 1.2855 -22.4434	19.0000 1.3683 -20.9700	19.0000 1.2905 -20.4479
21.0000 1.3166 -23.0782	21.0000 1.2446 -22.7243	21.0000 1.1357 -22.5883	21.0000 1.1944 -21.1201
23.0000 1.1167 -24.5083	23.0000 1.1791 -23.1937	23.0000 1.1791 -22.2626	23.0000 1.0412 -22.3130

con'd on following page

con'd from page 247

120.0000	125.0000	130.0000	135.0000
11.9142 12.3179	10.2932 10.5858	8.7421 8.9475	7.2768 7.4154
3.0000 9.0399 -2.3980	3.0000 8.1706 -2.0059	3.0000 7.2273 -1.6528	3.0000 6.2387 -1.3368
5.0000 4.9872 -7.5641	5.0000 4.9872 -6.2938	5.0000 4.8298 -5.1537	5.0000 4.5167 -4.1423
7.0000 2.5320 -13.4517	7.0000 2.5320 -12.1814	7.0000 2.6359 -10.4136	7.0000 2.7320 -8.5090
9.0000 2.4949 -13.5799	9.0000 2.1583 -13.5686	9.0000 1.8420 -13.5262	9.0000 1.6731 -12.7681
11.0000 2.0491 -15.2897	11.0000 2.0491 -14.0193	11.0000 1.8777 -13.3595	11.0000 1.5803 -13.2637
13.0000 1.5098 -17.9425	13.0000 1.5098 -16.6722	13.0000 1.5780 -14.8701	13.0000 1.5329 -13.5280
15.0000 1.5412 -17.7638	15.0000 1.3271 -17.7926	15.0000 1.1995 -17.2519	15.0000 1.2328 -15.4209
17.0000 1.3472 -18.9320	17.0000 1.3472 -17.6617	17.0000 1.1737 -17.4408	17.0000 1.0000 -17.2388
19.0000 1.1305 -20.4558	19.0000 1.1305 -19.1854	19.0000 1.1587 -17.5522	19.0000 1.0128 -17.1278
21.0000 1.1770 -20.1051	21.0000 1.0122 -20.1450	21.0000 0.9762 -19.0411	21.0000 1.0002 -17.2363
23.0000 1.0572 -21.0380	23.0000 1.0572 -19.7676	23.0000 0.8954 -19.7919	23.0000 0.8632 -18.5164

140.0000	145.0000	150.0000	155.0000
5.9120	4.6616	3.5382	2.5532
6.0012	4.7156	3.5685	2.5685
3.0000	3.0000	3.0000	3.0000
5.2349	4.2463	3.3030	2.4337
-1.0565	-0.8104	-0.5973	-0.4164
5.0000	5.0000	5.0000	5.0000
4.0648	3.5029	2.8691	2.2075
-3.2539	-2.4822	-1.8207	-1.2639
7.0000	7.0000	7.0000	7.0000
2.7320	2.5911	2.3047	1.8990
-6.7050	-5.1007	-3.7233	-2.5711
9.0000	9.0000	9.0000	9.0000
1.6731	1.7265	1.7039	1.5430
10.9640	-8.6274	-6.3468	-4.3741
11.0000	11.0000	11.0000	11.0000
1.2932	1.1655	1.1831	1.1831
13.2008	12.0402	-9.5153	-6.6815
13.0000	13.0000	13.0000	13.0000
1.3199	1.0372	0.8717	0.8717
13.0236	13.0532	-12.1682	-9.3343
15.0000	15.0000	15.0000	15.0000
1.2328	1.0653	0.8068	0.6680
-13.6169	-12.8206	-12.8401	-11.6462
17.0000	17.0000	17.0000	17.0000
1.0000	1.0000	0.8328	0.6012
-15.4347	-13.3707	-12.5644	-12.5607
19.0000	19.0000	19.0000	19.0000
0.8390	0.8390	0.8116	0.6165
-16.9585	-14.8945	-12.7885	-12.3432
21.0000	21.0000	21.0000	21.0000
0.8501	0.7047	0.7211	0.6323
-16.8446	-16.4095	-13.8147	-12.1231
23.0000	23.0000	23.0000	23.0000
0.8632	0.6897	0.6103	0.6103
-16.7123	-16.5971	-15.2636	-12.4298

con'd on following page

con'd from page 249

160.0000	165.0000	170.0000	175.0000
1.7167 1.7232	1.0370 1.0392	0.5211 0.5216	0.1743 0.1743
3.0000 1.6647 -0.2667	3.0000 1.0195 -0.1477	3.0000 0.5176 -0.0589	3.0000 0.1743 -0.0000
5.0000 1.5644 -0.8069	5.0000 0.9851 -0.4458	5.0000 0.5106 -0.1775	5.0000 0.1743 -0.0000
7.0000 1.4223 -1.6336	7.0000 0.9350 -0.8989	7.0000 0.5001 -0.3570	7.0000 0.1743 -0.0000
9.0000 1.2485 -2.7659	9.0000 0.8710 -1.5146	9.0000 0.4864 -0.5991	9.0000 0.1743 5.211533-12
11.0000 1.0555 -4.2245	11.0000 0.7955 -2.3025	11.0000 0.4695 -0.9065	11.0000 0.1743 -0.0000
13.0000 0.8587 -6.0162	13.0000 0.7113 -3.2738	13.0000 0.4496 -1.2823	13.0000 0.1743 -0.0000
15.0000 0.6768 -8.0837	15.0000 0.6221 -4.4381	15.0000 0.4270 -1.7307	15.0000 0.1743 -0.0000
17.0000 0.5320 -10.1742	17.0000 0.5320 -5.7962	17.0000 0.4019 -2.2564	17.0000 0.1743 -0.0000
19.0000 0.4464 -11.6980	19.0000 0.4464 -7.3199	19.0000 0.3747 -2.8644	19.0000 0.1743 -0.0000
21.0000 0.4243 -12.1401	21.0000 0.3718 -8.9086	21.0000 0.3459 -3.5599	21.0000 0.1743 -0.0000
23.0000 0.4397 -11.8295	23.0000 0.3159 -10.3233	23.0000 0.3159 -4.3468	23.0000 0.1743 -0.0000

reason, it is not unusual to find that a motor designed for 60 Hz operation will frequently rattle and heat somewhat when operated on a phase control at reduced power.

INVERTER OUTPUT WAVES

With the advent of solid state devices capable of switching and handling sizable amounts of power, a new type of motor control came into popularity. This is the *inverter control*. This type of control generally uses a dc supply with solid state switching devices to turn the dc into ac. We shall see a few inverter type circuits in subsequent sections, but here we are interested in the harmonic content and the analysis of the waveforms. In particular, we are interested in the *quasi-sine wave* inverter, which is a relatively recent development.

Figure 13-13 provides a set of definitions for the parameters of the quasi-sine wave. It may be seen that this is a wave formed by switching in one or more voltage levels for a given period of time. Note that when $E_1 = E_2$ the case resolves itself to a square wave. This is the simplest of inverter waves, the one to be found in most of the common inverters that plug into your car cigarette lighter to provide high voltage ac for operating your shaver etc. in the car.

The reason for operation of the simple inverter in the square-wave mode is a matter of efficiency and the power handling capacity of the transistors (or sometimes SCRs). As demonstrated in Chapter 2, a transistor can switch a great deal more power than it can

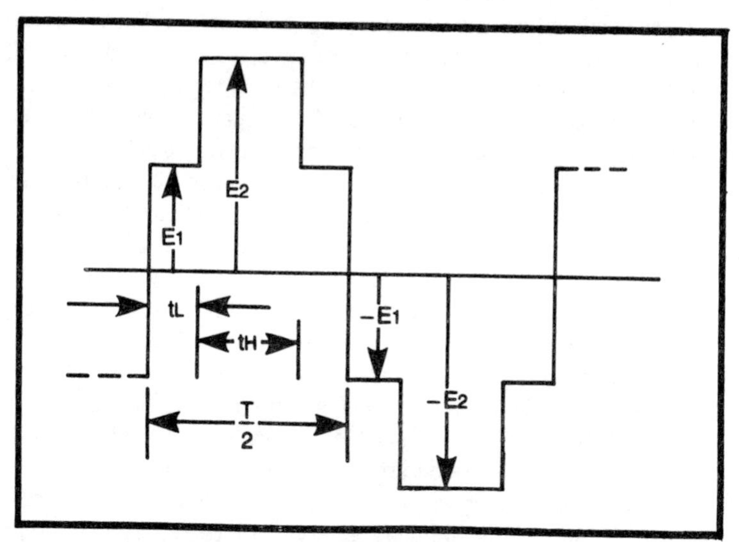

Fig. 13-13. The inverter output quasi-sine wave.

handle in a linear amplifying mode. Typically, a class B linear amplifier is only capable of efficiencies on the order of 50 to 60 percent, and a class C linear amplifier can scarcely exceed 75 to 80 percent, requiring careful tuning to achieve that. A true sine wave power supply with the amplifiers running in class B could scarcely achieve an efficiency in excess of 50 percent. This means that the transistors would be dissipating as much as they are supplying to the load. A transistor that can switch 7,000 watts was shown to be barely capable of dissipating 50 watts. Without any heat derating, therefore, a sine wave supply of 7,000 watts built with the same units would require about 140 of the transistors to handle the same power that could be switched by one.

To develop a couple of points, it is worthwhile to examine the case of the square wave switched supply. First of all we notice that a true square wave with 50 percent duty cycle satisfies both of the symmetry conditions; that is, it is symmetric about both the zero crossing and the midpoint between the zero crossings. As noted earlier, this means that it has neither even-order harmonics nor cosine (or A_n terms from equation 13.2). This is sometimes a little difficult to visualize. Thus the runouts of Table 13-2 are shown for a square wave—showing the individual A and B coefficients for the fundamental and the third harmonic. In looking at the readings in the A column you will find that there is a negative reading to cancel every positive reading, and to the limit of accuracy of the computer, the total of the cosine terms for all harmonics will in fact be zero. A very similar condition applies to the second harmonic. In looking at the table, it may be seen that the first term occupies just exactly one-half cycle of the sine function. The second harmonic would have had a full cycle, and all of the second half of the chart would have been numerically equal to and of opposite sign to the terms in the first half, so the even harmonics are also zero. Similarly, you will note that only the first half of the wave was considered. Since in this case the wave is possessed of the types of symmetry discussed earlier and is symmetrical about the zero axis as well, we may perform our analysis on only half of the wave. For the second half, both the sine expression and the voltage would have negative algebraic signs, so the column for the second half will be exactly like the column for the first half. It may be seen that these symmetries permit a reduction by a factor of four in the amount of calculation required.

Anyone who has tried to run an ac operated tape recorder from a square wave inverter will understand the reasons for the interest developed recently in the quasi-sine wave inverter. Among other things the induction motors, particularly shaded-pole types, do not

Table 13-2. Numerical Solutions For Equations 13.8 and 13.10 For a Square Wave With E = ± and T = 48.

	B A FUNDAMENTAL		B A THIRD HARMONIC	
#	B	A	B	A
1.	0.0654	0.9978	0.1950	0.9807
2.	0.1950	0.9807	0.5555	0.8314
3.	0.3214	0.9469	0.8314	0.5555
4.	0.4422	0.8968	0.9807	0.1950
5.	0.5555	0.8314	0.9807	-0.1950
6.	0.6593	0.7518	0.8314	-0.5555
7.	0.7518	0.6593	0.5555	-0.8314
8.	0.8314	0.5555	0.1950	-0.9807
9.	0.8968	0.4422	-0.1950	-0.9807
10.	0.9469	0.3214	-0.5555	-0.8314
11.	0.9807	0.1950	-0.8314	-0.5555
12.	0.9978	0.0654	-0.9807	-0.1950
13.	0.9978	-0.0654	-0.9807	0.1950
14.	0.9807	-0.1950	-0.8314	0.5555
15.	0.9469	-0.3214	-0.5555	0.8314
16.	0.8968	-0.4422	-0.1950	0.9807
17.	0.8314	-0.5555	0.1950	0.9807
18.	0.7518	-0.6593	0.5555	0.8314
19.	0.6593	-0.7518	0.8314	0.5555
20.	0.5555	-0.8314	0.9807	0.1950
21.	0.4422	-0.8968	0.9807	-0.1950
22.	0.3214	-0.9469	0.8314	-0.5555
23.	0.1950	-0.9807	0.5555	-0.8314
24.	0.0654	-0.9978	0.1950	-0.9807
	1.0000	1.0000	3.0000	3.0000
	15.2897	-1.000000-13	5.1258	3.000000-13

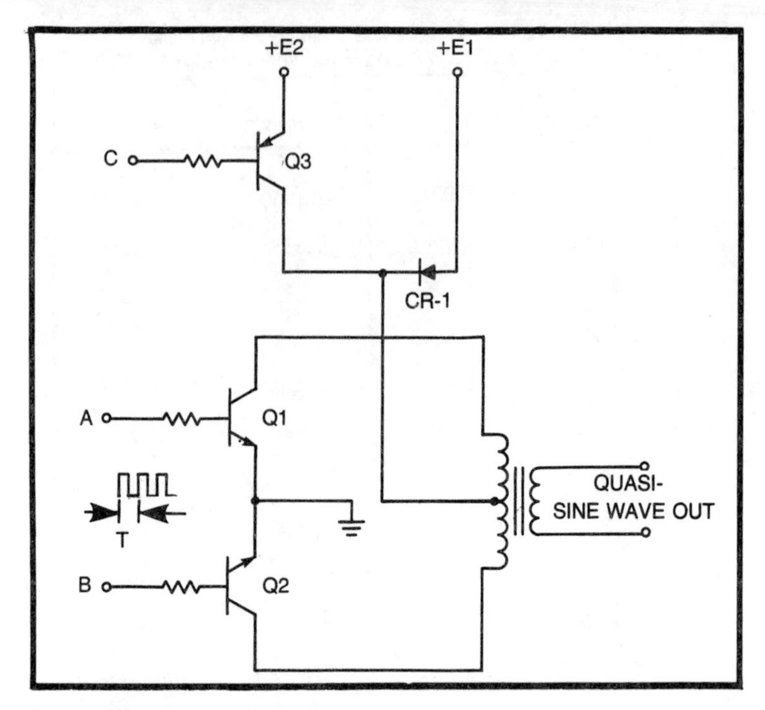

Fig. 13-14. The quasi-sine wave inverter.

take particularly kindly to square waves, and they may or may not run. And if they do run they have a tendency to heat even more than usual.

Figure 13-14 shows one type of simple mechanism for obtaining a waveform of the type shown in Fig. 13-13. A square wave applied between the bases A and B of the inverter switches the current through opposite halves of the transformer. When the base of $Q3$ is high, the unit draws its current from E_1, and when the base of $Q3$ goes low, the current is drawn from E_2. Diode CR-1 acts as a steering diode to keep current from flowing backward into E_1. There are other techniques which can be used to give the waveform with only a single supply voltage, but for rectified ac applications this one has the advantage of simplicity.

It seems likely that there ought to be some combinations of voltages and relative times that will give a more or less optimum distribution of the harmonics. The data of Table 13-3 illustrate some results. The left hand column is the familiar square wave. The third harmonic term compares reasonably well with the analytic value of −9.5424 dB with an error of 0.0497 dB. The central column was chosen to have the same rms value as a sine wave of the same peak voltage. It shows a rather remarkable suppression of the lower harmonics and the harmonics which are multiples of three. This would clearly be a favorable choice since the harmonics are spread out mostly in the higher frequencies where they are easier to filter out and absorb.

Perhaps the most remarkable part of the table is the result in the right-hand column which indicates that the same properties can be obtained with $E_1 = 0$; in other words, the harmonic structure of the simple inverter can be reduced by a judicious choice of duty factor. This result is the more unusual in the fact that the harmonic structure is identical to the $E_1 = 0.5$ volt wave. The identity holds

Table 13-3. A Quasi-Sine Wave Inverter Output With T = 24 and $E_2 = 1$ V For a 24-Sample Summation.

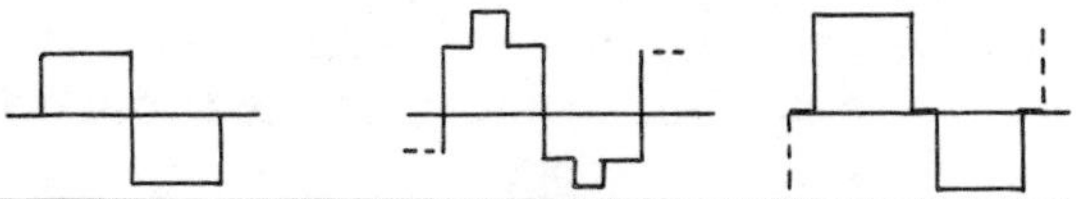

1.0000	0.5774	0.5000	0.4750	0.0000
0.	6.	8.	9.	4.
....	....	....	....	....
1.	1.	1.	1.	1.
15.2897	13.3972	11.4673	10.3345	13.2413
3.	3.	3.	3.	3.
5.1258	1.4279	-2.750000-12	-0.0514	-3.000000-13
-9.4927	-19.4461	-252.4026	-46.0586	-272.8962
5.	5.	5.	5.	5.
3.11110	0.8666	2.3332	2.9866	-2.6942
-13.8300	-23.7834	-13.8300	-10.7820	-13.8300
7.	7.	7.	7.	7.
2.2609	1.9811	1.6957	0.6197	-1.9580
-16.6021	-16.6021	-16.6021	-24.4420	-16.6021
9.	9.	9.	9.	9.
1.7999	1.5771	0.0000	0.4933	0.0000
-18.5828	-18.5828	-180.0000	-26.4226	-180.0000
11.	11.	11.	11.	11.
1.5166	0.4225	1.1374	1.4560	1.3134
-20.0702	-30.0236	-20.0702	-17.0222	-20.0702
13.	13.	13.	13.	13.
1.3300	0.3705	0.9975	-0.0133	1.518
-21.2105	-31.1639	-21.2105	-57.7764	-21.2105
15.	15.	15.	15.	15.
1.2026	1.0538	-4.000000-13	0.8129	0.0000
-22.0849	-22.0849	-269.1480	-22.0849	-180.0000
17.	17.	17.	17.	17.
1.1149	0.9769	0.8362	0.7536	-0.9656
-22.7426	-22.7426	-22.7426	-22.7426	-22.7426

Table 13-4. The Fundamental Plus the Second Harmonic.

Table 13-1.		Table 13-2.		Table 13-3.	
CUTOFF ANGLE —	15.0000	t —	1.	E_1 —	1.0000
FUNDAMENTAL VOLTAGE —	35.9282	$\frac{\sin 2\pi nt}{T}$	0.0654	t_L —	0.
$\Sigma\lvert E\rvert$ —	45.2859	≈			....
HARMONIC NO. —	3.0000		24.	n; HARMONIC —	1.
VOLTAGE $\lvert E_n\rvert$ —	0.5176	Harmonic	0.0654	$\frac{2\pi nt}{T}$ —	15.2897
dB —	-36.8281	n —	1.0000	n —	3.
		$\sum_{0}^{T/2} E$	15.2897	$\sum_{0}^{T} \sin$	5.1258
				dB	-9.4927

with a finer grained analysis out to the 17th harmonic. The advantages of using a simpler supply mechanism are, of course, obvious.

The projection of this form of analysis in too much detail to the higher harmonics is not very worthwhile unless a very fine grained and time consuming analysis is employed since the actual detail of the higher harmonic structure is determined by matters such as the rise time of the transistors and the ringing or resonances and absorption in the load. These items are not very amenable to prediction with any simple model and require a detailed measurement of the actual circuit. They can be altered by such factors as board layout and lead dress, etc.

THE INVERSE FOURIER TRANSFORM

Once the analytic expression for the wave is known and can be written as in equation 13.1, the currents for the fundamental and each of the harmonics can be calculated using ordinary ac algebra. The current summation can then be written. It may be seen that this is a relatively lengthy process suited to computer synthesis, more especially so since not only the amplitude of the harmonic currents but the phase angle must be accounted for. It should be noted that in real-world iron motors and transformers, the inductance varies with frequency and with excitation. For this reason the fundamental current can cause a change in the harmonic currents and can, in fact, modulate them at the fundamental rate. Except in the largest and most expensive iron equipment, this stage of the analysis is seldom carried out except for the first few harmonics. When only air-core transformers are involved this process is actually quite accurate and is, in fact, identically the technique used for predicting the patterns of array antennas. The latter process is regularly performed with great precision, and a large scale antenna is never constructed without a carefully predicted set of radiation patterns.

14
DC Drive Circuits

The actual drive circuits for dc motors can take the form of simple on-off switches, speed controls or position controls, in order of sophistication. We have already treated the various mechanisms for feedback sensing and shall be concerned in this chapter with the actual drive devices, that is, the last items before the motor proper. Since the simple on-off control actually represents only a subclass of the speed control it shall not be treated separately.

SPEED CONTROLS

In a great many devices it is necessary only to control the speed of the motor. For example, many forms of tape recorder, phonographs, and mixers, do not require any knowledge of the actual position of the load at any time but rather only require a control of the speed. As noted in Chapter 4 this can be accomplished by controlling the input voltage to a PM motor, or controlling the field or armature voltage to a shunt wound motor. Occasionally, the requirement exists in a speed control to provide dynamic braking to bring the motor to a rapid halt. This is usually accomplished by short-circuiting the motor.

As with all inductive loads one of the major problems lies in transient protection. This can take different forms:

- The transistor must be protected against overvoltage and breakdown as discussed in Chapter 2.
- The unit should not generate transients that will cause radio frequency interference (RFI), which disrupts operation of

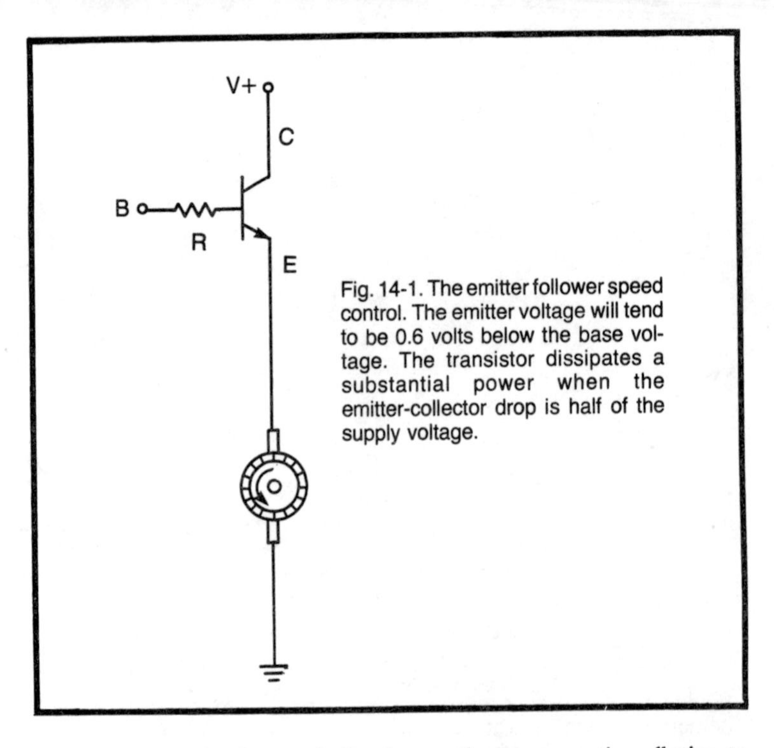

Fig. 14-1. The emitter follower speed control. The emitter voltage will tend to be 0.6 volts below the base voltage. The transistor dissipates a substantial power when the emitter-collector drop is half of the supply voltage.

other circuitry and also forms electromagnetic pollution to the environment.

Several techniques for dealing with the latter will be discussed.

A Simple Speed Control

The very simplest form of speed control for PM and shunt motors is shown in Fig. 14-1. In this case, the transistor is employed simply as a variable resistor to control the input voltage. It has the advantage over an old fashioned resistor control in the fact that the effective resistance of this "resistor" is a dynamically varying function, which employs the property of the emitter follower to keep the emitter voltage close to the base voltage without too much sensitivity to the emitter current. In an ordinary silicon transistor, however, the finite forward-current gain tends to limit the effectiveness of the compensation.

For example, let us consider that the transistor is a General Electric D42C type NPN silicon device and the motor is the little 2 watt unit discussed in Chapter 3 (see equations 3.11 and 3.12). The unit has a locked rotor resistance of 3.7 ohms. If V+ is 12 volts, the transistor is held at a base voltage of +6.8 volts, and the motor is operated with no external torque load we will find the motor current to be about 0.095 amperes and the emitter voltage to be 6.0 volts. Now, if we interrupt the operation and then let the motor try to start a heavy torque load so that, at least instantaneously, the motor CEMF is zero, we will find that the motor current rises to something like 1.5 amperes and the emitter voltage sags to about 5.7 volts due to the increased current. In the lightly loaded case, the transistor represented an equivalent resistance with a voltage drop of 6 volts for a current of 0.095 amperes, or 63.16 ohms. In the locked-rotor case, the transistor showed a 6.3 volt drop with a current of 1.5 amperes, or an equivalent resistance of 4.2 ohms.

Consider what would happen if we try the same regulation with a series resistor. In order to regulate the no-load speed we simply insert a 63.16 ohm resistor in series with the motor. Then in the locked-rotor, or starting case, the maximum rotor current would be 12 V/63.16 ohms, or 0.189 amperes. The motor would have only 0.189/1.5 = 0.127 times as much starting torque. In essence, the transistor control will give eight times the starting torque that a fixed resistor controlling the no-load speed would allow. If the supply voltage had been higher, the advantage would be even greater.

A somewhat improved version of this simple control is shown in Fig. 14-2. Here the immense gain of the op amp is employed to hold the emitter voltage essentially constant. With this arrangement it is

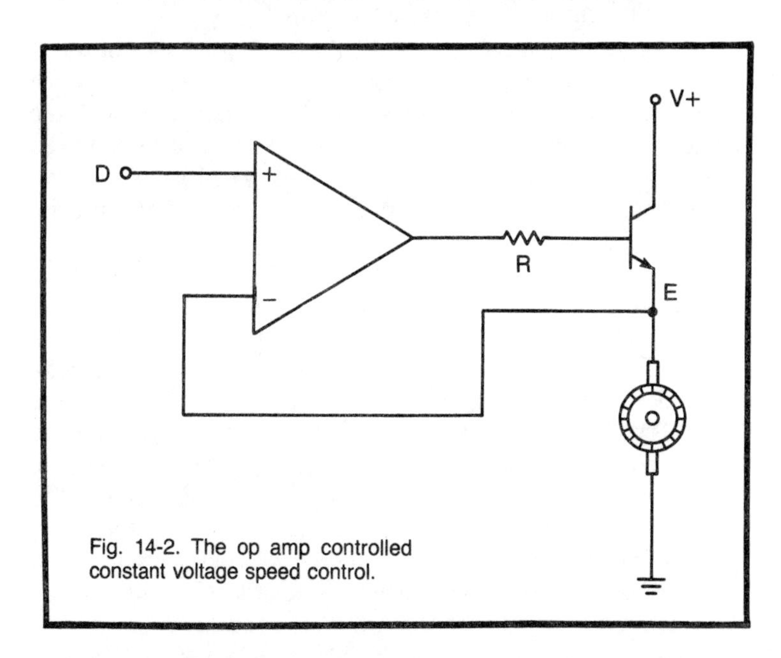

Fig. 14-2. The op amp controlled constant voltage speed control.

not unusual to find that the emitter voltage can be held constant to within a few millivolts from zero load to the locked-rotor condition. This unit also removes nearly all of the variation due to the base-emitter temperature of the driving transistor. In essence, the speed regulation of the system is controlled only by the resistance of the motor itself. With the motor described in Chapter 3, as the torque load was increased from zero to 6 in.-oz the speed would sag from 60 rpm to about 57.8 rpm, or about 4 percent. This is about as good as you are going to get with an open-loop system. If the unit were equipped with one of the various closed-loop feedback sensors described earlier, even better performance could be obtained.

One of the relatively obvious drawbacks to this system is the fact that the transistor control is operating in a linear mode. Under the no-load condition the transistor will be dissipating 6 V × 0.095 amp = 0.57 watt. With the increase in torque load to 6 in.-oz, the motor current rises to 0.15 amperes, and the transistor dissipation rises to 0.9 watt. In the locked-rotor condition, the dissipation rises to 6 V × amp = 9 watts. This is about the maximum rating on this power-tab transistor with the tab heat-sunk at 50°C. The technique generally ceases to be applicable to motors of any significant size due to the size of the transistors required or, alternatively, due to the number of transistors required.

It should be noted that the locked-rotor condition also tends to exceed the capabilities of the motor as well.

The motor can be protected in any of a number of ways. Perhaps the cheapest is the supply of a simple thermostat switch which simply removes the voltage from the system when either the motor or the transistor temperature exceeds some safe limit. The second alternative is to equip the system with a thermistor or silicon diode bridge to sense the temperature of the motor and/or transistor. (See Fig. 10-6 and text.)

The thermal sensing is perhaps the most effective and direct way of protecting these units since it responds directly to the threat to the system. Both the motor and the transistor tend to burn up as a result of excess heat. When the ambient is low, the units will stand a great deal more of this type of overload than when the ambient is high. The thermal overload system will permit the motor to try as long as possible without damage and then will shut down the system for as long as it takes to get back to a safe try-again position. The thermal sensor should be equipped with a substantial amount of "toggle," or hysteresis, so that the system will not try again until the temperature has fallen well below the cutout level.

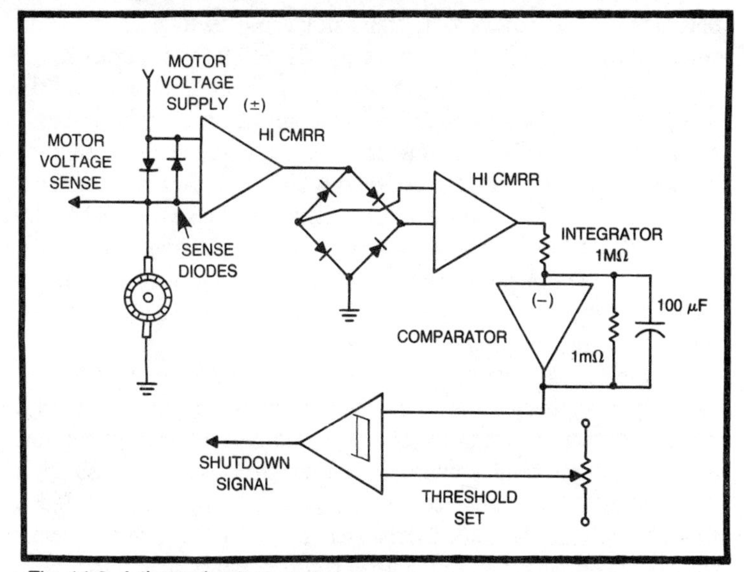

Fig. 14-3. A thermal cutout control.

An alternative but perhaps less effective system is shown in Fig. 14-3. The unit is arranged to accept both polarities of excitation as in a positioning system or even an ac system. The first hi CMRR op amp is used to sense the forward drop across the sense diodes. This is a function of both the current in the system and the junction temperature of the diodes. The sensed signal is then applied to the bridge rectifier to make the signal polarity constant regardless of the input polarity. The second hi CMRR op amp drives a comparator. The threshold on the comparator can be set so that the circuit will trip only when the diodes are both hot and handling a heavy current. The integrator serves to prevent premature tripping and to hold the circuit off for a period of time once the trip has taken place.

With circuits of this nature, it should be remembered that the diodes will generally cool much faster than the motor, so some form of integrating holdoff is generally required. This can take the form of a long one-shot delay or the form of an integrator discharge delay. With an adequate hysteresis, the integrator values will hold the circuit off for something on the order of a minute.

DUTY FACTOR CONTROL

The circuit of Fig. 14-1 can also be made to operate in a duty factor modulated mode in which the transistor functions only as a switch. This has a considerable advantage in terms of transistor dissipation. The same D42C transistor can effectively switch currents up to 5 amperes. Assuming a little derating to 4 amperes, it would have a saturated forward drop of about 1.25 volts so the dissipation in the saturated condition is only about 5 watts. Depending upon the unit designation within the family, the unit can stand-off between 30 and 80 volts (D42C1 and D42C10), so a transistor of the same physical size can handle something between 30 V × 4 amp = 120 watts, and 80 × 4 = 320 watts in the switching mode. Of course, the voltage rating should be derated also to account for transient handling. However, the fact of the very large advantage remains for the switching regulator.

Figure 14-4 illustrates a form of switching regulator in which the switching frequency is held constant and the pulse width is varied in order to obtain control of the average voltage applied to the motor. In this circuit an astable multivibrator constructed from a 555 timer establishes a constant frequency. The voltage across timing capacitor C2 is a stable sawtooth. This is coupled through a capacitor and combined with a speed control reference bias at B. Voltage A represents an integrated average of voltage sample D which goes progressively more negative as the average positive value of D rises. Whenever the instantaneous value of B goes more positive than the value of A, the comparator output goes low and the control transistor shuts off. As A goes lower and lower, the control transistor is on for progressively shorter periods as shown in the waveforms at the bottom of the illustration.

The system also possesses a form of self-regulation based upon the motor CEMF. As noted in the waveforms, when the control transistor goes off, the CEMF of the motor remains and discharges through CR-1, which also serves as a free-wheeling diode to protect the control transistor from the inductive kick of the motor. The exaggerated waveforms at the bottom of the page show that a heavy load or a low CEMF will tend to decrease the value of A and thereby summon a longer duty factor. This will, in turn, cause the motor to speed up. Unfortunately, the effectiveness of this feedback mechanism tends to decrease as the duty factor approaches one, since the length of time in which the CEMF is sampled becomes progressively less important. This can be improved by using a sample-and-hold circuit to measure voltage D only immediately after the cutoff of the control transistor.

Figure 14-5 illustrates such a gated CEMF measuring scheme. Transistors Q1 and Q2 turn on simultaneously and Q3 is cut off. When Q1 and Q2 turn off, the base of Q3 instantaneously goes to ground and Q3 saturates for about 80 microseconds thereby charging C1 to the CEMF level. Transistor Q3 acts as a switch connecting the integrator to the CEMF. It then cuts off and will not close again until the next falling edge of the switch signal for Q1. Diode CR-2 prevents the base of Q3 from rising above its own emitter. The remaining portion of the circuit can be configured as in Fig. 14-4.

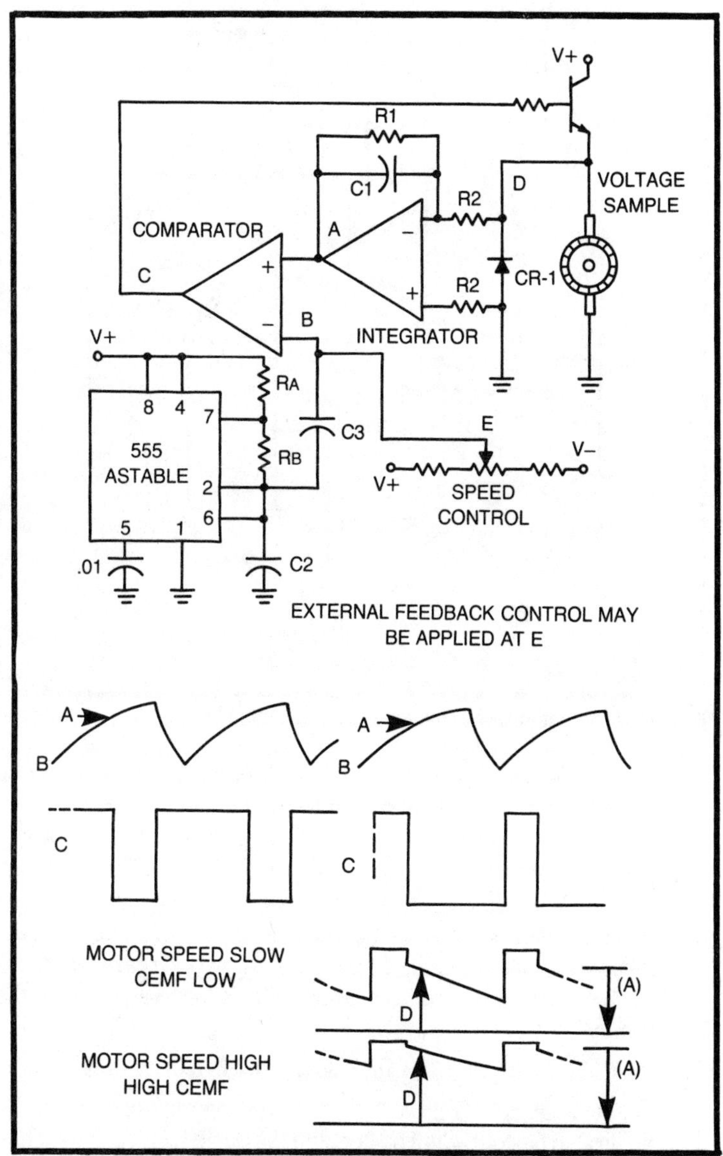

Fig. 14-4. The constant-frequency switched speed regulator.

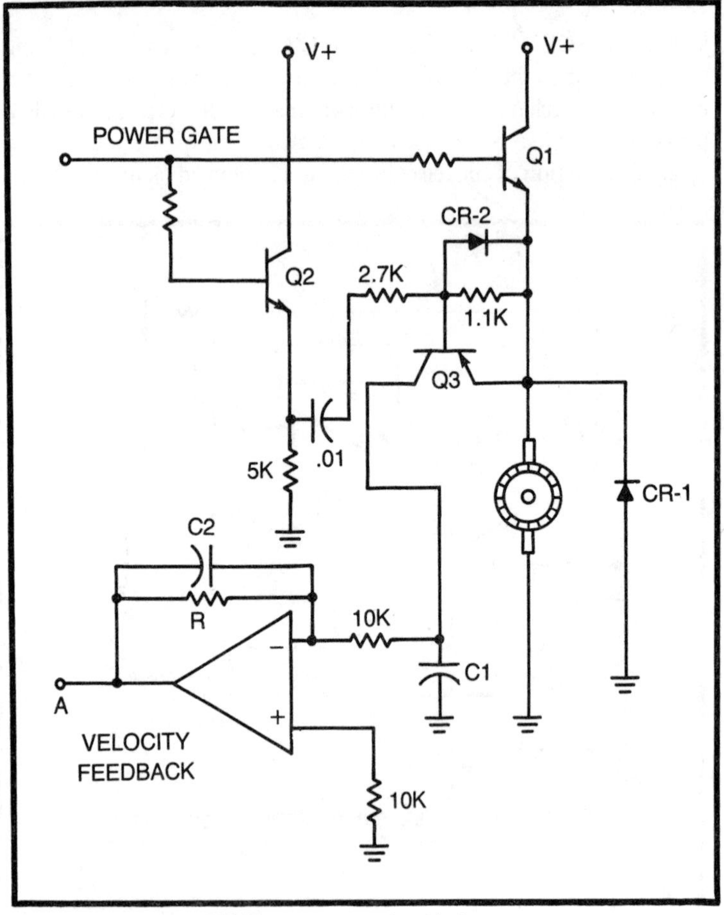

Fig. 14-5. The gated CEMF measurement control.

The ability to sense the CEMF directly during nonexcited portions of the drive cycle is one of the advantages of the switched speed regulator. This unit will always apply full power to the motor at the instant of turn-on and bring the motor up to speed nearly as fast as possible. For this circuit, an op amp with a relatively high slew rate is required in order to obtain a reasonable rectified output from the relatively short sample. The circuit will work reasonably well with a 741 provided that too much gain is not required in the integrator.

THE VARIABLE FREQUENCY CONSTANT PULSE WIDTH CONTROL

It should be relatively obvious that the average current to the motor can also be controlled by keeping the pulse width constant and varying the pulse rate. Figure 14-6 illustrates one such configuration. The 556 is configured as a VCO for the low-numbered pin section. The output of this section at pin number 5 is a train of negative going narrow pulses whose repetition rate is determined by R1, C1, and the voltage on pin 3, with the frequency increasing with decreasing control voltage. The output of pin 9 is a train of pulses with constant width determined by R3 and C2. With the values shown the VCO can be modulated between about 10 kHz and 1 kHz with the input modulation signal. The output pulse width must naturally stay somewhat smaller than the repetition frequency of the monostable.

The 555 and 556 devices are exceptionally useful since they can operate over a relatively wide range of supply voltages, and the output can source or sink as much as 150 milliamperes, which means they can directly drive a fairly good sized control transistor.

The two techniques have about the same characteristics for the range of speed control that can be obtained. Most dc motors will not run over a much wider speed range than about 10 : 1 and do any useful external work. This range of excitation is available from either circuit. The constant frequency-variable pulse width circuit has the advantage that the harmonics tend to be stationary in frequency and to fluctuate in amplitude with control variation, whereas the variable frequency circuit tends to vary both the frequency and the amplitude of the harmonics. There is an argument which goes to the effect that

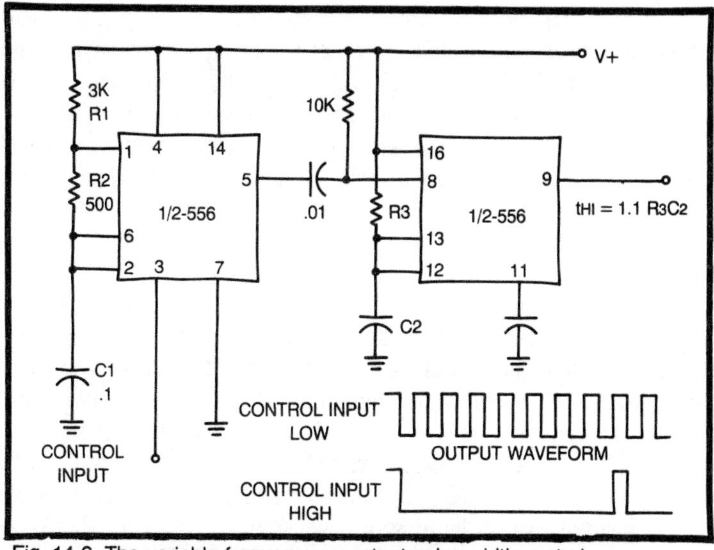

Fig. 14-6. The variable frequency, constant pulse width control.

the stationary frequency distribution of the harmonics makes the job of filtering and "sanitizing" the output somewhat easier since the harmonics at least stay put during the measurement. The advantage to the constant frequency-variable pulse width system is relatively slight, but worthy of note if a new design is contemplated.

The various circuits have been shown with packaged ICs since the continuing fall of the prices of these items has tended to make them more economical to use than discrete designs. Any of the functions shown can be realized entirely with discrete components at the expense of some addition in parts count. The sample-and-hold circuit is shown with discrete components because most of the ICs will not operate in the voltage range usually employed in motors; however, that is true only at the time of this writing. By the time you read the discussion, sample-and-hold ICs with a 400 volt input range and power ICs with a 10 ampere output rating and built-in thermal overload protection may well be available.

THE POSITION CONTROL

In many applications it is necessary to control not the speed of the motor but rather the position of the load, which usually requires a drive capable of reversing the direction of rotation of the motor. This is frequently done with a bridge-type circuit for dc motors. Fig. 14-7 shows a circuit of this type with some unusual drive features. Among others, it incorporates a 7446 BCD-to-7-segment converter!

The control bridge consists of transistors Q1 through Q4, with the motor across the balanced arm of the bridge. Suppose that the bases of Q1 and Q4 are taken low while Q2 and Q3 are high. In this case A is connected to V+ and B is connected to ground, and the motor will run, for example, clockwise. If the situation is reversed and the bases of Q2 and Q3 are taken low while Q1 and Q4 are high, then B is connected to *V*+ and A is connected to ground. Note that it is important that Q1 and Q2 **never** be taken low simultaneously since this shorts the V+ to ground! The 3 ohm resistor serves an important function in limiting the current upon instantaneous reversal. If the motor is running full speed in one direction and instantaneously reverses, the CEMF is in the direction to add to the supply voltage and the motor will instantaneously draw nearly twice the normal locked-rotor current. This is not only a bit hard on the brushes but can also demagnetize the stator on a PM motor. With the weaker field, the unit will run faster and faster (see Chapter 3). Some form of current limiting or reverse delay, or both, should be incorporated as a matter of good engineering practice. The resistor

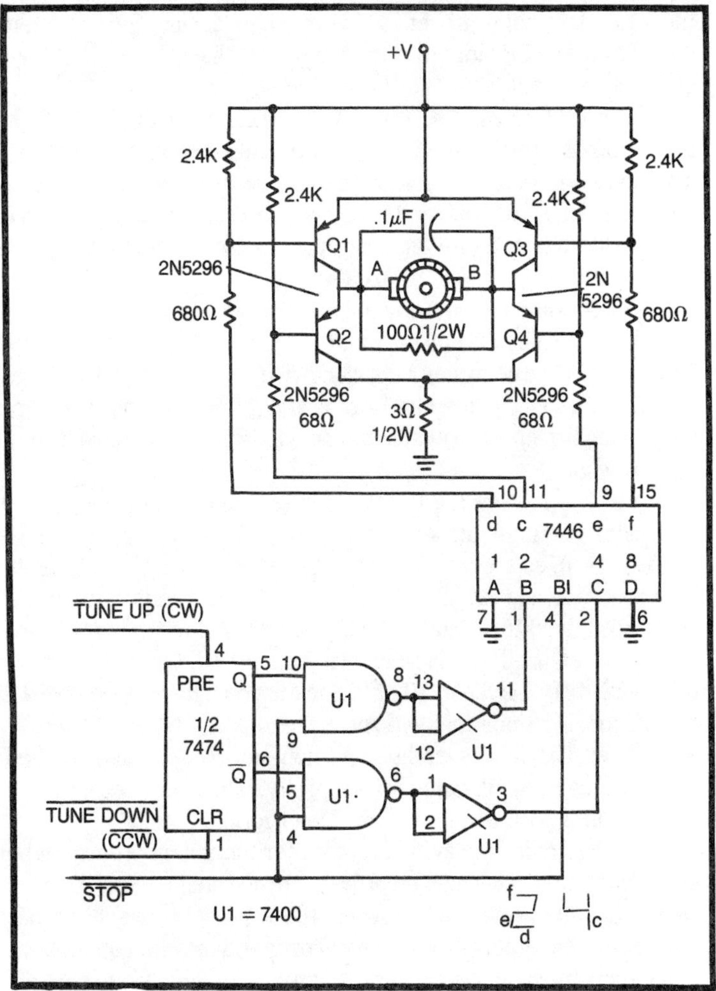

Fig. 14-7. Servo bridge drive.

and condenser across motor terminals A and B serve to limit the commutating transients and the shutoff transients and to reduce RFI due to brush noise.

A set of four D42C1 transistors with V+ = 12 V will drive the little Siemens motor earlier discussed in very snappy fashion for a limited-duty-cycle positioning servo. The actual voltage drop across the motor will be about 10 volts, which exceeds the nominal rating of the motor somewhat, but neither excessive heating nor demagnetization will be experienced if the duty cycle does not exceed about 65 percent. Since a good positioning servo should not be running all the time, this does not represent a severe under-design. Servo systems very often employ limited duty ratings on the motors in order to obtain an adequate "zip" in the positioning performance.

In order to obtain the requisite base drive for the bridge using the TTL logic, the 7446 BCD-to-7-segment converter is employed. This unit was designed for driving lamps and has open-collector outputs rated at 30 volts and is equipped to sink 20 milliamperes on each of the segment outputs. Thus it is capable of directly driving the transistors for V+ levels up to 30 volts.

When connected as shown, with the pin 1 input high and pin 4 input low, the unit will pull segments a, b, d, and e low. This excites Q1 and Q4, and the motor runs clockwise. With pin 4 high and pin 1 low, the unit pulls segments b, c, f, and g low. The blanking input must be high in order to enable the device, since a low on BI holds all outputs high.

A low trigger on the CW line presets the flip-flop and the unit runs clockwise until further notice. A negative or low trigger on the $\overline{\text{CCW}}$ line causes counterclockwise operation. Taking the STOP line low causes the unit to halt. This unit functions very successfully in an automated radio transmitter tuning system. If the triggered operation is not required, the flip-flop may be omitted, but it is necessary to ensure that pins 5 and 10 of U1 never go high (or open on TTL) simultaneously since this will short the supply.

Note: The limit switches and non-jam diodes have not been incorporated in the drawing for simplicity. These are nearly always required in a positioning system. Don't forget them!

A speed control may be required for the unit in order to attain stable operation. This may be provided by including one of the speed control circuits previously shown in the V+ line. However, a more elegant means of duty-factor control of speed may be obtained with this circuit by simply applying the duty factor modulation ANDed onto the $\overline{\text{STOP}}$ line. This technique is actually employed on the radio tuner mentioned.

BACKUP CONTROL

A technique that helps greatly in obtaining both a reasonably zippy performance and good stability is to reduce the duty factor on the successive reversals. For example the motor can be triggered into full speed operation initially. After the first deadband crossing the duty factor is cut in half etc. This has a tendency to discourage overshoot in the system bounceback.

RFI CONTROL

Most brush type motors are notorious generators of RFI because of the extremely fast rise times generated on the mechanical commutation. Fast rise-time repetitive switching is also a common

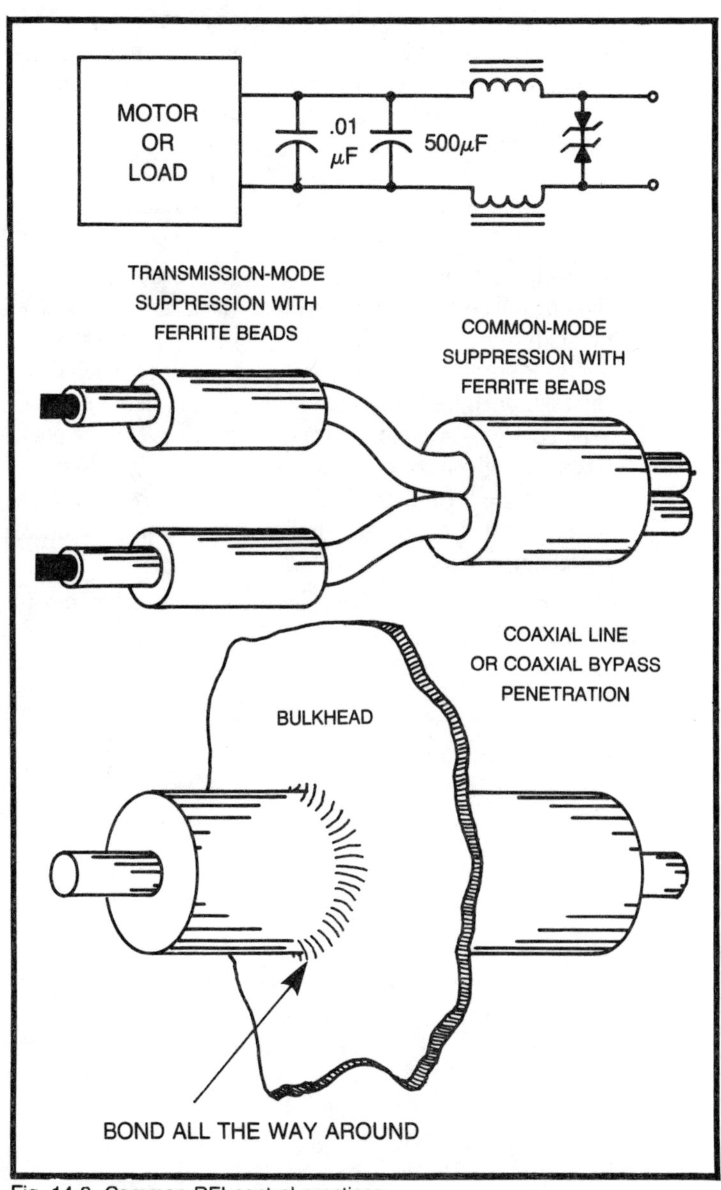

Fig. 14-8. Common RFI control practices.

source of RFI. The detailed treatment of RFI control is beyond the scope of this text, but a few basic rules are well worth reiteration.

1. Keep ground leads as short and as heavy gauge as possible. Try to avoid using the power return ground as part of the signal path.
2. Bypass all lines with both large and small capacitors. A 500 μF electrolytic may be essentially an open circuit at radio frequencies. If fast rise times are involved, always include something like a 0.001 μF in the bypass scheme. Bypass as close to the noise source as possible.
3. When possible, include very high permeability ferrite beads around supply leads. These are generally good RFI absorbers.
4. If economically feasible, enclose the entire unit up in an RFI tight box. This means that all joints which cannot be welded or soldered shut should have "fingers" or mesh at the joints. Riveted and bolted joints with spacings much larger than one inch between fasteners will leak like a sieve.
5. When passing through the walls of an RFI enclosure put all leads through on coaxial bypass capacitors. When passing a coaxial cable through an RFI bulkhead, be sure to bond the outer conductor to the bulkhead around the entire periphery.

Figure 14-8 shows a few of the more common techniques for RFI suppression.

15
AC Drive Circuits

The ac drive circuits may be subdivided into roughly three main groups: *phase control, duty factor zero crossing*, and *inverter* controls. There are also a large number of tap-changing and *cycloconverter* types which are rather specialized and usually confined to items like traction motors in electric locomotives and large rolling mill motors. We will deal with the latter only briefly.

ZERO CROSSING CONTROLS

The zero crossing type controls are perhaps the easiest to deal with since they are usually the simplest. As noted earlier, the SCR was one of the first solid-state devices to be incorporated into motor controls because of the availability in the early 1960s of devices capable of switching large voltages and currents at a time when transistors were still pretty expensive and feeble. The SCR is also at a marked advantage in ac applications in that it always tends naturally to interrupt the circuit when the current is zero. This tends to make for very smooth circuit breaks with minimum transient stress and RFI. In ordinary on-off switching service, the circuit shown in Fig. 2-4 also turns on at the zero crossing of the voltage since the transistor does not gate on until the anode voltage is slightly positive. These switching characteristics make for extremely quiet and smooth operation of the switch. It can be shown that the harmonics of a wave train with the sine wave power turned on and off only at the zero crossing fall beneath a sin X/X curve spaced above and below the fundamental sine wave frequency, with the harmonics spaced in

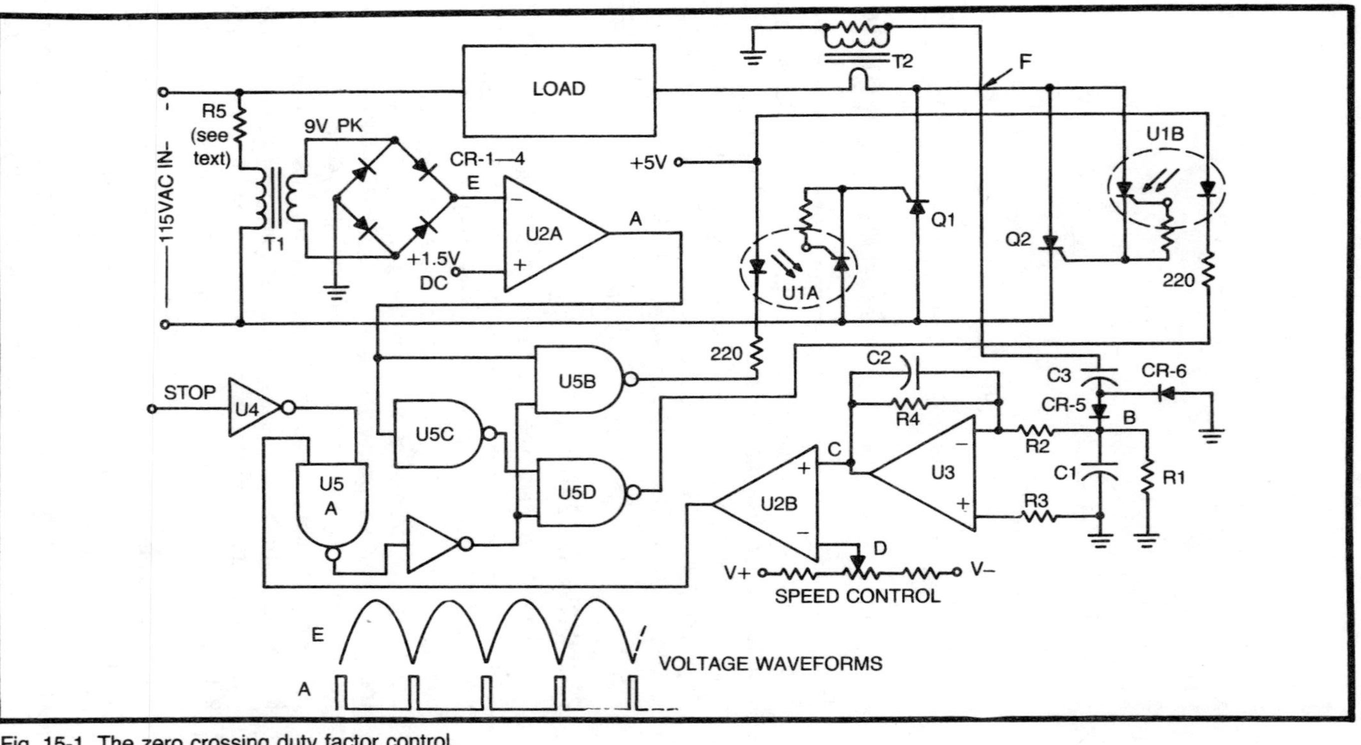

Fig. 15-1. The zero crossing duty factor control.

frequency by the repetition frequency of the pulse train. These harmonics have attenuated to negligible values just a relatively short distance from the fundamental sine wave frequency.

Figure 15-1 illustrates a *zero crossing duty factor control*. This unit employs a pair of large SCRs back-to-back to handle the main load current. These units are driven by photo SCR isolator U1A and U1B. These units provide a very large isolation from the line, typically 1500 volts or more, and permit the unit to be driven directly from TTL logic levels. If the load current does not exceed 300 milliamperes, the large SCRs, Q1 and Q2 may be eliminated. A typical photo isolator unit is the General Electric H211C2.

The assembly consisting of the transformer T1 and bridge rectifier CR-1 through CR-4 generates a full-wave rectified sine wave, clamped to ground by one-diode forward drop. Comparator U2A goes high only during the period when this wave is less positive than the 1.5 Vdc reference. Allowing for the forward drop of the diodes in the bridge, this is the period within ±5.7° of the zero crossing, or 266 μsec each side of the zero crossing.

Note: This depends upon the current in the primary of T1 being in phase with the line voltage. If the sense signal is taken from a second winding from a loaded power transformer, this will probably be the case. Otherwise, resistor R5, which is much larger than the inductive reactance of T1, should be included. In the power transformer case the slight lag due to the leakage inductance will serve to shift the trigger slightly in the right direction.

The output of this comparator will be a series of short spikes at a 120 Hz repetition rate.

The unit is shown in a current-regulating configuration. If a voltage regulating function is desired, T2 may be eliminated and the sense line connected at F. In either case, the voltage doubler rectifier made up of CR-5, CR-6, R1, and C1 rectifies a voltage proportional to the parameter being controlled; op amp U3 is again our familiar integrator, which produces a negative going dc output proportional to the sample at B. Comparator U2B output goes high whenever the integrator output is higher than the reference voltage at D. This enables U5B and U5D to pass the trigger train to the two photoisolators which, in turn, permits Q1 and Q2 to fire, as appropriate. The trigger is actually supplied to both SCR gates simultaneously, but only the one whose anode is + will fire. The circuit break with the zero-current crossing is inherent with the SCRs. The feedback loop from B tends to stabilize the output of the device.

In general, most motors prefer to be stabilized with a constant voltage drive to provide for the self-regulating action of the motor whereby it draws more current with increasing torque load, so the voltage sensing action should be used. However, with certain types of loads, such as ovens and lamps, it is sometimes desireable to limit the current at turn on, because of the low resistance under the cold condition. The current control is shown here for that reason.

The principal shortcoming of the duty factor control of a zero crossing regulator is the fact that at low outputs, it is on so little of the time. When operated at anything less than half-setting, this control will give an objectionable flicker to lamps and at a duty factor of a third, or less, motors will begin to surge noticeably. Obviously, if the unit were set for a duty factor of 1/60, the output would consist of 4 millisecond bursts spaced a second apart and the motor operation would be noticeably jerky. This system is chiefly advantageous for oven and heater controls and other systems where a limited speed range and a high inertia are present.

THE SCR PHASE CONTROL

The first significant solid-state speed controls for motors and appliances were of the *SCR phase control* variety. This is really somewhat of a misnomer since what the device does is to multilate the sine wave to obtain a form of duty factor control. This does not control the phase in the ordinary sense of the word. However the title has stuck so we shall use it.

The phase control operates by delaying the closure of an SCR switch after the zero crossing. This effect is shown in Fig. 15-2. The control may be employed with a single SCR to obtain a 0 to 50 percent control. If a diode is placed across the SCR, with reversed polarity, the device can be operated from a 50 to 100 percent duty factor. In both of these cases the halves of the current cycle are not balanced, and a considerable dc magnetizing current flows in the load. Transformers and inductive devices have a tendency to be unhappy about this and will often saturate during a portion of the cycle and draw excessive current, thus experiencing excessive heating. A superior arrangement is the full-wave, or symmetrical, SCR control shown in Fig. 15-3. If the halves are truly symmetrical, no dc current flows in the load unless the load itself is unilateral, i.e. contains diodes or other rectifying devices.

The phase-control SCR has the advantage over the zero-crossing control in that it can be made to deliver flicker-free power even at very low levels. With the symmetrical system, the load always receives 120 impulses per second, even at low duty factors, whereas the zero crossing control could give no more than two per

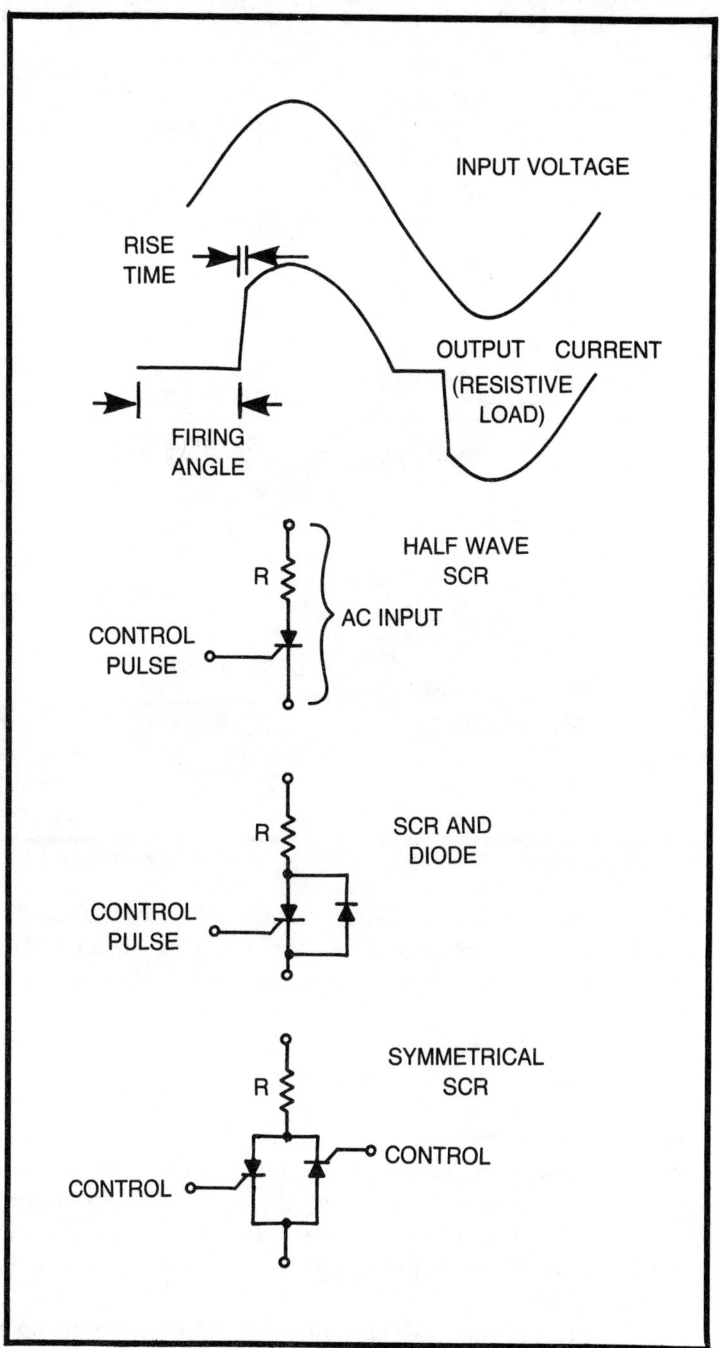

Fig. 15-2. Typical SCR phase control waveforms.

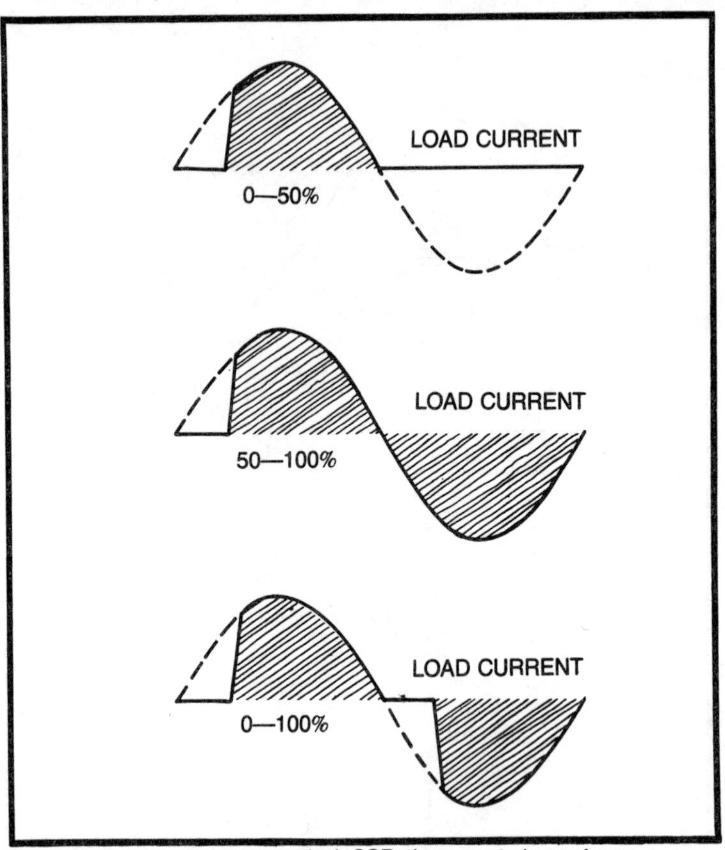

Fig. 15-3. Full wave, or symmetrical, SCR phase control waveforms.

second when operation at 1/60 power level. For this reason the phase control is still used in lamp dimmers and small motor controls where there is not enough inertia in the response to smooth out the roughness.

The principal objection to phase control systems is the radio frequency interference (RFI) generated by the very fast turn-on time. SCRs with a rise time of less than a microsecond are not at all uncommon. Since the high order harmonic content of a wave is determined largely by the rise time, this tells us that there will be a considerable harmonic component at 1 MHz. There components can be difficult to filter out of the load current since the SCR and the load impedance at the harmonic frequencies must be many times lower in order to suppress this effect.

Another significant problem with the SCR phase control concerns operation with inductive loads. The SCR will commutate, or turn itself off, when the *current* goes to zero. With a highly inductive load, such as a transformer or an induction motor, the current waveform has a tendency to lag by nearly 90° for an unmutilated sine wave. In the phase control circuit, the current may not have decayed to zero before the other SCR fires, thus leaving both SCRs in the fired condition for a short time. Fig. 15-4 presents a somewhat simplified view of the resulting waveform. It may be seen that the trailing edge of the waveform is the section most mutilated. This action tends to be a bit hard on SCRs and death to Triacs. It is

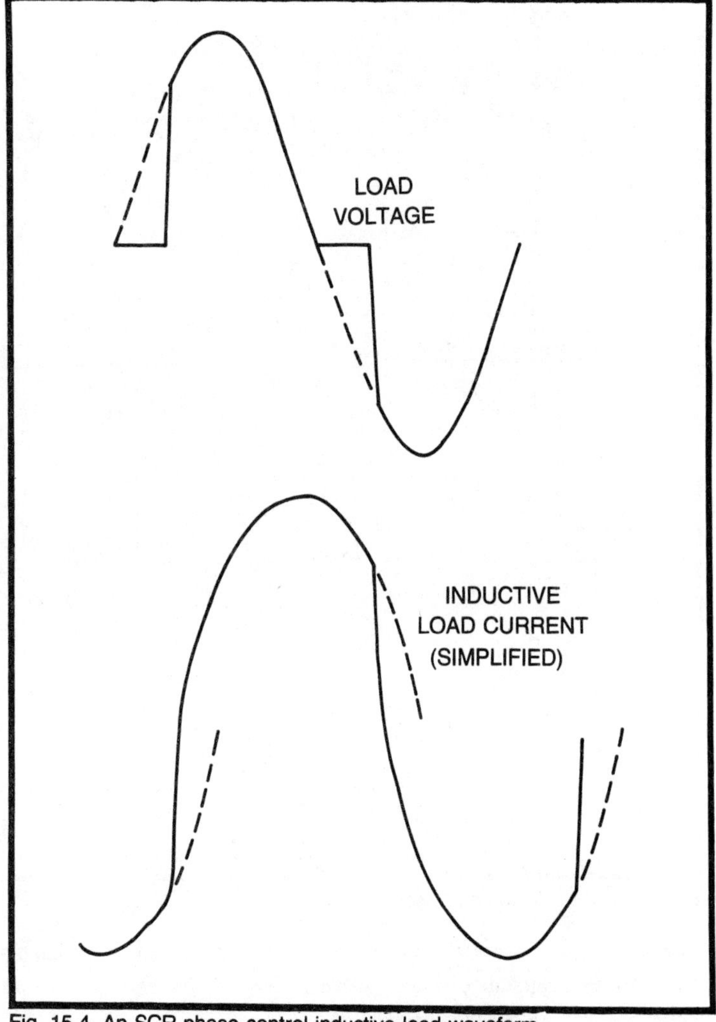

Fig. 15-4. An SCR phase control inductive load waveform.

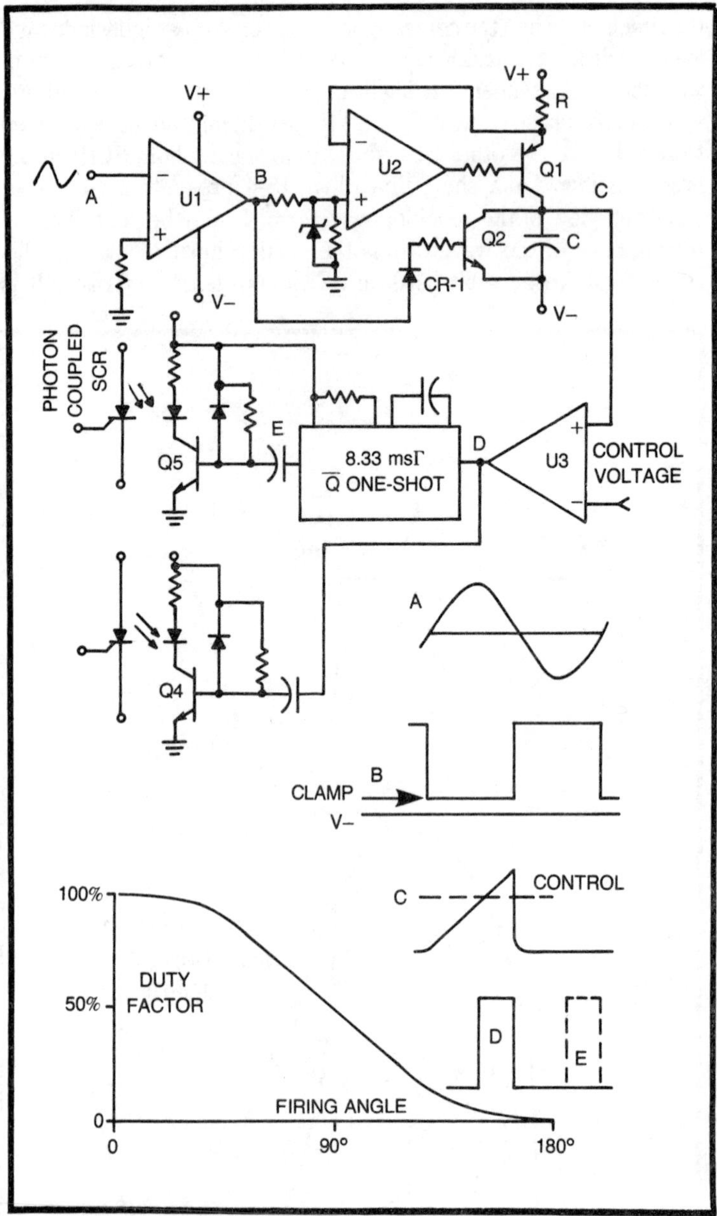

Fig. 15-5. Linear firing angle control.

generally a good practice to provide some power factor correction in the form of a capacitor to help prevent this. The use of the snubber capacitor was described in Chapter 2.

TIMING CIRCUITS

The SCR phase control has been around for a long time and a large number of timing circuits have been developed. Many of these employ other trigger type devices such as unijunction transistors (UJTs), Programmable UJTs (PUTs), breakdown diodes.etc. These devices have mainly been employed in low cost drive circuits for lamp dimmers, power tools, etc. The variety of these devices is so large and the operation so different from most of the circuitry used here that space does not permit covering them. In addition, many of these circuits are not too stable with temperature and line voltage variation. Many are prone to have hooks, flats, or hysteresis in the output curve. These problems are not too serious in hand-tool or lamp dimmer operations but are very hard to deal with in servo controls, where stable operation is necessary. Accordingly, we shall concern ourselves here with circuits intended primarily for electronic control. The General Electric text referenced in Chapter 2 is recommended for those wishing to pursue the matter further.

Figure 15-5 illustrates a technique capable of yielding an accurately linear relationship between the control voltage and the firing angle. Op amp U1 is a comparator with a zener clamp on the output. This unit acts as a zero crossing detector. U2 and Q1 form a voltage-to-current converter. This linearly charges capacitor C when the wave at B goes to the clamp level. When B goes high, transistor Q2 rapidly discharges C. Comparator U3 compares the voltage on *C* with the control voltage. When this voltage exceeds the control voltage, Q4 briefly fires the photoisolator. The 8.33 msec one-shot also triggers, and one-half cycle later Q5 fires. The photon isolated SCRs may be connected in anti-parallel as in Fig. 15-7 or may be used to trigger other larger SCRs.

Although this circuit does give a linear relationship between firing angle and control voltage, the relationship between duty factor and firing angle is reasonably linear only between about 45° and 135°. At smaller angles there is not much lost, and at larger angles there is not much left.

The circuit has the advantage of being nearly independent of line voltage fluctuations since the line is employed only to obtain a zero crossing reference (the circuit power supply should naturally be regulated). This feature is quite important in some cases since a good sized motor will often cause a considerable sag in the line. Many of the simpler controls will change the firing angle with a line transient and can wind up by fighting themselves since the motor under control may provide the transient.

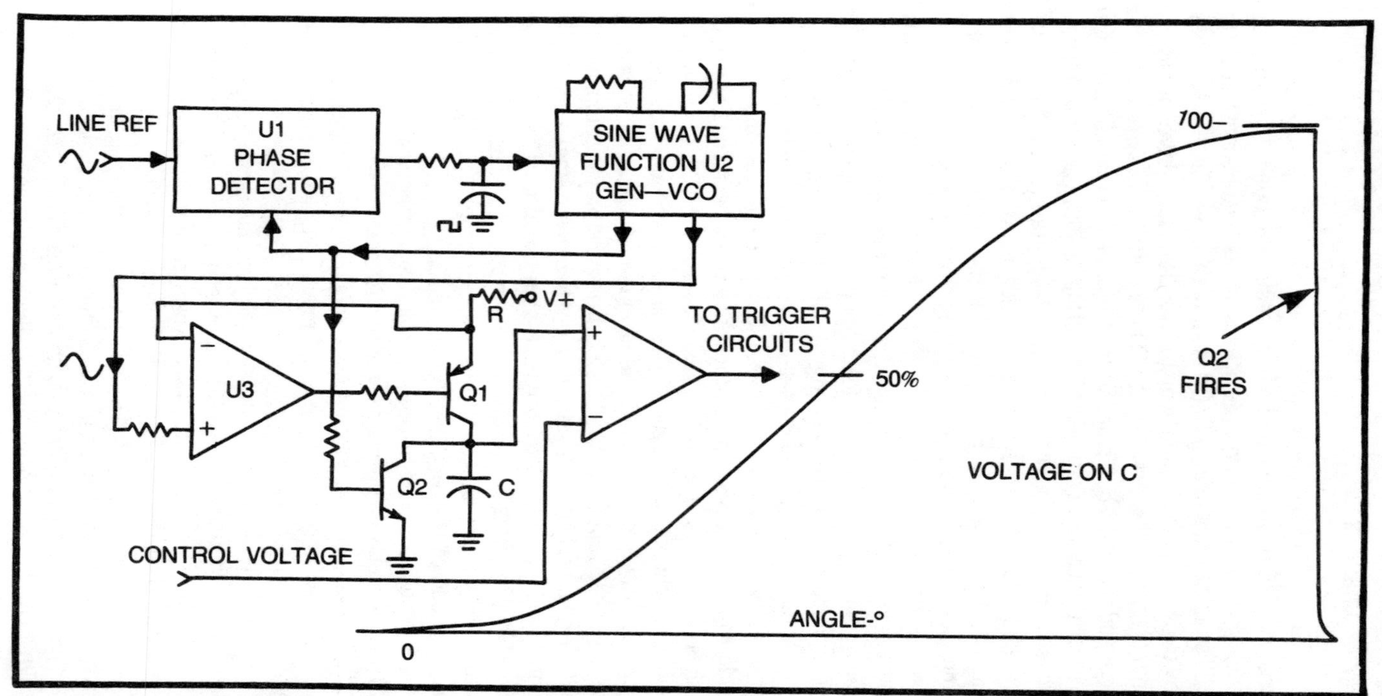

Fig. 15-6. Linear duty factor phase control SCR timing generator.

A circuit which will supply a highly linear relationship between duty factor and control voltage is shown in Fig. 15-6. Here, a sine wave from a function generator is integrated in capacitor C by the voltage-to-current converter. The voltage on the capacitor is actually directly proportional to the fraction of the cycle which is past. If all of the elements were perfect, the circuit would yield a perfect linearity between 0° and 180°. As a practical matter, the circuit can be made linear within about two percent between 25° and 160°.

The reason for the elaborate phase locked loop scheme is the fact that this circuit is extremely sensitive to variation in the sine wave reference voltage when operation at less than 45° or more than 135° is attempted. The phase locked sine wave generator chips will provide a sine wave output with less than one percent error, and the output is quite stable. Some temperature compensation may be required to account for a slight variation in the sine wave. This could be built into the control signal line with a temperature compensating amplifier. If circuit operation is intended to take place mainly below 45°, the linear angle circuit would be the better choice.

RFI SUPPRESSION

The supression of radio frequency interference is one of the most difficult parts of phase control SCR design. On a one-horsepower motor, the impedance is typically on the order of 15 ohms at the line frequency and may be a good deal less at radio frequencies. Large SCRs with rise times on the order of one microsecond are not unusual, and one could expect current jumps from zero to seven amperes, or more, to be present on the output line. For components of this speed, the inductance and capacitance of the line wire itself is not at all negligible, and the stray leakage to earth can cause the currents in the return line to be unequal to the outgoing current. Some of the sources of this unbalance are shown in Fig. 15-7 for a common circuit arrangement. The nonzero impedance in the return line and leakage impedance Z2 combine to divert a portion of the current designated as i_3, thus making $i_1 > i_2$. If i_1 were precisely equal to i_2, the external magnetic field would nearly cancel. The "nearly" is because the slight difference in distance from the wires to a distant point keeps the external field from cancelling perfectly. This effect is very small with respect to the radiation which is caused by the common-mode current. For the common-mode current, the entire line length becomes an effective transmitting antenna. The resulting radiation is not only an illegal annoyance to others, but can also very easily disrupt the operation of the control circuitry and other low-level circuits within the system.

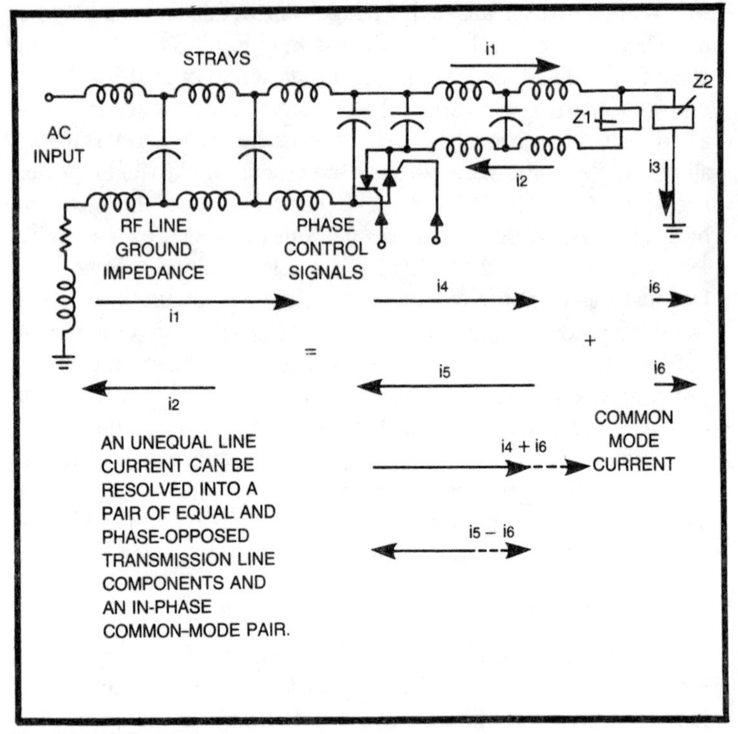

Fig. 15-7. RFI unbalance.

The most effective single step toward elimination of this effect is to completely enclose the system within a continuous, highly conductive shield. Run all of the wiring within a shield braid or conduit which is tightly sealed to the control enclosure etc. This cannot always be done since the load may be an electric appliance supplied by others. In any event the system will usually enter the electric mains via a wall plug or some other connection not under control of the designer. For this reason, the inclusion of common-mode suppression circuitry on both sides of the SCR set is good engineering practice.

Figure 15-8 illustrates the basic features of a good noise suppression system. The SCR set is enclosed within a shielded enclosure. All power, load, and control lines exit through a coaxial capacitor with the outer conductor securely grounded. Each line is arranged with two such capacitors, and an inductor in the form of a Pi-section low-pass filter. The outboard capacitor ensures that the outboard impedance is low at the high frequencies. The inductor ensures that the impedance seen by the inboard capacitors is high, thereby permitting them to effectively bypass the high frequency components.

Coaxial bypass capacitors especially designed for RFI service are available from a number of sources. The use of these units is highly recommended over the use of two-lead conventional units because of the very low inductance and high self-resonance frequencies designed into the coaxial units.

The filters illustrated in Fig. 15-9 show some means of obtaining the very high impedance to common-mode rf currents required to permit good bypassing action by the capacitors. The bead style is easier to install but generally does not offer high common-mode impedance to the lower frequencies unless several units are strung on the line. The bifilar toroid filters offer the highest impedance but are more of a problem to wind and mount. Cup-core types offer a compromise in that they are easier to wind and mount.

It should be noted that these inductors do not tend to be magnetized by the transmission-mode currents, which induce cancelling fields. The single-ended unit shown on the unbalanced control line does, and this effect must be considered. Those of the bifilar type will generally function effectively from about 1/5 of the self-resonant frequency upward. The fact that the very high microferrite is lossy at higher frequencies actually is helpful in suppressing noise. If the RFI content and average energy is high enough, there may be

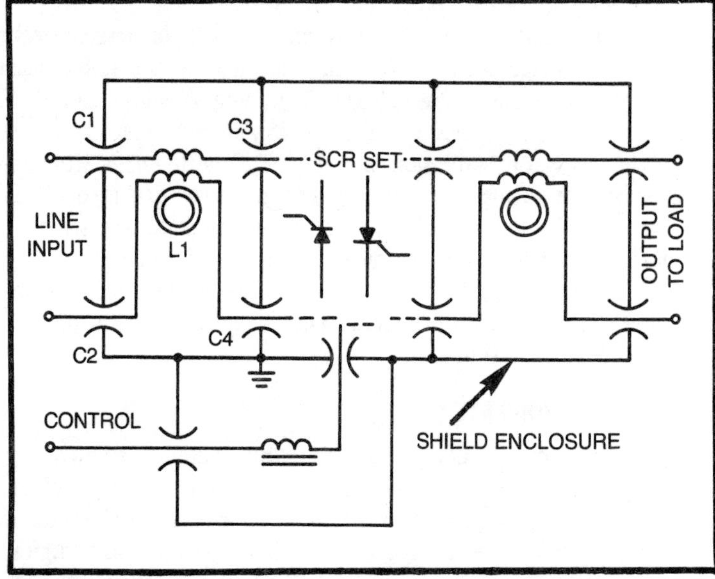

Fig. 15-8. RFI control.

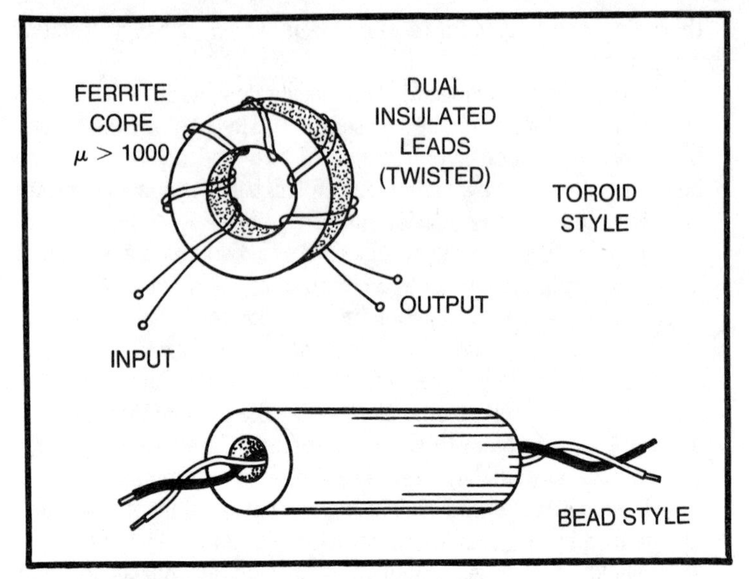

Fig. 15-9. The bifilar common-mode choke (L1 in Fig. 15-8).

some heating in the core, which may control the minimum usable size. In general, this is not the case. A core which is large enough to permit winding a suitable inductance with wire large enough to handle the main line current can seldom be overheated by RFI absorbtion in a 60 Hz phase control circuit.

The various precautions shown in Fig. 15-8 do not really represent over-design in any equipment that must function in the vicinity of, or from the same power line used by, any sensitive circuit. The malfunctioning of digital control circuitry in particular can be very nearly assured by omission of just one of the features. Radio and television performance can be blighted and the error-rate on a RTTY circuit can be dramatically increased by the same omission. The maximum radiated noise for different classes of controls are subject to MIL specs and FCC regulations. Anyone designing a phase control SCR circuit for production should familiarize himself with the appropriate requirements.

THE CYCLOCONVERTER

The term cycloconverter is sometimes applied to a variety of devices in which the frequency is changed in order to effect an efficient voltage or power conversion. The simplest of these devices simply rectifies the ac into dc and then employs an inverter to produce the new frequency. This technique is usually employed on single-phase low to medium power installations. On modern units of this type, either SCRs or transistors may be employed. Very large units are generally SCR types.

The frequency conversion is nearly always upward. The principal advantage in an upward frequency conversion is in the weight of motors and transformers for a given power rating. As noted earlier, the output power of a motor is essentially proportional to shaft speed, hence a high frequency motor is generally capable of much higher shaft output than a low frequency motor of the same physical size.

A similar argument applies to transformers. The weight of a power transformer of a given rating decreases about as the reciprocal of the square root of frequency. For this reason we find that many inverters are run at the highest possible frequency, particularly where it is desired to produce rectified dc as the final product. When a motor is to be driven with the ac output, the frequency is usually held below 1000 Hz because of the core losses. The motor armature and stator are less easy to fabricate from the very thin laminations required for higher frequencies than the stationary transformer parts.

Figure 15-10 illustrates one of the simpler forms of SCR inverter. This is a resonant type device which delivers a reasonably clean sine wave output. The operation of the device is controlled by alternately triggering SCR-1 and SCR-2. Presume that SCR-1 is first turned on with a short trigger. Current will flow through L1 to charge C3. Since C2 is ten times larger than C3 the charge on C2 changes only slightly. Because of the inductive overshoot effect C3 will charge to nearly twice the dc input voltage. The current will then reverse and tend to flow back to the battery through CR-1. This turns off SCR-1. After C3 is discharged, the control circuit triggers SCR-2 and the process repeats itself with C3 charging to the opposite polarity.

The SCRs are commutated by the resonant swing and the diodes which pass the reverse current swing. Obviously, the control drive must operate at some frequency related to the resonant frequency of $L1$ (or $L2$) and $C3$. For the values shown, this is about 13,400 Hz. The distortion of the waveform seems to work out to be minimal when the drive trigger frequency is only about 3/4 of this frequency, or 10 kHz in this case. The frequency can be lowered by increasing the value of either $L1$ or $C1$, or both.

There are actually some 35 or more classes and configurations of inverters based upon such qualifiers as whether the SCRs are self-commutating or externally commutated, whether the unit changes frequency to accommodate itself to load changes etc. The discussion of these devices is beyond the scope of this text, and the reader is referred to the General Electric text again.

An important class of inverters in modern work is the series which simulate a sine wave by using pulse width modualtion. Figure 15-11 shows the waveform generated by such a unit. It may be seen that the output consists of a series of pulses of varying width, arranged so that the output can be relatively easily filtered into a sine wave. This class of devices is increasingly finding use for variable frequency motor drives using transistors. One circuit for realizing this form of control is shown in Fig. 14-4 in the chapter on dc motor controls. A sine wave applied to E in this circuit will produce the output waveform of Fig. 15-11.

The same end may be accomplished by employing pulses of constant width and varying the frequency of repetition. Figure 14-6

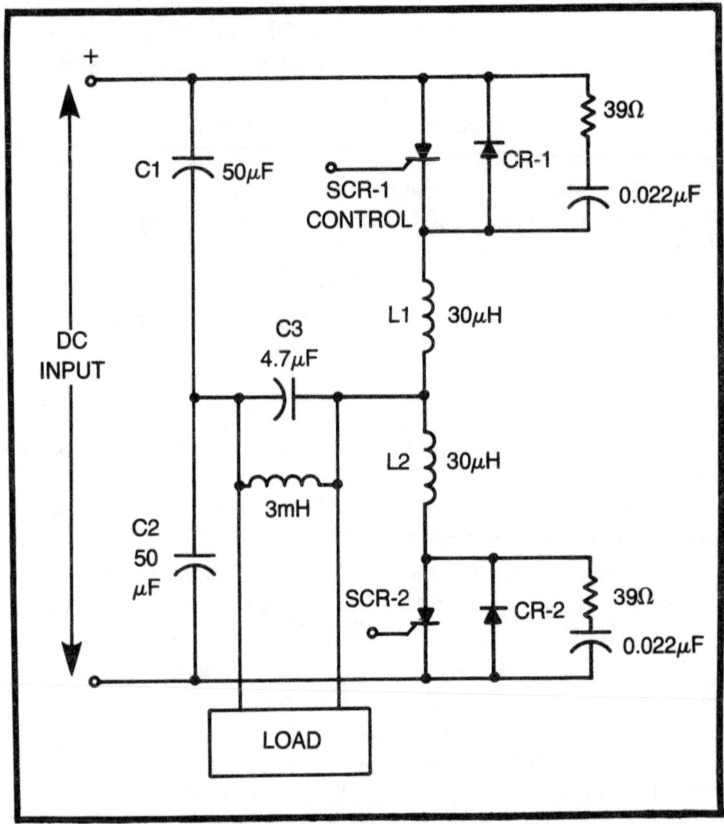

Fig. 15-10. A class A SCR inverter.

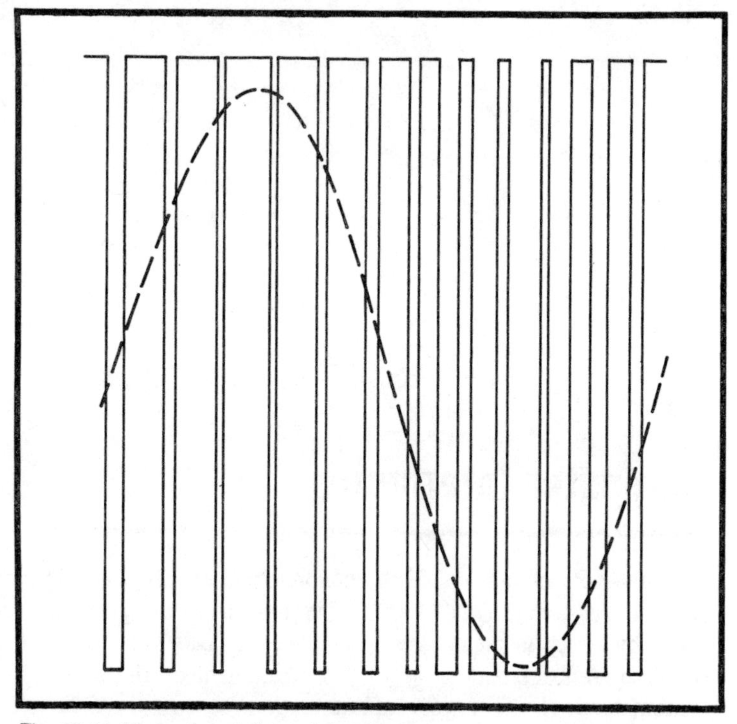

Fig. 15-11. The pulse-width-modulated inverter.

illustrates a mechanism for obtaining the requisite frequency modulation of a train of constant width pulses.

There are advantages and disadvantages to each of these techniques. The constant pulse-width technique tends to have a lower envelope of high order harmonics whereas the constant repetition period train has stationary components which vary only in amplitude.

Both of these techniques can be filtered with a resonant series circuit in order to smooth out the sine waves. Both have the advantage of operating the transistors in the switching mode for minimum losses.

16
Computer Interfacing

For a variety of reasons, both economic and sociological, the past decade has seen the rise of what I will refer to as *intelligent automation*, by which is meant the automation of tasks which require a definite feedback and some level of judgement on the part of the automation. The first wave of automation (at least the first to be called such) came in the 1950s and started largely in the automotive industry. It had its ancestry in the Jacquard Looms, which would weave cloth with a pattern determined by punched cards. These were followed by lathes that would automatically turn decorative furniture legs and pieces, and later still by automatic screw machines that could produce complete turned parts from information stored in the machine in the form of special tools, cams and stops.

THE MODERN AUTOMATION

The 1950s wave of automation was different in the sense that not just one but many machines were coupled together to produce not single, simple parts, but complex items which required a number of different machines and operations. Chrysler Motor Car Co. probably pioneered this move in about 1954 when they automated the complete operations required to produce a V-8 block with a strung-together super-machine which stretched out nearly a mile and spanned Jefferson avenue in Detroit in the process.

This automation was largely done with hydraulic valves, pistions, and sensors along with microswitches, relays, and cams. The logic was "hard wired" in the sense that it was capable only of producing the part it had been programmed for. The establishment of these automated lines was a difficult and time consuming process since plumbing had to be run in and new cams machined whenever a change in the product was required. Also, in the ordinary sense of the word, there was no adaptive decision making on the part of the machine. Once started, it simply ground out more of the parts until the operator told it to stop.

This type of automation is very effective in situations where a large number of identical parts are required, as in the case of automobile engines, refrigerator compressors, etc., but it is far too expensive for operations where only a few or a few hundred of the items are required. Accordingly, the next wave of automation was to be found in the form of *numerical control* machines.

As the craftsmen who could set the stops and adjust the cams and valves of automatic machines became more scarce, the numeric control machine that could be programmed and sequenced from a punched paper or Mylar tape appeared on the scene. With these machines, a technician could create a program with which the machine could fabricate one or a dozen or a few hundred parts economically on a general purpose machine. When a different part was called for, the machine could be switched over with just the change of the tape and perhaps a few tools. However, these machines are still principally just *sequencers* in that they simply follow the tape instructions about what to do with the tools. They pass no judgement about the quality of the work they are turning out or the sharpness of their own tools, etc.

Intelligent automaton is the most recent development in this parade and generally the most sophisticated. It generally will not operate with "hard wired" logic because of the relative complexity of the task. It should be noted that the people who design and use these machines would probably not recognize them by the name of "intelligent automaton." Instead the machine is generally known as an "automatic sorting whatzis" or a "computer controlled thingamabob" etc. Because the people who use the machine are used to buying a "whatzis" or a "thingamabob" and have little interest in generic classifications.

To pick out an example of an intelligent automaton, consider the lumber industry. Tree trunks do not grow very uniformly and they contain knots, blemishes, and various types of inclusions. When the finished lumber is sold, however, it is sold on the basis of a classification, depending on whether it is "clear," No. 1, with small fast knots less than an inch in diameter; No. 2, with larger knots and a few loose knots; or "construction" grade, which is rough and not planed. The price of course depends upon the grade, given here in descending order of value. When the tree is first slashed open, the actual state of that piece of wood comes to light and a decision must be made about the handling. A given board might be judiciously cut to yield three premium "clear" boards with some level of waste or two No. 1 boards with less waste or a single No. 2 board with still less waste, or simply sold as construction grade (for concrete forms etc.) with no waste and no further planing or treatment. The decision about what to do with the board is generally made by a highly skilled inspector or lumber grader. Very recently, however, several groups have been working upon an automated machine which is capable of assessing the board and calculating the net financial return for each of the alternatives. The machine then issues the commands required to execute the optimized decision and to perform the entire operation at a speed competitive with the somewhat intuitive and experienced judgment of the skilled lumber grader. While the machine may not be exercising "intelligence" in the usual sense of the word, it is equipped to evaluate a series of alternatives on the basis of economic merit and to initiate action based upon that judgement.

The semiconductor industry has long operated automated sorting operations in which the devices are tested and marked for type as a function of the test results. This is the principle reason for the existence of "family" lines of transistors and SCRs with different voltage ratings etc. Those units that show a higher beta, a lower leakage, or a higher breakdown or zener voltage are differently numbered and sold at a premium price.

In the initial generation of machines of this type, the *aerospace ground environment* (AGE) type were the most sophisticated and comprehensive. These machines were evolved in order to test the working of each of the various subsystems in modern jet aircraft. A stimulus was applied and a response measured for each subsystem to determine whether operation was within tolerance limits. In the case of marginal or incorrect response, a branch routine was employed to diagnose the source of the failure. Volkswagen has pioneered the use of this type of automated testing for automobile service operations, and the practice is rapidly spreading to other makes. There seems little doubt that intelligent automatons will form a major part of the electronics industry in the next generation.

SENSES AND MUSCLES

In order to accomplish some of these goals, it is necessary that the computer be provided with both "senses" to determine the condition of things or the status of the operation and "muscles" to permit it to do something corrective or constructive. The senses may be any of a variety of transducers, ranging from something as simple as microswitches or photointerruptors to something as sophisticated as a mass spectrometer. The muscles may be as simple as an electric heater or as sophisticated as a remote anthropomorphic manipulator which has shoulder, elbow, wrist, and grasping motions. In each case, however, the devices must be adapted to the interface requirements of the computer or controller.

With the exception of IEEE Standard 488-1975, which will be treated in the next chapter, there is very little in the way of standardization of computer interfaces. The computer output may be serial or it may be parallel, or it may be arrangeable to be either, and the same is true of the input configuration. The logic may be either active low or active high and may be mixed on a given machine with some functions only active high and others only active low. Hence, it is difficult to generalize on any of these matters, so the following discussions are non-specific in nature and provide a few general guidelines for interfacing. Some or all may, and again may not, be applicable to a particular situation.

LEVELS

By far and away the most common interface level used for connection between a completely packaged computer and the outside world for nearby control and sensing application is the TTL set using negative logic. This is defined in 448-1975 as:

Logic State 0 (also called high state)

This may be derived from either an open collector device or a three state device. For a transmitting device of the three state type this represents a voltage equal to or higher than +2.5 volts with a current of −5.2 milliamperes. For an open collector transmitter, a more detailed set of specifications is required. Fig. 16-1 shows the envelope of voltage and current that is within the specification tolerance.

For a receiver, any voltage greater than 2.0 V must be perceived as a high state. It should draw no more than −40 μA leakage current at +2.4 V and −1 mA at +5.25 V.

Logic State 1 (also called low state)

The output logic voltage for either a three state device or an open collector device shall not be greater than +0.4 V at +48 mA

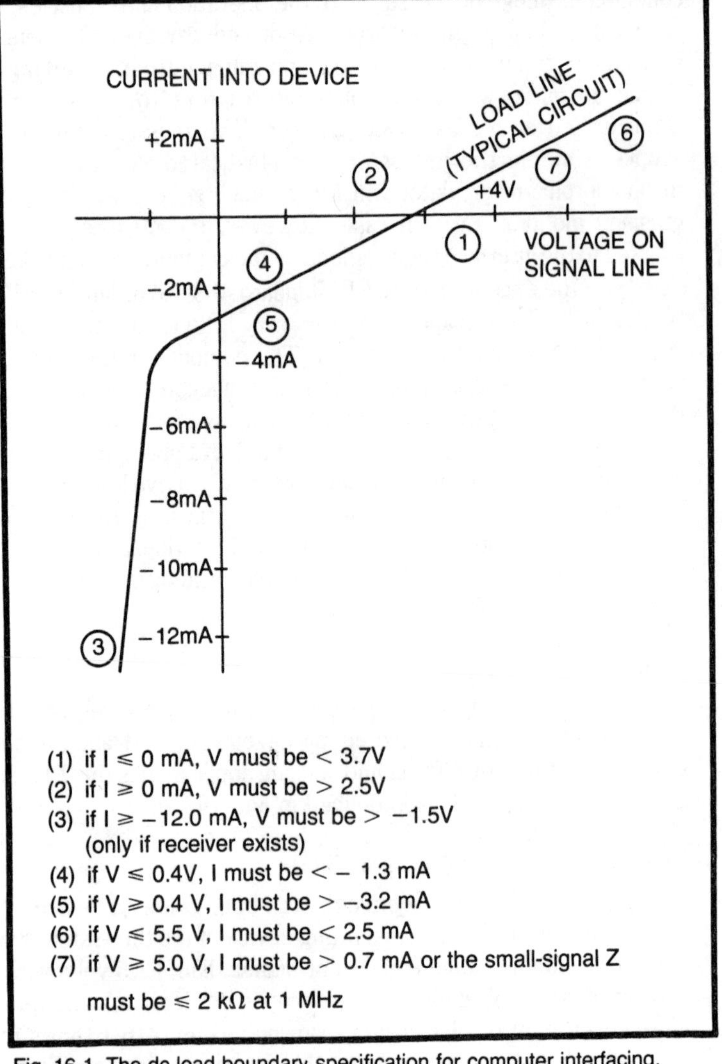

Fig. 16-1. The dc load boundary specification for computer interfacing.

sink current. For the receiver, any voltage less than 0.8 V shall be perceived as a low state and the device shall not draw more than −1.6 mA at +0.4 V.

Capacitance

The internal capacitance load shall not exceed 100 pF on any one line.

Negative Clamping

Each signal line to which a receiver is connected must contain circuitry to limit the negative voltage excursions. This is typically a diode clamp.

Other Level Considerations

In general, the interface requirements will be met by any good quality TTL package, with the transmitters being of the open collector type. It is usually a good practice never to place more than one transmitter and one receiver on any single line. Fig. 16-2 illustrates an interface transmitter and receiver circuit that will meet these requirements.

The TTL level signaling is generally used in cases where the computer and the device under control are separated by no more than a few feet and generally where parallel data bus operation is employed. Most often, the data bus is bidirectional, and the device under control must be capable of either transmitting or receiving on the line. Furthermore, the line is a party line and the device must be prepared to share the operation of the line with other devices which are also under control of the computer.

The use of the active-low type transmission has the advantage that it inherently permits this type of operation. In effect, each of the tri state or open collector devices is in a wired OR configuration since any one can pull the line low, and an inactive transmitter simply goes for a free ride. Moreover, any one can be added to the line or unplugged from the line without disturbing the operation of the remainder of the system. An active-high system would not permit this simple parallel operation.

For noisy industrial applications where the control lines get to be longer, the use of higher-level logic signals, or HINIL type levels, is frequently advantageous. In this case special ICs with higher operating and threshold voltages are usually employed. Relatively high-level signals may also be used with discreet device interface circuits. The RS-232C and 20 mA or 60 mA teletype interfaces are generally not used directly for motor control applications because of the amount of decoding logic involved. When this form of transmission is employed, a local terminal, perhaps containing a microcomputer or microprocessor, is also present, and the interface between that device and the devices under its control is often at TTL level. It should be noted that a great many of the microprocessors do not of themselves operate at TTL levels, and a major number of the devices on the microcomputer card may be involved in the level

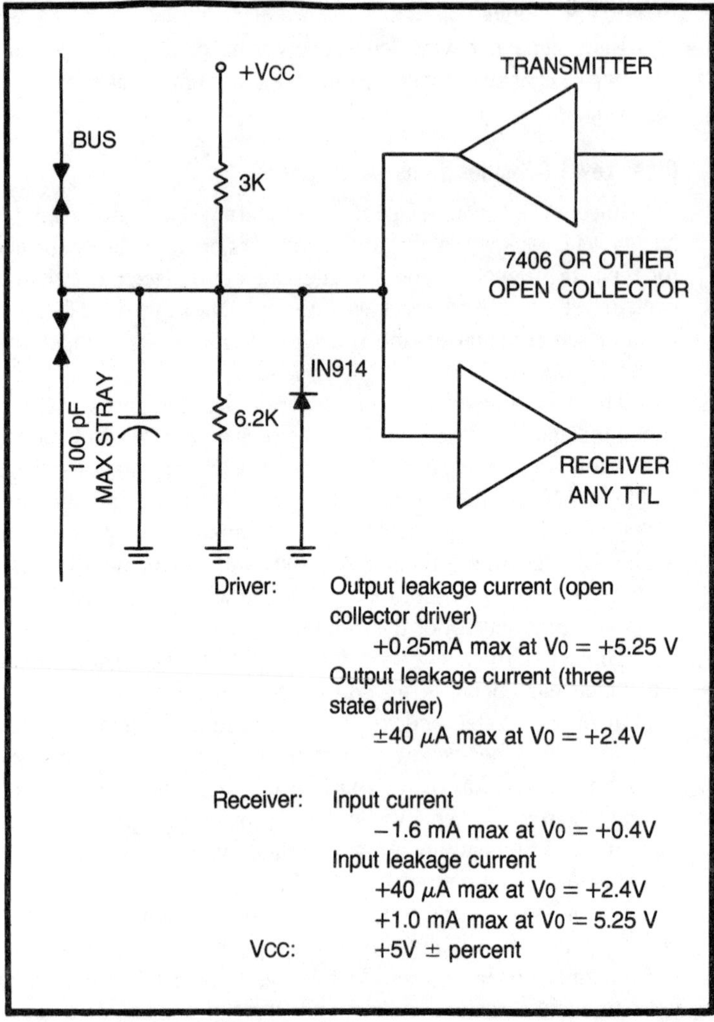

Fig. 16-2. Typical transmit-receive, or bidirectional, data terminal. Only a single driver and receiver may be connected to each signal line in the typical configuration.

translation. If either of these drive a printer, plotter, or tape cassette, the transfer is usually accomplished at TTL level.

LATCHING

Most microcomputers and a great many of the mini's have a rather elaborate protocol of input and output. Many will only input data and output data at some very tightly defined portion of the machine cycle. Input and output may only be valid or available between bits 8 and 15 of a repetitive 20 clock cycle period, for example. Furthermore, the number of clock cycles within a machine cycle may be variable depending upon the nature of the instruction being executed. For example, a subtract might take more time than an EXCLUSIVE-OR. In a calculator-configured machine, a square root or a cosine can take as much as 100 times as long as an add.

For this reason, it is usual that a given control be capable of latching the instruction between successive machine cycles and holding that state until it receives a different instruction. You will note, on review, that a number of the control circuits in previous chapters had this feature.

When a latched control is installed in any system that has limited travel, the latching must, in most places, be provided with end-of-travel interlcoks and at least the equivalent of the non-jam diodes.

Perhaps the most persuasive reason for latching is the fact that computers, even of the micro variety, are expensive. They are seldom used to control only a single function. The computer bus is, thus, a "party line," and each device to be controlled should answer only its own "ring." Also, it should not "talk" when the call is not addressed to it. Fig. 16-3 shows a rather typical computer bus arrangement whereby a number of devices may be addressed.

For example, let us consider an eight-wide bus as shown in the figure. Suppose we arrange the controlled devices so that they would accept line h as "I am talking" in the low condition and "I am listening" in the high condition.

Let us further assume that lines g, f, and e are reserved for address. With a three-bit address we can encode up to eight devices with serial numbers 0 thru 7 thus:

g f e	Device No.
H H H	0
H H L	1
H L H	2
H L L	3
↓	↓
L L L	7

H=High
L=Low

An appropriate code on lines g, f, and e will permit a unique answer or device address selection. If we forgo the use of address 0 we obtain one of the advantages of the inverted logic. It is the nature of TTL devices in general to consider an open circuit on the input as

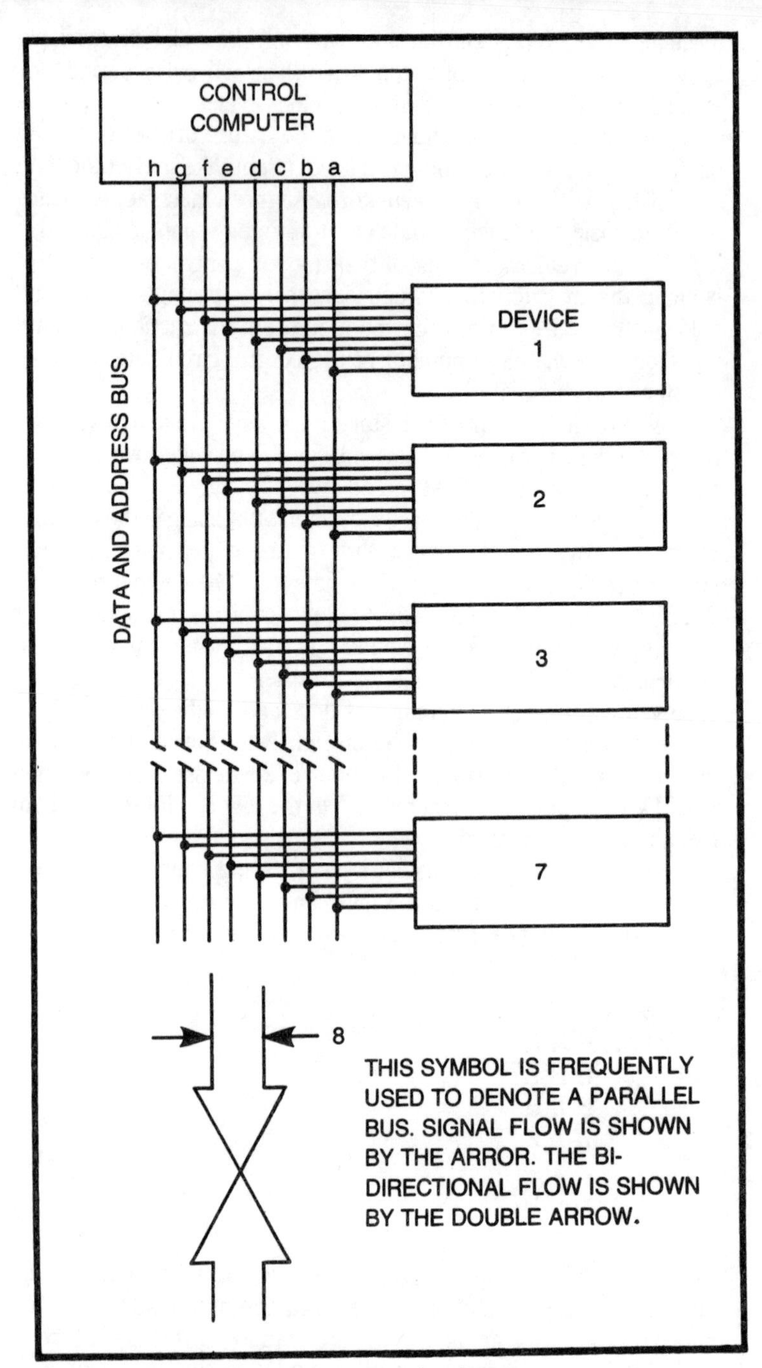

Fig. 16-3. A parallel-bus computer control system.

high. So *any* of the devices, 1 through 7, equipped with input/output circuits similar to Fig. 16-2 would simply consider themselves not addressed and would do nothing if unplugged from the data bus. This is an invaluable aid in troubleshooting and maintainance. The ability to operate the remainder of the system without one of the members is often most important, and the ability to remove a member from the system without causing a violent runaway is more important still!

CAUTION

Never disconnect an energized unit from the data bus unless you can be sure that a violent runaway will not result.

The next four bits, d through a, may be used to provide any one of sixteen commands; thus*:

d c b a	Example Command	No.
H H H H	Stop	0
H H H L	Run CW	1
H H L H	Run CCW	2
All others	Stop	
L L L L	Stop	15

It is usually a good practice to program a motor system to stop on all but valid coded commands, and particularly to stop on a null or all-high code which can result from a disconnect or a broken cable. It is also sometimes advantageous to have an all-high bus state clear all latches.

PARALLEL AND SERIAL DATA HANDLING AND MULTIPLEXING

Multiplex operation is one of the most common features of large scale integration (LSI) devices. The reasons for this are largely economic. At a given point on the "learning curve" the cost of an IC is nearly directly proportional to the number of pins on the package, almost without regard to the content or complexity of the device. An even more persuasive argument can be made for the wiring.

Consider a relatively simple pocket calculator, with only +, −, ×, and ÷ functions. It must also have a clear entry, clear, +, decimal, and digits 0 through 9, 18 keys in all. An eight digit unit would require eight times the seven segments used to display the digits plus eight decimal points, for a total of 64. As a bare minimum, the calculator chip will also require + and − supply connections.

---

*This limited example uses only a fraction of the available commands.

With no multiplexing, the calculator chip would require an 18 + 64 + 2= 84 pin package, which remains to be perfected! The low-price "four banger" depends for economic success upon a multiplexing scheme that wires all of the display segments in parallel (7 + (.) = 8 lines) and uses eight digit-select lines to strobe the data into the correct slot. The eight digit-select lines are also used to strobe the keyboard onto two or three keylines. Including the + and – supply this can be done with 8 + 8 + 3 + 2 = 21 pins. This is a considerable saving. In addition, a 64 conductor cable to the display is cut to 16 and an 18 conductor cable to the keyboard is cut to 11. A six digit calculator has similarly been implemented in an 18 pin package with a bit of doubling up.

It is fairly easy to see from the above example that some measure of multiple use of wire and pins is nearly always required in an intelligent automation. There is a legitimate question, however, as to how much. For example, it is entirely feasible from a technical viewpoint to multiplex the entire display message for a calculator onto a single two conductor line. But the display LED array would then require considerable on-chip decoding to sort out which digit segments and which decimal point should be illuminated from the high-speed serial bit stream. Only in cases where the display was quite distant from the calculator chip would the wire and pin savings offset the added complexity over the self-sorting segment/digit scheme.

For control applications where the distance between the controller and the items under control is relatively small—for example, 15 feet or so—the overwhelming choice is for bit parallel/byte serial organization with eight or sixteen bit wide data bus. The IEEE 448-1965 standard employs an eight bit bus for control and handshaking and an eight bit bus for data; both are bidirectional.

There are many applications where the device or devices under control may be seperated from the processor by distances measured in hundreds of feet or more. In such cases, the provision for an eight or sixteen bit wide bus would be inconvenient and expensive. Even longer links spanning tens and hundreds of miles using commercial telephone and telegraph lines are not uncommon. These devices generally employ some form of universal asynchronous receiver-transmitter (UART) and modulator/demodulator (MODEM). The UART marshals a 5, 6, 7 or 8 bit wide bus into "single file" or a single bit-stream that can be sent over a two-wire pair. The MODEM modulates the bit stream into a two tone frequency shift keyed (FSK) carrier for transmission on telephone facilities that will not handle dc signals, or it keys the signal in an RS-232 or a 20 mA or 60 mA teletype signal for telegraph system transmission where continuous currents can be handled. At the far end the process is reversed and the original data bus is reassembled. This process is depicted in Fig. 16-4. A discussion of UARTs and MODEMs is well

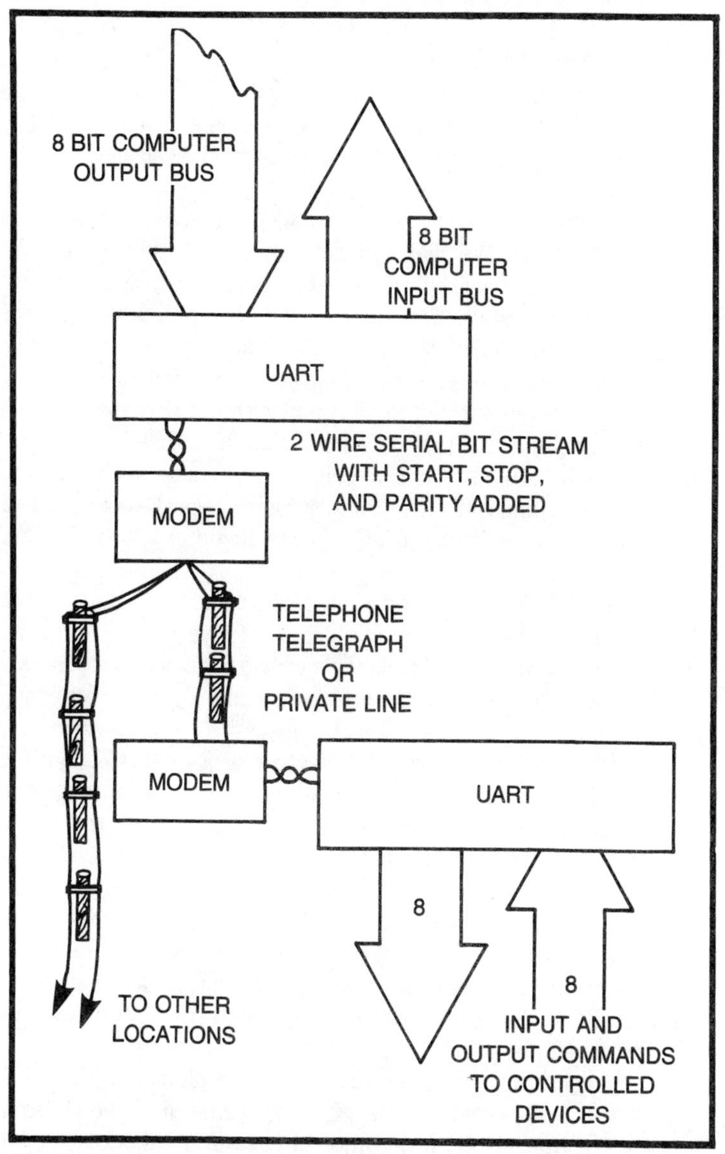

Fig. 16-4. Basic serial links for long distances.

beyond the scope of this text, and in fact, each easily warrants a substantial text for explanation by itself. However, at the motor control level, the system is often completely transparent and the motor control operates from the UART exactly as if it were on the original data bus. The UART and MODEM functions are usually purchased separately from the motor control.

A SYSTEM EXAMPLE

For our system example we shall consider a vehicle control for a hypothetical electrically propelled and braked, and generally computer-managed, high speed truck. The truck shall have an electric motor on each wheel for both propulsion and braking. These are independently controlled for antispin starting and antiskid braking by a central processor. We shall not elaborate on other features of the control. The motors will be controlled by a single central processor. Properly implemented, the system would represent an advance over the no-skid control now required for new highway trucks and tractor-trailers, and offered by a few non-U.S. cars.

We shall equip our vehicle with an "accelerator pedal" or "foot throttle" that acts like that of a conventional automobile on level ground with no wind, i.e., the velocity of the vehicle will be controlled to be proportional to the pedal depression regardless of grade or wind. Each wheel will have an individual tachometer to permit wheel speed readings.

Figure 16-5 shows the principle features of the motor control. The motor is a shunt wound type using field weakening to control speed between 25 and 100 mph. This provides an energy conservation feature in the form of regenerative braking. Suppose, for example, that the truck is going 75 mph and wishes to slow down. He eases off on the foot throttle and the system senses that V_{ave} and V_n, (the speed of the Nth wheel) are above $V_{command}$, (the throttle setting). This causes the motor field to increment until the motor CEMF equals the battery voltage and the motor ceases to draw current, and the vehicle coasts down to the slower speed. If a more vigorous slowdown is commanded, the field will increment to the point where motor CEMF exceeds battery voltage and the motor will actually charge the battery through the appropriate units CR-1 through CR-4. Diode CR-5 prevents this current from backing up into the prime power source.

The algorithm for this control system is shown in Fig. 16-6, written in a form of calculator language for simplicity. The program forms a single large loop with a number of smaller loops based upon conditional jumps shown in the diamond blocks.

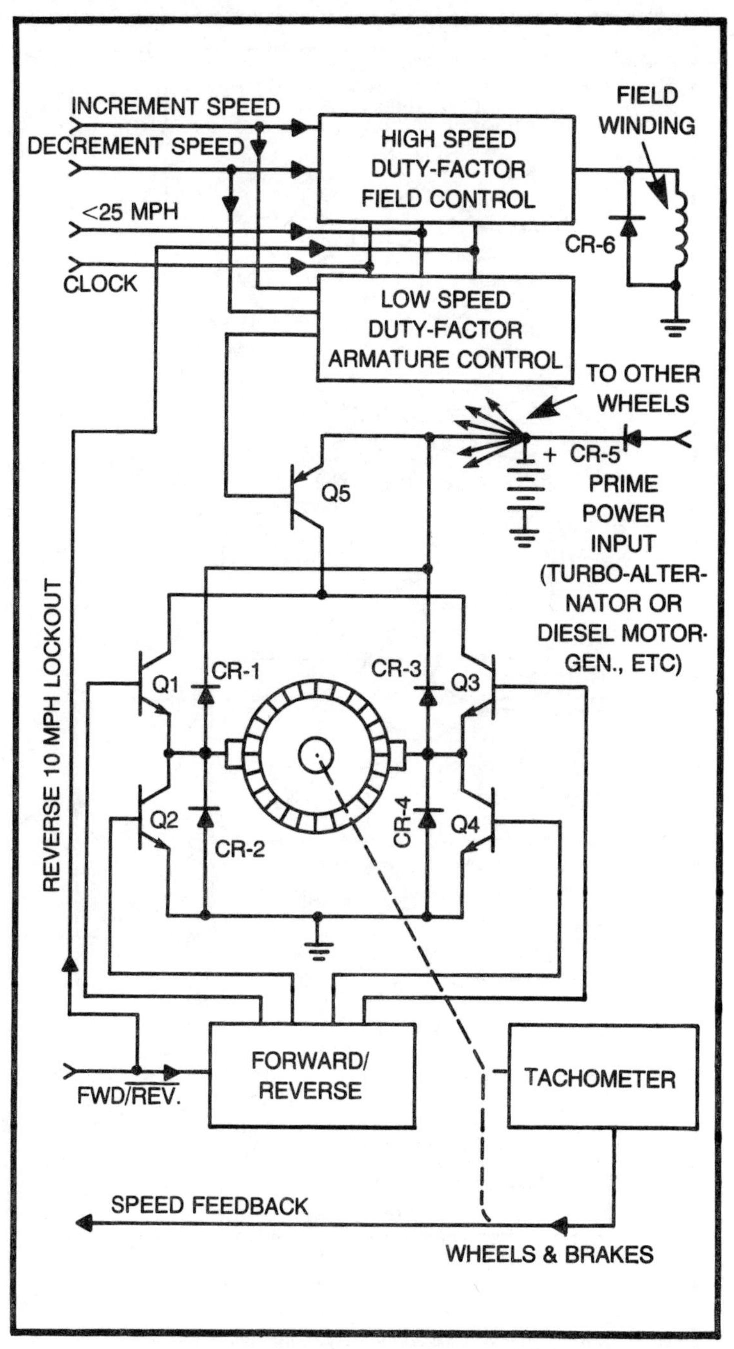

Fig. 16-5. Wheel drive and control.

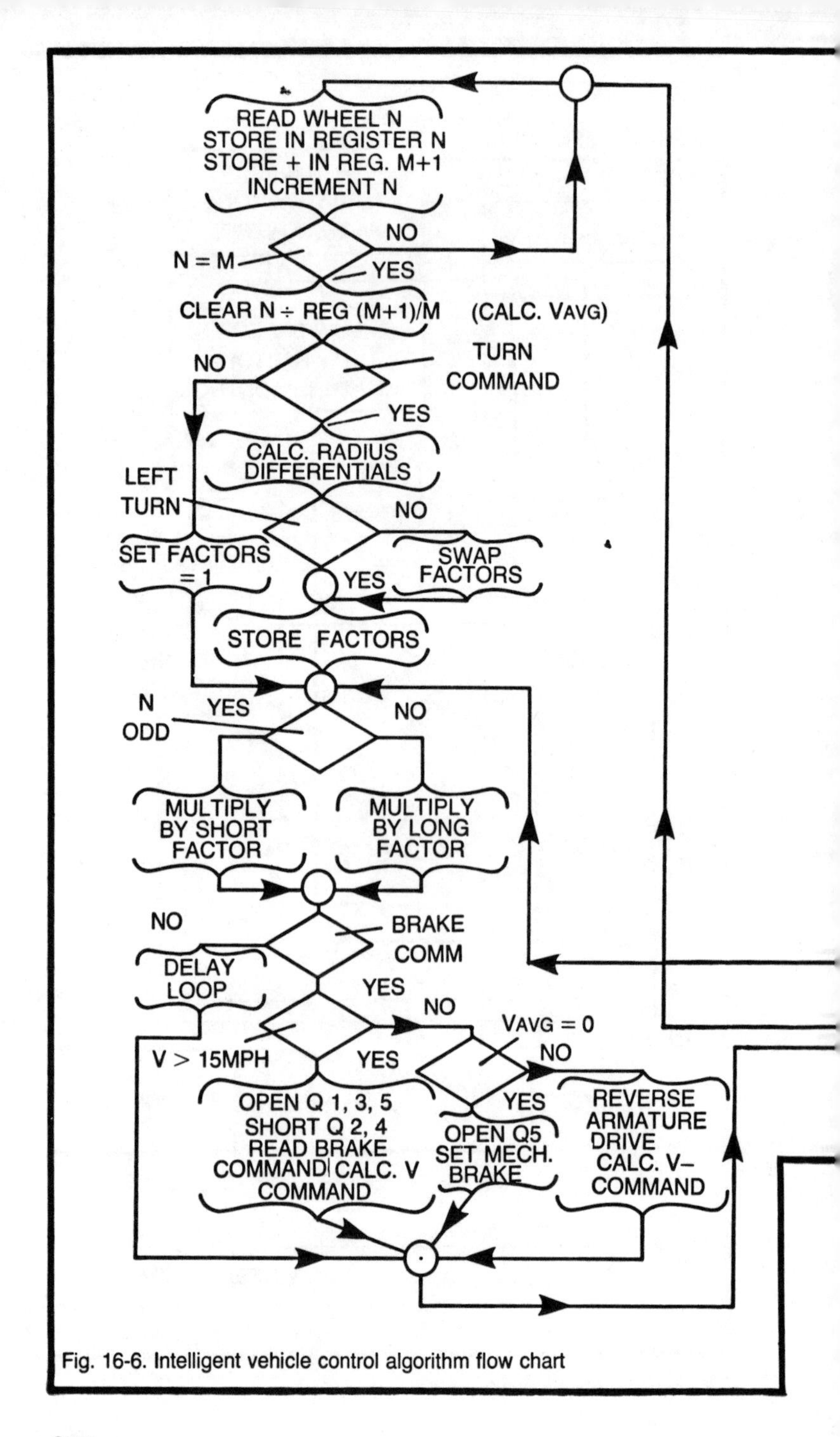

Fig. 16-6. Intelligent vehicle control algorithm flow chart

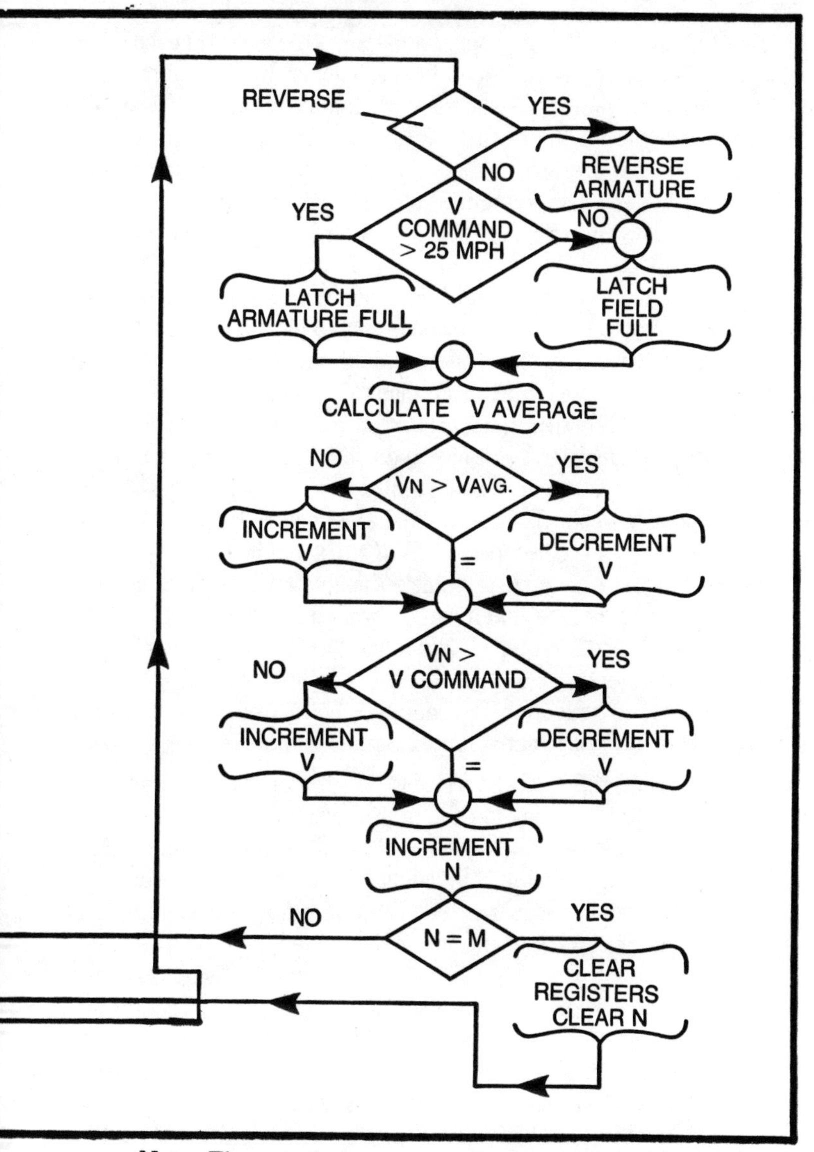

Note: The actual program required to perform this algorithm would be a series of several hundred mnemonic or hexadecimal steps. It would be completely incomprehensible to anyone not intimately familiar with the specific microcomputer employed. The generalized algorithm in calculator language is used for instructive purposes even though few, if any, calculators would be fast enough to actually perform the task.

The algorithm first measures and stores the speed of each wheel and also totals the speed of all of the wheels. V_{ave} is computed by dividing by the number of wheels. Note that the number of wheels to be controlled is not "hard wired" into the logic but can be sensed as a data item. One or more trailers with powered and braked wheels could be added. The system is also prepared to accept steering geometry data and calculate wheel speed corrections from steering wheel deflection to allow for inside and outside turn speeds. This effect should be slightly underdone since the vehicle should have a mild tendency to straighten out on a turn to permit stable handling. When properly done, the system would have an effective built-in power steering whenever it was rolling. The algorithm assumes that wheels with odd serial numbers (N = odd) are on the left-hand side.

The next decision point has to do with braking. If no braking is required, the system enters a slight delay loop which is software programmable. This provides a means of coontrolling "jerk" or the rate-of-change of acceleration ($d^2 V/dt^2$). Except in a racing car, it is seldom advisable to have a vehicle capable of accelerating up to speed at the same rate that it can be braked to a "panic" stop. This system is intended to behave like a fine "heavy" car with all minor speed adjustments issuing from the foot throttle. When a more vigorous speed reduction is required, the system operates on a three-phase program; activated by depression of the brake pedal:

1. While the average velocity is greater than 15 mph, the system disconnects the motor armature, short circuits it, and calculates a sequence of fictitious V commands which give each wheel an individual deceleration proportional to the pedal pressure. The braking effort of each wheel is under control of the motor field.
2. When V_{ave} falls below 15 mph, the armature is reconnected in the reverse direction and a new sequence of fictitious V commands is generated to bring the vehicle to a smooth stop.
3. When V_{ave} falls to zero, the armature disconnects and the mechanical brakes are set. This decision point should be programmed to test the recent history of V_{ave} to avoid setting the brakes on an all-wheels-locked slide.

It will be noted later that this system should, given sufficient torque reserve, permit the fastest possible stop that the road surface will allow in a "panic" stop condition since each wheel is controlled to prevent its departure from V_{ave}, thus a too-vigorous braking command will require all of the wheels to slide. An illogical, too sudden all-wheels-locked slide can be programmed to override the manual braking command and keep the wheels rolling.

The next decision point and the following one test for a normal reverse and an over 25 mph command. Below 25 mph, the field is locked full-on so that minimum torque is available. Speed control comes from armature current duty-factor control. The diodes in the bridge smooth the control impulse and limit the transients.

The next decision point is the one that actually performs the no-slide start and the no-skid stop control. The speed of each wheel, as modified by the turn command, is tested against V_{ave} and controlled to be neither faster nor slower. A sixteen-wheeled vehicle with one side entirely on ice and the other entirely on dry pavement would have all wheels rotating in perfect synchronism.

The following decision point considers whether the velocity is equal to the command velocity, and issues speed-up or slow-down signals accordingly. The vehicle will start and stop smoothly with limited jerk since the acceleration will climb at a constant rate from zero to maximum and then accelerate at maximum rate until $V_{command}$, derived from the foot throttle or the brake command, is reached. A limited-jerk approach to $V_{command}$ could be included, though most drivers would rather perform this function for themselves by not depressing the foot throttle to its final destination at the instant of starting.

The final decision point considers whether all wheels have been serviced, and either loops to service the next or loops to the start of the program, and begins the sequence anew.

If we consider that the whole program might be 200 instructions long and that average instruction time is 25 clock cycles, each of which is 0.5 microseconds in the minicomputer, we come up with a total average main loop time of 2.5 milliseconds to service all of the wheels. Each wheel would have its speed adjusted 400 times per second. This should be adequate to prevent wheelspin and skid since the wheels will travel less than 4 inches between successive readings. The 400 Hz modulation of the motor currents is probably about as high as we would care to go on a large motor because of hysteresis losses in the iron, and the requirement for thin laminations in a motor that must otherwise be very robust and tolerant of large overloads.

The transistors, motors, and other components capable of realizing a system such as this for a large high-speed truck are not available at this writing. The example was selected to provide the reader with a glimpse of the sophistication available through the use of microcomputers in motor control applications.

17
The IEEE Interface

At the time of this writing, the only de facto standard covering the complete range of digital instrumentation interfacing is IEEE Standard 488-1975 entitled:

IEEE Standard Digital Interface for Programmable Instrumentation
Published by:
The Institute of Electrical and Electronics Engineers, Inc.
345 East 47th St.
New York, N.Y. 10017

The incorporation of this standard into control equipment wherever feasible is highly recommended, for a number of very significant reasons which will be discussed.

The standard was first employed by the Hewlett-Packard Company and is represented in a number of their products which include calculators and instruments. Since the standard is very inclusive and covers nearly the full spectrum of such matters, including the plugs and sockets to be employed, a device built to this standard will be directly plug-compatible with an ever-growing number of instruments and programmable calculators available from Hewlett-Packard and others. Most of the firms in these fields are actively implementing the standard. The advantages of being able to plug-in and operate your device under computer software control and to automate performance measurements are very great.

Because of the easy availability of the standard itself and the need to incorporate any revisions, no attempt will be made to directly quote or reproduce the standard except for the discussion of TTL interface levels included in Chapter 16. Anyone attempting to design a compatible interface should have a copy of the standard itself. This chapter is intended only to discuss the basic features of the interface as they apply to motor controls.

BASIC STRUCTURE

The illustration of Fig. 17-1 shows some of the most basic features of the interface. You will note that the various types of devices have been divided into just four categories:

1. Devices able to talk, listen, and control. This includes calculators and certain control devices.
2. Devices able to talk and listen only. This includes instruments such as digital voltmeters, digital thermometers, etc. which may have to be told what to measure, and on which scale, and would then report the measurement results.
3. Devices able only to listen, such as signal generators, power supplies, and motors.
4. Devices able only to talk, such as single scale single point instruments, digital panel meters, tachometers, etc.

It is fairly easy to see that nearly anything one might want to operate or control falls into one of these classes. The motor control example of the previous chapter belongs in the second category since it listened to velocity commands and reported speed data.

Data and control transfer are carried on a set of two segregated 8-bit-wide buses. These buses are, generally speaking, bidirectional. A complete system may contain a variety of combinations of devices in any mix. Obviously, a system containing only two talkers or only two listeners is trivial. Also, if the system contains only one talker, such as a frequency counter, or a tachometer and one listener, such as a printer, the data bus will be unidirectional in action even though it is bidirectional in principle. The one-talker, one-listener system is definitely not trivial and is very likely to be used.

CONNECTION

The standard permits the use of cable lengths of 2 meters times the number of devices or 20 meters, whichever is less. It permits cabling in either a star or linear circuit or combination thereof. A group of devices in close proximity to one another might form a star through stacked connectors at the end of a linear routing back to the computer or controller.

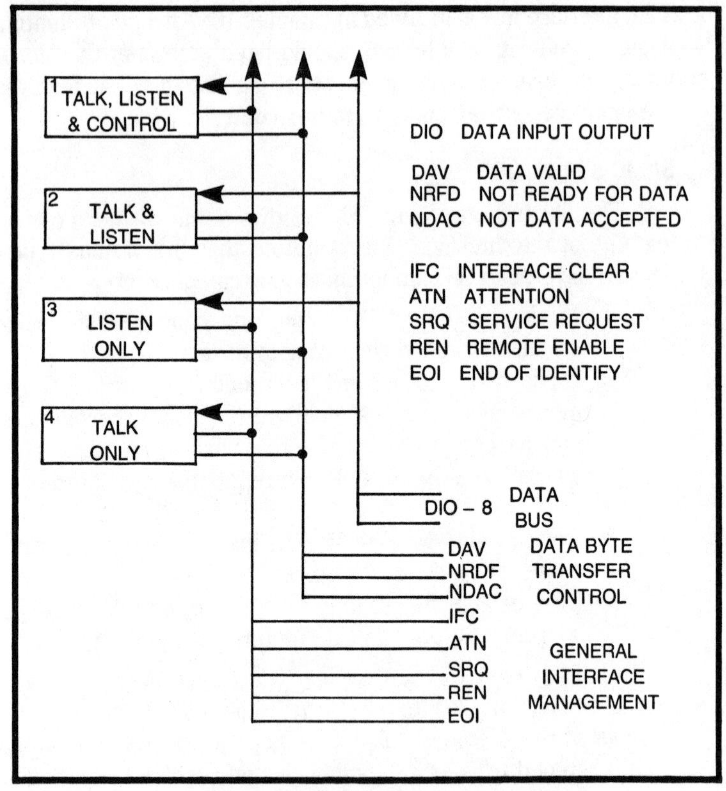

Fig. 17-1. Four types of interface features.

SPEED

The speed of data transfer is also a matter of interest in some cases, especially where rapid data transfer is required. The data bus by itself will handle 250,000 bytes per second with an equivalent standard load every 2 meters out to 20 meters if 48 mA open-collector drivers are used. This speed increases to 500,000 bytes per second with standard three-state drivers. Faster data rates may be obtained by reducing the cable length. For example, the three-state drivers could be used to 1,000,000 bytes per second with one standard load per meter out to 10 meters.

The actual error-free data rates achievable are generally somewhat more dependent upon the delays generated within the devices than they are upon the cable. With most programmable type calculators, for example, the machine itself does not complete an internal cycle in much less than 100 microseconds. If the machine must furnish an address, ask for data, recognize the data, do sometning useful with the data (like store or test it), and finally loop or branch, you are likely to find only four data transactions taking place in a millisecond. We shall also see later that the slowest device to acknowledge sets the speed of the entire system. This is often the calculator when one is used.

Some of the newer generation of microcomputer chips will do considerably better. The Fairchild F-8 is particularly intended for control applications and has 16 input/output latches per chip. With a device of this sort, simple transactions might wind up by achieving data rates on the order of 100,000 bytes/second. Other micro's, such as the 8080, will handle arithmetic faster but must generally shift the data one bit at a time into a shift register latch. In cases where there is little arithmetic to be done, the control-oriented chips will generally be faster. On the other hand, if considerable number crunching is required one of the general purpose or arithmetic oriented chips might be faster. It is very difficult to guess or generalize about which chip will offer the best speed because of the considerable differences in the programming. Until the program is developed and debugged for that particular machine/chip combination, the designer can only guess at the achievable transaction rate. Even at that point, the designer cannot be sure that another algorithm or a "tighter code" could not be written to improve the speed.

One of the full-scale minicomputers such as the PDP-11 can generally be written with tight, machine-language code to perform fast enough to avoid being the limiting item on simple operations. However, if the control algorithm should require something like dividing the sum of a pair of three digit, two exponent complex numbers by their difference (as required in impedance control applications), then rest assured that only the biggest and fastest computers will get up to 250,000 bytes per second.

In cases where a large amount of number-crunching is involved, a very good case can be made for the use of a calculator or a microcomputer/calculator chip multiprocessor in which the micro "punches the keys" and "reads the 7-segment display" which must of course be decoded. The calculator chip will sometimes have a debounce disable pinout for factory testing which will considerably speed the operation. The micro/calculator combination also offers some ease of programming complicated arithmetic, but this is the area where the calculator shines. The software savings in a calculator system are usually very significant, particularly where algorithms involve trigonometric or logarithmic functions.

THE DATA BUS

Data input output lines DIO-1 through DIO-8 form the data bus, which operates in the negative logic framework described in Chapter 16. The various devices on the bus are configured as a wired-OR, and any device can pull the line down. The Standard permits the use of any commonly understood binary, BCD, or alphanumeric code on this bus after a talker and listener have been addressed via interface messages and while ATN is in the high state. The use of the ISO-7 bit code sent with ATN tow is recommended. This goes on lines DIO-1 through DIO-7. This is an alphanumeric code with capital letters only and provides great flexibility. When used for numbers, this code used the three most significant digits to identify the column, and the last four are BCD.

DIO 8	7	6	5	4	3	2	1	Number
H	H	L	L	H	H	H	H	0
H	H	L	L	H	H	H	L	1
H	H	L	L	H	H	L	H	2
H	H	L	L	H	H	L	L	3
↓					↓			
H	H	L	L	L	H	H	L	9

The principal disadvantage in the use of this code is the fact that it transmits only one decimal digit per data byte. Any given byte will handle only the numbers 0 through 9. If more precision is required, subsequent messages must be sent.

The BCD code provides a net advantage in this matter in that any number from zero to 99 may be sent in a single byte. There are a great many control applications where the 1.1 percent increments are adequate in accuracy, and the data exchange can be reduced to a single data byte transaction. This is a considerable time saver.

Even greater resolution may be obtained, of course, in a straight binary data code in which any number from 0 to 255 may be sent. For both binary and BCD, the most significant bit goes on DIO-8 and the least on DIO-1. ATN (attention) must be left high if other listeners are on the line in order to denote the nonstandard code. It should be noted that the generation and handling of a straight binary code is not necessarily possible in all calculators. A good many of these are constrained to BCD.

CONTROL BUS

The control bus, like the data bus, consists of a series of eight lines. These, unlike the data bus, are not all bidirectional. The eight lines are actually broken into two subgroups which shall be discussed separately. The first of these is used to effect the *Three Wire Handshake*.

NOTICE

> The three wire handshake is the subject of one U.S. patent and corresponding patents in foreign countries. These patents are held by Hewlett-Packard. Neither the IEEE nor this writer takes any position on this subject. See the foreword of IEEE Standard 488-75 or contact Hewlett-Packard for license information.

The three wire handshake system is, in the humble opinion of this writer, a very neat way of permitting a number of devices with miscellaneous response characteristics to accept asynchronous data. The idea is good enough in its own right to warrant the use of the data bus in a great many situations because of the assurance that all of the devices will be accepting stable, settled data. This is one of the biggest problems involved with the transfer of miscellaneous, unsynchronized data at unspecified time intervals.

The system consists of the three lines labled DAV (for data valid) NRFD (not ready for data) and NDAC (not data accepted). The meanings of these somewhat awkward sounding terms will, hopefully, become a little clearer through the example.

The timing diagram of Fig. 17-2 shows the sequence of events making up the "handshake." This takes place only after an "address" and "acknowledgement of address" procedure. In the latter, the designated talker does the equivalent of assuming the podium and shaking the hand of the chairman, thereby indicating his willingness to speak to all designated listeners.

From this point on, the process gets a little more complicated and a little less familiar. However, it does bear a marked resemblence to a revival meeting, with slides (data).

The speaker starts to insert a slide in the projector and focus it (Puts data on DLI-1 through DIN-8) and asks "Do you want to see what SIN is doing?" (Sets DAV to high, or false, for data not yet valid). The audience responds with:

yes!
yes!
yes!
yes!
Show it to us brother!
We want to see it!

This is the equivalent of each member individually setting NRFD to high, or false, (indicating that no one is NOT Ready For Data). At

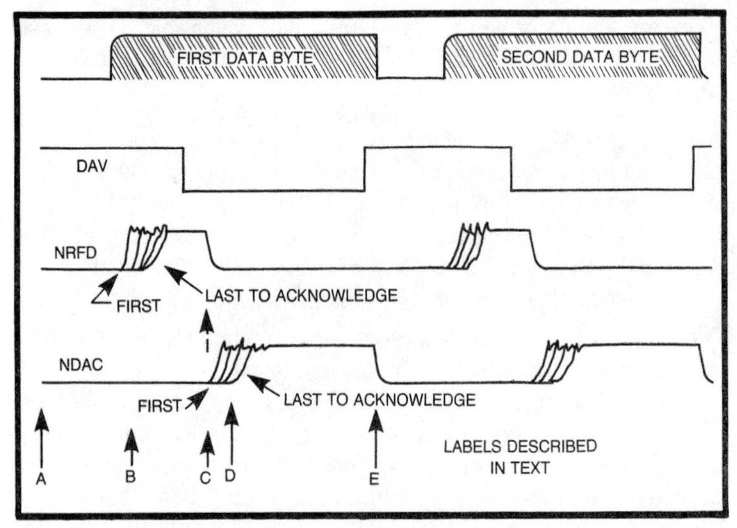

Fig. 17-2. Three wire handshake timing diagram.

this point the slide is focused, showing a row of porno shops etc., (data) and the speaker, sensing the silence (all NRFD HIGH, all DAV low, or true), says "Look at this!" (sets DAV true, or low). The audience responds with:

Shocking!
Terrible!
Abomination!
Horrors!

Individually, each sets NDAC to high, or false, (no one did Not Accept the Data) and the speaker starts to insert another slide (new data not yet settled, thereby setting DAV to high, or false, again). The cycle is ready to begin anew.

Without straining our simile further, the system has the following features:

1. It permits the talker to get the data on to the data lines without extremely tight control of the rise and fall time of each individual line. No one will accept the data until the talker declares it to be true.
2. The talker has the advantage of knowing when the slowest device is ready for data. As long as any device holds NRFD low, or true, nothing more happens.
3. The talker has the advantage of knowing when the slowest device has accepted data. As long as NDAC is low, or true, held there by any listener device, the process halts there.

4. Each receiver has the advantage of being able to reset NRFD low at its own rate.
5. Each receiver has the advantage of being able to reset NDAC low at its own rate after DAV goes high. It can internally constrain NRFD from going high before NDAC goes low again.

The object of this entire scheme is to permit the widest possible range of devices to function together with minimal constraints upon rise time, response time of individual circuits, etc. The beauty of the thing is the flexibility offered. The individual items functioning as transmitter and receiver can be internally synchronous machines capable of changing data only at times set by their individual internal clocks and with no synchronism to the outside world or the other members of the system, but they can be made to communicate precisely over the system.

The three wire handshake requires the use of a form of *state machine* in that the output must be related to the previous operational sequence. The machine shown in Fig. 17-3 is intended as an example only, since it shows only a portion of the circuitry. Drivers have been omitted for simplicity and certain other signals relating to general interface functions have been left off for simplicity.

The block diagram may be seen to have an 8-bit-wide data latch for capturing data. This unit is indicated as latching on the rising edge of $\overline{\text{DAV}}$ (which is inverted by U1). NAND U2 is shown as a 4-bit-wide NAND gate equipped with three signals not previously mentioned, which are internal device functions.

Input VLA is shown as the valid listen address. This will be latched high, or true, by all valid listen addresses to the device. Note that this is an overriding function. Whenever VLA = LOW, FF-1 and FF-2 will be set with Q = high, which represents a false, or "0," output on NRFD and NADC.

Note: The use of inverted logic on the output and normal logic on the other part of the circuit lends a certain amount of difficulty to understanding.

Input RFDI represents the device internal readiness to accept data. If the machine using this interface is internally synchronous and can only accept data at certain times, RFDI can provide the clock strobe. Let us assume that the device has been validly addressed and VLA goes from low to high. This action clears FF-1 and FF-2 and sets Q = L (true) as at point A on Fig. 17-2. The interlock from FF-2 goes high. At the first instant when DAV and RFDI are both high, the output of U2 goes low, thereby presetting FF-1 and rendering

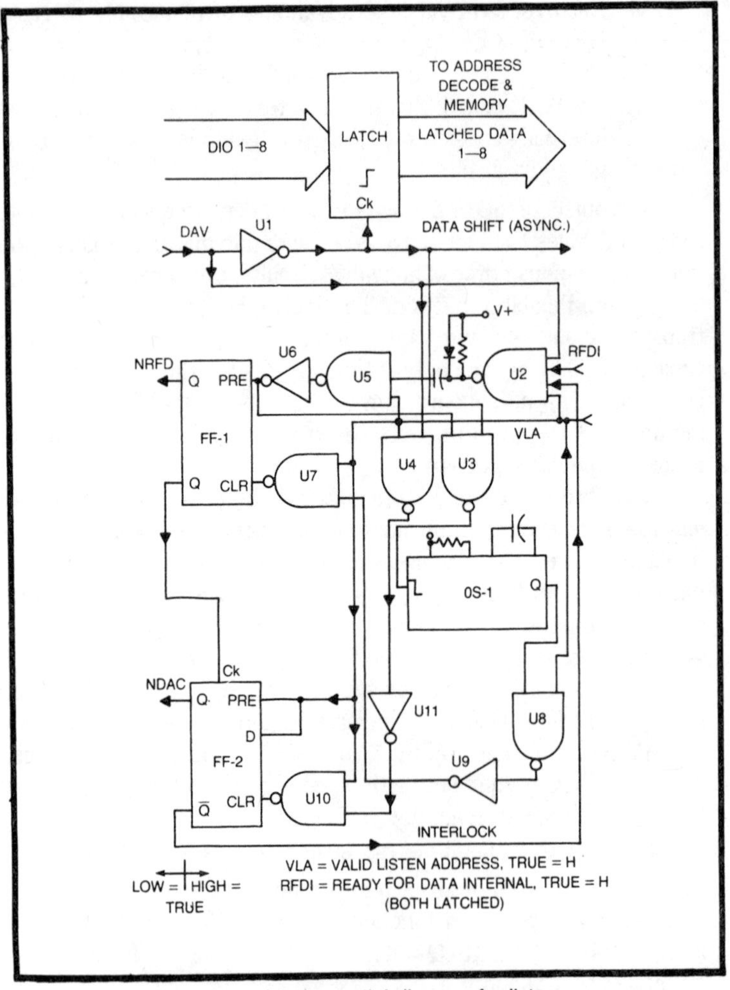

Fig. 17-3. A three wire handshake partial diagram for listener.

NRFD = high, or false, as at point B. The RC network behind U2 converts this to a short spike.

As soon as the present spike is over and when $\overline{\text{DAV}}$ goes low, U3 triggers one-shot OS-1. If VLA is still high, the $\overline{\text{Q}}$ output pulse triggers U8, U9, and $\overline{\text{U7}}$, and clears FF-1 as at C.

The rising edge of $\overline{\text{Q}}$ from FF-1 clocks a high through on NDAC, rendering it false, as shown at D. The next rising edge of DAV triggers U4, U11, and U10, thereby restoring NDAC to the true condition as shown at E. The data byte exchange is now complete, and the system ready for the next exchange.

The purpose of the interlock is to lock out NRFD until NDAC has cleared. In some cases, it might prove necessary to delay the clearing of FF-2 also.

In synchronous devices capable of accepting data only at certain times it may be necessary to include a new-byte flag to prevent reading of the same data byte into memory more than once.

The specification requires the use of open-collector transmitters on NRFD, NDAC, and SRQ. The remaining drivers can be either open collector or three state. There is one exception in that DIO-1 through DIO-8 must use open collector drivers if parallel polling is to be permitted.

ATN AND SRQ

The ATN (attention) is used to specify the way in which the data on the DIO lines is to be interpreted. As long as ATN is low, or true, each device will interpret the content of DIO-1 through DIO-8 as an address or an instruction. A typical sequence for a data transfer illustrates this usage in Table 17-1.

If the talker runs out of things to say and wishes to end the speech it may set line EOL = low during the last data byte. This tells the control device that the transaction is completed.

The SRQ function permits one of the devices to request service from the control device. The actual transaction involves a poll-taking action on the part of the control device. There are a number of applications where this type of operation is advantageous in terms of

Table 17-1. A Typical Data Transfer Sequence.

ATN Status	Control Mnemonic	Comment
L	UNL	UNL = "unlisten." This command cancels all previous listen commands
L	LAD_1	$LAD_{1\text{-}n}$ = Listen Addresses.
L	LAD_2	Any number of devices up to 15,
	LAD_n	the system capacity, may be addressed.
L	TAD	TAD = Talk Address. This command
L		specifies which device will do the talking. The device will not talk until ATN = H
H	DAB_1	$DAB_{1\text{-}n}$ = Successive data bytes
H	DAB_2	sent by the talker to the listeners. Each data byte transaction proceeds
H	DAB_n	through the three-wire-handshake procedure.

time saving. For example, a system might be configured to control and monitor a chemical reaction. The temperature of one of the reactor vessels might be of no interest unless if falls below some level. The use of the SRQ would permit this function to be omitted in the normal control monitoring while still permitting an out-of-range response to obtain service.

THE CONTROL LANGUAGE

The IEEE interface operates by turning the ensemble of devices into a large state-machine. Each transaction on the control and data bus represents a specific state of the device. Most of these states have been given a specific name and mnemonic designation, and a considerable effort was devoted toward making these descriptive of the state or function. For example TADS represents "talker addressed state," and IFC represents "interface clear state." The number of states which are available is so large that it is not practical to attempt to cover them here. Attention is again directed toward the published specification itself.

Another thing which tends to complicate the situation is the fact that nearly all of the states are restricted to a limited subset tf states into which they may exit. This sounds more complicated than it actually is. For example, consider SRQ. In the middle of some transaction the control device cannot simply drop everything and jump to service the device. This would leave all other devices in some intermediate state and chaos would result.

The SRQ request is a great deal like the bell cord which used to run the length of a bus or trolley car. When one wanted to get off he pulled the cord, and the bell rang. The driver or motorman had no way of knowing who wanted to get off since anyone in the car could have pulled the cord. He simply arranged to stop at the next convenient (and designated) stopping place and then waited to see who got off. He could have asked "Who wants to get off?" and waited for a show of hands. This would be described as a *parallel poll*. Alternatively, he could have gone down the aisle and asked each individual "Do you want to get off?" The latter choice represents a *serial poll*.

Some further insight into the mnemonic usage may be obtained from our earlier three-wire-handshake discussion. After an UNL (unlisten) command all of the devices are *not* listening and the system (from a listener viewpoint) is in the AIDS state, or acceptor idle state. In this state no device in the system pulls the NRFD or NDAC lines, or true. Upon receipt of an ATN signal or a ($\overline{LADS}$) (listener addressed state) or a ($\overline{LACS}$) (listener active state) message within a prescribed time interval, the listener enters the ANRS (acceptor not ready) state. This corresponds to point A on the timing diagram of Fig. 17-2. This also corresponds to VLA, NRFD, and NDAC all in the true condition.

When all of the devices commanded to listen acknowledge, DAV goes low, or true, and the system is in the ACRS (acceptor ready) state. In this state, shown at C on the timing diagram, NRFD is high, or False, and NDAC is low, or true.

When NRFD clears and goes LOW, or true, the unit enters the ACDS (accept data) state.

After a delay NDAC goes high, or false, and the device enters the AWNS (await new cycle) state, shown at D on the timing diagram.

From AWNS the device may reenter ANRS when DAV goes high, or false, to begin a new data byte transaction. This sequence will normally be repeated until the entire data exchange of N bytes is complete.

After completion of the exchange, the device could exit the loop to enter the AIDS (acceptor idle) state upon receipt of an UNL (unlisten) command.

As with any high-level computer language, the syntax and definitions take a while to learn and are rather taxing. However, the advantages of standardization of the digital automation are enormous. These include:

Interchangeability

At the present writing a sizeable spectrum ot plug-compatible talkers, listeners, and control devices is available. Within a few years, nearly any type of measurement or active device should be available. The advantage of being able to purchase "sense," "muscle," and "control" devices from diverse manufacturers and have them simply plug together and function with no soldering or black box construction represents an overwhelming economic advantage. The cost increment in the individual devices will, in most cases, be completely offset by the reduction or elimination of design and fabrication time.

Flexibility

Since the system reduces all data and commands to a single digital format, the construction of nearly any automated or interactive function can be done by simply plugging together the available "sense," "muscle," and "control" units. Where a given function is too specialized to permit the use of standard items, only these need be designed and fabricated. An example of this was our driven truck wheel of Chapter 16.

For most instrument and process control applications, the largest part of the operation is reduced to the problem of developing the proper sense and control algorithm, or the "software." Large changes in the operating procedure can be obtained in many cases without changing any of the physical hardware items, by simply rewriting the program.

An additional advantage is the ability to perform extensive mathematical manipulation of the data. The simplest example of this is the linearization of sensor data. Many commonly used devices, like thermocouples, have a nonlinear input/output curve. For years people have tricked-up operational amplifiers to more or less linearize the response. In a computer operated system the linearization can be done mathematically directly from the nonlinear output.

In a similar vein, many "muscle" devices tend toward a response that is variable with temperature or some other parameter. For example, the armature and field resistances of electric motors rise with increasing temperature. This limits the torque and thereby the response of the device. Rather than making the effort to compensate the device itself, the computer controlled system could simply read the temperature and include a mathematical compensation in the algorithm that determines the input commands. It could simply supply a higher voltage or a longer current pulse to a hot motor.

Accuracy

Compared to an analog system, the digital system has nearly unlimited accuracy, paid for only in the time required to transfer data. An 8-bit binary code can describe a voltage, force, or other parameter to within one part in 255, or ± 0.20 percent, in a single byte. At 250,000 bytes per second a single such transaction might require 20 microseconds. If the data word were upped to four bytes the accuracy would be one in 4.29×10^9, and the full transaction time would rise to something like 32 microseconds. At the level of human perception, it is an analog world that we live in and many of the things we measure are done in an analog way. In a digital system, the accuracy of the initial A/D conversion can be made to be the controlling factor with no further errors added in the data handling.

Recent years have seen the developement of a new breed of sensors in which the measurement is made directly digitally. The Laser Interferometer will measure distance to an accuracy of a few parts in 10^9. The photointerruptor/tachometer will measure speed to a few parts in 10^7. Several types of pressure and force gauges have been developed which convert the parameter to frequency and can measure to one part in 10^7. The quartz thermometer can measure temperature to one part in 10^4. Only in a digital system can this data be handled easily and without loss of accuracy.

Interconnectability

The use of the bus architecture can greatly reduce the requirement for interconnection as shown in Chapter 16. For larger and more widely disbursed systems it is entirely feasible to distribute microprocessor/interface/sense/muscle systems which could funnel into a large central processor serving in only a supervisory capacity. The micro could marshall the pertinent data onto a single co-axial cable, in addition to its duties as "foreman" to the local devices.

CONCLUSION

It seems to the writer that the next few years will witness an ever rising level of automated control in all walks of life. Most of these controls will be developed using solid-state semiconductor devices. The capabilities of these devices are increasing at a dizzying rate. An example of this is the fact that between the time of writing of Chapter 2 on switching devices and the completion of the final chapter two semiconductor houses have announced op amps with output capabilities in the 150 volt, 1 ampere range (data was not available to permit inclusion). Nothing short of a complete reversal of world economy seems likely to turn the tide.

Index